Lecture Notes in Computer Science 1105

Edited by G. Goos, J. Hartmanis and J. van Leeuwen

Advisory Board: W. Brauer   D. Gries   J. Stoer

Springer
Berlin
Heidelberg
New York
Barcelona
Budapest
Hong Kong
London
Milan
Paris
Santa Clara
Singapore
Tokyo

Tuncer I. Ören   George J. Klir   (Eds.)

Computer Aided Systems Theory – CAST '94

4th International Workshop
Ottawa, Ontario, Canada, May 16-20, 1994
Selected Papers

 Springer

Series Editors

Gerhard Goos, Karlsruhe University, Germany

Juris Hartmanis, Cornell University, NY, USA

Jan van Leeuwen, Utrecht University, The Netherlands

Volume Editors

Tuncer I. Ören
Ottawa Center of the McLeod Institute of Simulation Sciences
University of Ottawa
Ottawa, Ontario, Canada K1N 6N5

George J. Klir
Thomas J. Watson School of Engineering and Applied Science
State University of New York
Binghamton, NY 13902-6000, USA

Cataloging-in-Publication data applied for

Die Deutsche Bibliothek - CIP-Einheitsaufnahme

Computer aided systems theory : 4th international workshop ;
selected papers / CAST '94, Ontario, Canada, May 16 - 20,
1994. Tuncer I. Ören ; George J. Klir (ed.). - Berlin ;
Heidelberg ; New York ; Barcelona ; Budapest ; Hong Kong ;
London ; Milan ; Paris ; Santa Clara ; Singapore ; Tokyo :
Springer, 1996
  (Lecture notes in computer science ; Vol. 1105)
  ISBN 3-540-61478-8
NE: Ören, Tuncer I. [Hrsg.]; CAST <4, 1994, Ottawa>; GT

CR Subject Classification (1991): J.6, I.6, I.2, J.7, J.3, C.1.m, C.3

ISSN 0302-9743
ISBN 3-540-61478-8 Springer-Verlag Berlin Heidelberg New York

© Springer-Verlag Berlin Heidelberg 1996
Printed in Germany

Typesetting: Camera-ready by author
SPIN 10513380   06/3142 – 5 4 3 2 1 0   Printed on acid-free paper

Preface

This volume consists of a selection of papers presented at the Fourth International Workshop on Computer Aided Systems Theory, CAST'94.

CAST'94 was organized by the Ottawa Center of the McLeod Institute of Simulation Sciences, the Department of Systems Science of the State University of New York at Binghamton, and the Master's Programme in Systems Science at the University of Ottawa. It was held on the campus of the University of Ottawa, May 16-20, 1994.

Out of the 82 abstracts/papers submitted, 58 have been included in the workshop program. Fully refereed, revised, and previously unpublished, 31 articles directly relevant to CAST are included in the book.

This volume, as a good representative of the state of the art of the CAST movement, provides in-depth knowledge on all three aspects of CAST, i.e., its foundations, methods, and tools and environments. Accordingly, the book is divided into three sections:

1. Foundations of CAST: Theory and Methodology (10 articles)
2. CAST Methods (7 articles)
3. CAST Tools and Environments (14 articles)

The efforts of the referees and the cooperation and patience of the authors are very much appreciated. We would also like to express our appreciation to Profs. F. Pichler and R. Moreno-Díaz for their contributions as the founders of the CAST movement as well as for their support of this CAST Workshop organized for the first time in North America.

May 1996   Tuncer I. Ören , George J. Klir
                                                             Canada; NY, USA

Contents

1 Foundations of CAST: Theory and Methodology

2 CAST Methods

3 CAST Tools and Environments

1 Foundations of CAST: Theory and Methodology

Systems Science and Systems Technology: From Conceptual Frameworks to Applicable Solutions

Franz Pichler [1)]
Heinz Schwärtzel [2)]
Roberto Moreno-Diaz [3)]

Abstract. The paper points out the importance of systems technology for dealing with complex design tasks in real life problem solving especially in the field of engineering.

After an introduction, where Systems Science and Systems Technology are generally discussed the general framework of Design Science and related Design Technologies are considered.

For the definition of "Systems Technology" the concept of the Design Propeller is used. This is followed by the discussion of CAD tools for dealing with problems on systems level. The specific roles of "Computer Aided Systems Theory", "Computer Aided Engineering" and "Computer Aided Manufacturing" in the context of "Computer Aided Systems Technology " (CAST) are pointed out.

Future perspectives of necessary CAST research for applications in the "Engineering of Computer Based Systems" (ECBS) and in the development of "Application Specific Intelligent Micro Systems" (ASIMS) are given.

1   Introduction

The logistic need in praxis to organize complex systems to meet certain optimal goals led in the past to the founding of new scientific fields such as Operations Research, Systems Engineering, General Systems Theory and Cybernetics.

Today we can consider these fields under the common umbrella provided by *Systems Science*. As part of the design methodologies Systems Science is responsible for a scientifically based treatment of problems which appear on the architectural level of the design process. Its main concern is to put an emphasis on the treatment of systems complexity and to relate complex systems specifications to the given design requirements and at the same time also to possible decomposed structures which are feasible models for the final realization. In that sense Systems

---

[1)] Institute of Systems Science, Johannes Kepler University Linz, Austria

[2)] F A S T - Research Institute for Applied Software Technology, Munich, Germany

[3)] Instituto de Informatica y Sistemas, Universidad de Las Palmas, Las Palmas, Canary Islands, Spain

Science is a field, which interfaces on one ("above") side with the field of specification science and on the other ("below") side with the different engineering fields (in the narrow sense) which deal with the final realization, testing and maintenance by application of the different appropriate technologies.

In engineering less complex systems the "systems approach" (that is to make explicit use of the different frameworks of Systems Science in the design process) is usually avoided. However the experienced designer makes implicit use of it. In engineering complex systems, however, the application of the "systems approach" in design is today a necessity. The professional engineering community takes this into account. Of course, there has been an "up and down" concerning the acceptance of Systems Science. After an "up", which offered a lot of potentials in the early sixties there was a "down" until recently. Today, by different reasons, where the "computer" and the reached maturity of the field of Computer Science plays certainly an important role, we have again an "up" and it is the duty of the systems scientists to keep this level of acceptance by proving the excellency of the systems approach.

2   Systems Technology

By *Systems Technology* we understand the collection of all systems knowledge together with the existing tools to apply this knowledge in the engineering praxis. For these tools the computer has a major role as an universal "realization machine". Computer Aided Design tools (CAD tools) are common in many engineering disciplines.

However, we consider them only then as part of Systems Technology, if they strongly support the systems level in design. Over the last years several efforts have been undertaken to promote research and development of Systems Technology. Such efforts have been in the past the international workshops on Computer Aided Systems Theory (EUROCAST'89, 91, 93, published in [1], [2], [3]) which emphasized the investigation of formal systems specification, modelling and simulation, and the development of appropriate tools (CAST tools) for it. The main tasks of *Systems Theory* are the following:
- selection of proper formal models and formal coupling concepts (formal model building)
- provision of methods for formal model composition and formal model decomposition (formal systems architecture building)
- optimization of formal models which are components of the formal systems architecture (formal systems architecture optimization)

In the past CAST research was devoted to tools which support systems theory tasks in this sense and which can be integrated to common CAD-tools.

Another contribution in the past has been the work on research and development of the "Working Group on System Engineering" of the ESPRIT project ATMOSPHERE - an European project in information technology which has found careful documentation by three voluminous books [4], [5], [6]. A third activity which should be mentioned here is the ongoing work of the IEEE Task Force on the *Engineering of Computer Based Systems* (ECBS) in which participants from industry and university discuss the problems of design, implementation, and maintenance of large scale engineering systems which contain computers as essential components for their practical realization [7], [8].

3   Design Technology

Systems Technology, in the sense as we defined it above can be considered as part of *Design Technology*. To support the initial phases of engineering design we need technological means to deal with the client/designer cooperation in problem specification. We consider these means as part of *Specification Technology*. The phase of specification is followed in design by the phase of modelling which is performed on systems level in which Systems Technology provides the basic means. Modelling, in which an appropriate architecture of the final systems has to be found and evaluated, is followed by implementation. By *Systems Architecture* we understand a model which meets the required specification and is a hierarchical multi strate model (in the sense of M. Mesarovic). Its final node components are considered by the systems designer as feasible for implementation and/or realization. In addition the coupling systems of the final node components are also considered to be feasible for implementation and/or realization. Implementation is defined by the determination of the final structure which can be realized using Hardware Technology or Software Technology.

In discussing Design Technology we like to distinguish implementation from realization, which is the actual production (building) of the real system which follows the implementation step.

In the engineering community our given definition of different parts of design technology, is not standard. However, for the purpose to emphasize the concept of systems technology, we prefer to introduce this form of partition of design technology.

Specification/designer/client document

Systems Design/systems architecture

Implementation/implementation model

Realization/process model

Figure 1: Design phases and results to achieve

4   Design-Automation and CAST

It is a commonly accepted goal in design science to automate the different steps of the design process as much as appropriate. The ratio behind is to give support to a human designer by a (computerized) *design assistant*, which does all kind of work which is for the human designer either
- computational too complex, or
- needs a lot of efforts in searching (in a data base), or
- has to be done straight forward with no need in searching and evaluating alternatives.

By the existence of a design assistant of this kind, the design process gets partially automated and the designer can concentrate on innovative tasks and make more effective use of all his experience and imagination.

On the systems level the designer and also the design assistant apply the CAD tools provided by the available Systems Technology. We consider such a CAD tool as part of *Computer Aided Systems Technology* (CAST). The development of a design assistant (which takes care of the different design tasks which we mentioned above) requires the existence of formal mathematical/systems-theoretical models with deductive power. This specific development process on a meta-level of design has to be supported by specific CAD tools which we call *Computer Aided Systems Theory-tools* (CAST tools). To sum up, in order to automate parts of the design process by an (computerized) design assistant we propose to apply Systems Theory and related CAST tools.

For an example of an application of a CAST tool we refer to our former research ([9], [10], [11]) where the application of Finite State Machine Theory to optimize controllers in VLSI circuits was the goal. There the CAST tool CAST.FSM, a method base system for dealing with finite state machines [12], has been successfully applied. The developed algorithms can be used as part of logic synthesis methods in high level CAD tools to realize a design assistant for VLSI design.

5   Design Chart

For the representation of the embedding of CAST tools which support Systems Theory in engineering design we have introduced in former publications the so called *design propeller* [12], [13], which is shown in Figure 2.

The design propeller, which we call here also a *design chart*, is a generalization of the Y-chart of Gajski and Kuhn well-known in VLSI design [14]. It represents in pictorial form the integration of CAST tools (which support formal modelling activities) with common CAD tools and CAM tools for implementation and realization, respectively. The three sections of the propeller can be considered as the modelling part of the overall state space of the design process. After reaching the state of a agreeable specification a design trajectory enters the design chart and evolves by the activities of the designer by non deterministic state transitions from outer sections of the propeller to the inner regions until it reaches the center circle.

The states at the center circle represents the result of the modelling part of design, this means in our framework, that the step of realization is finished and production can be started.

Systems design activities can be located in the outer parts of the design chart. There, Systems Science should provide the conceptual framework for the formulation of the different design states. Furthermore it should provide methods for performing transformations (state transitions) between design states.

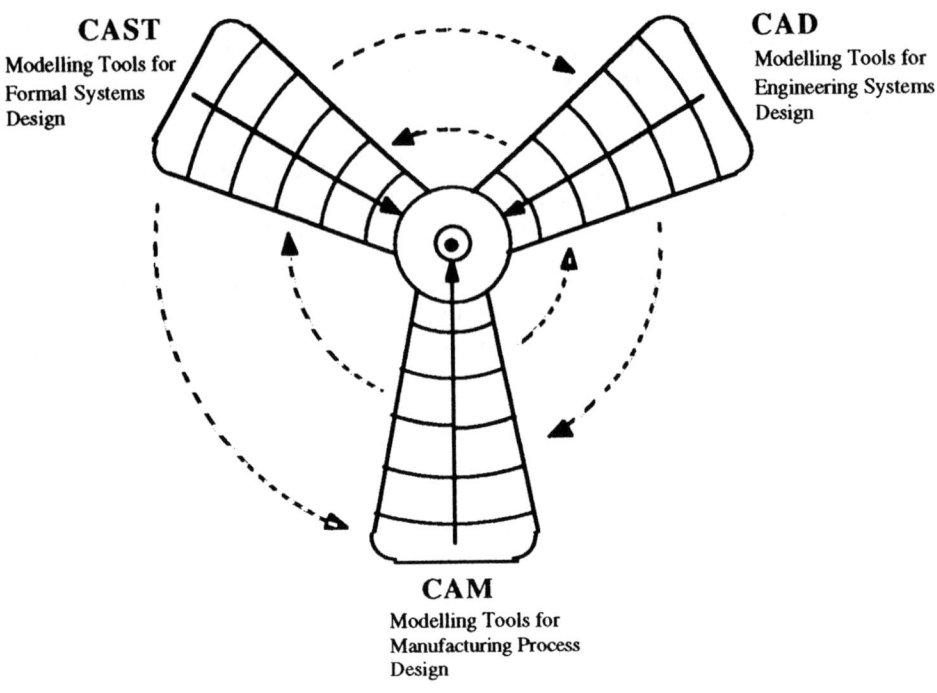

CAST
Modelling Tools for
Formal Systems
Design

CAD
Modelling Tools for
Engineering Systems
Design

CAM
Modelling Tools for
Manufacturing Process
Design

Figure 2: Design propeller (design chart)

In our interpretation of the design chart we associate with the design states and design state transformations also the related tools for giving computer assistance the definition and computation of design trajectories. We distinguish such tools in two ways: The three sections of the design chart divide the tools into CAST tools which support the application of Systems Theory and Mathematics), into CAD tools (which support the usual activities of engineering systems design) and CAM tools (which support the part of engineering design which deals with the modelling of the production process). The second way, in which we like to distinguish these tools is, that we consider the outer parts of the design chart as part of Systems Technology and where the inner parts represent tools of Implementation Technology. Refinements in graphical detail of the design chart can be used to indicate the degree of integration of such tools into each other.

The design chart can be considered as a suitable pictorial representation of Systems Technology and its role as part of the design process.

6   Design Complexity

In the following we discuss general ideas for measuring the complexity of the design process. The presented considerations might be useful as a basis for defining more specific complexity measures.

By the different phases of the design process, namely the phases
-        specification
-        systems design
-        implementation
-        realization

we have already a decomposition of the design process in sequential order. Based on this order an overall complexity measure C can be considered by $C = C_{spec} + C_{sys} + C_{imp} + C_{real}$, where the individual parts relate to the distinguished phases of the design process.

What are the factors which influence the complexity C and what quality or quantity should C measure. A generally accepted quantity is certainly given by the required "manpower" m and the required "time" t, usually measured by manpower times years. We denote this specific complexity measures by $C(m \times t)$, $C_{spec}(m \times t)$, $C_{sys}(m \times t)$ and $C_{imp}(m \times t)$, respectively.

In many examples of complex systems design we might agree that the following inequality is valid:

$$C_{spec}(m \times t) + C_{sys}(m \times t) >> C_{imp}(m \times t) + C_{real}(m \times t)$$

Figure 3 which is taken from the field of VLSI design shows a similar relation between cost factors [17].

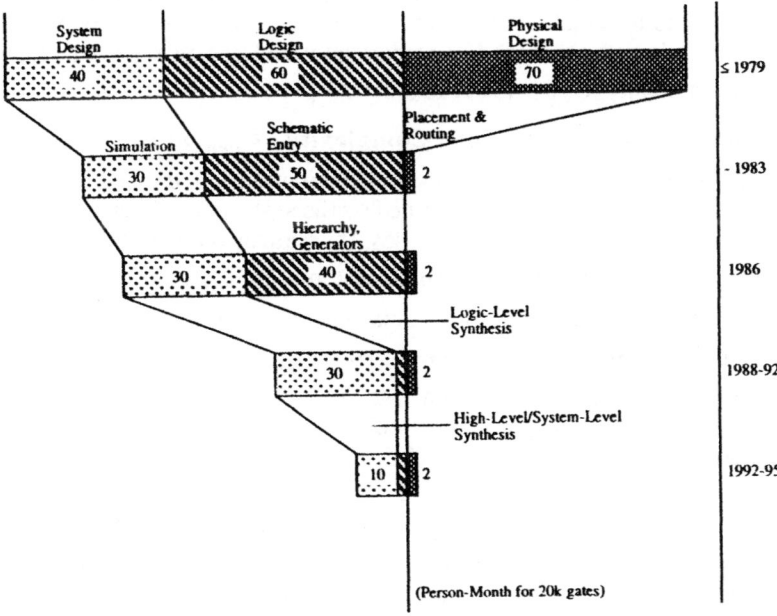

Figure 3: Development of effort spent in various VLSI design steps over the years
[17]

Let us discuss some factors which influence the given different complexity measures:

(1)   Specification Complexity $C_{spec}(m \times t)$

Specification is based, as we know, on the interaction of the client and the designer. Therefor, the effort necessary to specify the system to be implemented and realized depends strongly on the means for reaching a model of the overall model which meets semantically the objectives of the client and which is, at the same time, operational for the designer to serve as basis for the subsequent steps of design. To keep $C_{spec}(m \times t)$ as low as possible from both parties some maturity based on experience is urgently required. The client should have a praxis in dealing with similar problems; the designer is expected to have accumulated experience in the design of comparable systems. From a certain point on specification has to be done by using computer assistance. Therefor, the availability and applicability of suitable specification tools will also be a major factor to reduce $C_{spec}(m \times t)$.

(2)   Systems Design Complexity $C_{sys}(m \times t)$

Systems design can be considered as the "art part" of the design process. It needs from the designer an invention, to select the proper type of systems specification (usually of network-type) which can serve as the architecture of the real system which has to be designed. The systems architecture, as we define it, ensures (by looking "up") that the requirements as given by the final specification are met and ensures the feasibility of implementation and realization (by looking "down").

   For proving the feasibility the systems architecture is given by a multi-strata specification of the goal system, where the "atomic" components and the related interfaces are considered as feasible for implementation and realization, for example being known building blocks of a library of engineering models. Again, as in the case of specification, the availability of computerized design tools will influence strongly the complexity $C_{sys}(m \times t)$ of a design process.

(3)   Systems Implementation Complexity $C_{imp}(m \times t)$

Based on the results of systems design implementation has the task to refine the components and the interfaces, as determined by the systems architecture until a level of engineering description such that design realization can be started. The effectiveness of implementation, measured in terms of $C_{imp}(m \times t)$ is certainly depending upon the skills of the designer in using computerized tools.

(4)   Systems Realization Complexity $C_{real}(m \times t)$

The phase of design realization has to provide the logistic for the production process. Depending on the kind of the production technology (from handmade fabrication to CIM installations) different steps have to be considered. They range from finding the right position of components and "wiring" them together to the determination of quality tests for production quality control. The complexity $C_{real}(m \times t)$ depends strongly on the built-up experience of the individual design team in dealing with production planning. Inclusion of production issues in early phases of systems design (e.g. paying attention to "Design for Testability" can considerably reduce $C_{real}(m \times t)$.

7   Design Tools

The design process in complex systems design has to be supported by using appropriate *design tools*. In the past, many tools for dealing with the implementation and realization level have been developed in the different branches of engineering. They are usually known under the name CAD/CAM-tools. In the last years by the needs of the industry, however, also specification tools (SPEC-tools) have been strongly developed. We wish the same could be said for tools dealing with the systems level of design. Unfortunately, this is currently not the case. Only for specific branches of engineering (such as for example control engineering, communication engineering, signal processing, VLSI design) we have today elaborated tools for systems design. Usually they are integrated into common CAD tools for realization. In the context of the design chart they can also be considered as CAST tools if they support also systems theory. Another example of systems design tools which are available today are tools for dealing with systems engineering, especially with software engineering (CASE-tools). It is an important task in Computer Aided Systems Technology to provide in the near future additional systems design tools which have a common a conceptual basis for tool integration. Different ongoing initiatives e.g., CFI (CAD Framework Initiative) or ECSI (European CAD Standardization Initiative) address this problem area. We should have CAD-tools (in the more general sense) which cover all phases of the design process and which contribute to the reduction of the design complexity C. The already mentioned efforts of the IEEE task-force on ECBS, dealing with large scale systems and the current efforts to extend VLSI design tools to CAD tools for microsystems design reflect the importance of this research line.

8   Conclusion

The automation of the design of complex engineering systems is one of the most important goals in Engineering. It requires the existence of well founded concepts, methods, and tools for specification and modelling. To support modelling the availability of an appropriate Systems Technology is strongly required. To avoid complexity in applications it is necessary to base the design tools on a common framework. Systems Science and Mathematics can provide the appropriate systems concepts and related methods to allow a high degree of tool integration. The ultimate (academic) goal should be to realize the means Design Technology by a well-defined state machine and to develop autonomous (intelligent) behavior for it to compute automatically design trajectories.

Different ongoing activities in engineering show that there is a common awareness of the importance of this type of research.

One field of activity is given by the current efforts to promote the field of computer based systems (ECBS). Another field which should be promoted are highly integrated microsystems, where, in addition to the existing tools for the development of Application Specific Integrated Circuits (ASICS), the development of proper CAD tools for the design of *Application Specific Intelligent Micro Systems* (ASIMS) is very much desired.

References

1. Pichler, F., R. Moreno-Díaz (eds.): Computer Aided Systems Theory: EUROCAST'89, Lecture Notes in Computer Science 410, Springer Verlag, Berlin, 1990
2. Pichler, F., R. Moreno-Díaz (eds.): Computer Aided Systems Theory: EUROCAST'91, Lecture Notes in Computer Science 585, Springer Verlag, Berlin, 1992
3. Pichler, F., R. Moreno-Díaz (eds.): Computer Aided Systems Theory: EUROCAST'93, Lecture Notes in Computer Science 763, Springer Verlag, Berlin, 1994
4. Thomé, Bernhard (ed.): Systems Engineering: Principles and Practice of Computer based Systems Engineering
John Wiley & Sons, Chichester 1993
5. Kronlöf, Klaus (ed.): Method Integration: Concepts and Case Studies
John Wiley & Sons, Chichester 1993
6. Schefström, D. and G. van den Broek (eds.): Tool Integration: Environments and Frameworks
John Wiley & Sons, Chichester 1993
7. Thomé, Bernhard (ed.): Task Force on Computer-Based Systems Engineering-Newsletter
IEEE Computer Society, Vol 1, No 1, Summer 1993
8. Thomé, Bernhard (ed.): Task Force on Engineering of Computer-Based Systems -Newsletter
IEEE Computer Society, Vol 1, No 2, Winter 1993
9. Geiger, M.: CAST.FSM Applied to VLSI Synthesis: Experimental Results and Requirements for Industrial Use
in: Computer Aided Systems Theory: EUROCAST'91, Lecture Notes in Computer Science 585, Springer Verlag, Berlin, 1992, pp. 422-441
10. Müller-Wipperfürth, T., J. Scharinger, F. Pichler: FSM Shift Register Realization for Improved Testability
in: Computer Aided Systems Theory: EUROCAST'93, Lecture Notes in Computer Science 763, Springer Verlag, Berlin, 1994, pp. 254-267
11. Müller-Wipperfürth, T.: On the integration of CAST.FSM into the VLSI design process
Lecture presented at CAST'94: Fourth International Workshop on Computer Aided Systems Technology, University of Ottawa, May 16-20, 1994 (General Chairman: Prof. Tuncer I. Ören)
12. Pichler, F., H. Schwärtzel (eds.): CAST Methods in Modelling
Springer Verlag, Berlin, 1992
13. Pichler, F., H. Schwärtzel: CAST - Computerunterstützte Systemtheorie
Springer Verlag Berlin, 1990
14. Gajski, D., H. Kuhn: Guest Editors Introduction, New VLSI Tools
IEEE Computer, vol. 16, no.12, pp. 11-14, 1983
15. De Man, Hugo: Microsystems: A challenge for CAD Development. in: MICROSYSTEM Technologies 90 (H. Reichl (ed.).) Springer Verlag Berlin, 1990, pp. 3-8.

16.   Müller-Glaser, K.D., H. Rauch, W. Wolz, J. Bortolazzi, C. Kuntzsch, R. Zippelius: A Specification Environment for Microsystems Design. in: MICROSYSTEM Technologies 90 (H. Reichl (ed.).) Springer Verlag Berlin, 1990, pp. 9-16

17.   Michel, P., U. Lauther, R. Duzy (eds.): The Synthesis Approach to Digital System Design
Kluwer Academic Publishers, Boston, 1992

Soft Computer-Aided System Theory and Technology (SCAST)

George J. Klir
Department of Systems Science
T.J. Watson School of Engineering and Applied Science
Binghamton University
Binghamton, NY 13902-6000 USA

Abstract. *Soft Computer-Aided System Theory and Technology* or SCAST is introduced as a branch of CAST whose focus is on computing, system theory, and technology that exploit the tolerance for imprecision and uncertainty to achieve tractability, robustness, and low cost. *Soft computing* is currently viewed as a junction of fuzzy logic, neural computing, probabilistic reasoning, and genetic algorithms. *Soft system theory* is based on fuzzy set theory, fuzzy measure theory, rough   set theory, and their combinations. *Soft technology* plays a dual role in SCAST. Its first role is to develop supporting software and hardware for soft computing, while its second role is to develop applications of soft computing and soft systems theory in other areas. These various components of SCAST and their relationship are overviewed.

1   Systems Modeling and CAST

As stated by Franz Pichler, the originator of the CAST movement, "CAST has as its goal the supplementation of current and future CAD tools by system theory software (CAST tools) which can be applied in model building and in model application" [31, p.6]. That is, the aim of the CAST movement is to encourage the development of software tools by which methods resulting from system-theoretic research become accessible to practitioners in the areas of model building and model application.

In my opinion, the *construction of systems models* is a fundamental activity of both the sciences of the natural and the sciences of the artificial [35]. In the *sciences of the natural*, we attempt to construct systems that are adequate models of some aspects of reality; the purposes for constructing systems models in this domain include understanding, prediction, retrodiction, control, decision making, or policy making. In the *sciences of artificial*, we attempt to construct systems that are adequate models of desirable man-made objects; the purpose is to prescribe operations by which a desirable artificial object can be constructed so that appropriate objective criteria are satisfied within given constraints.

This broad view of systems modeling is also taken by Pichler in formulating the CAST mission: "Problems arise contextually in connection with open answers to questions concerning a given reality. A reality can already exist or be planned to exist. Existing realities are typical for problems of science, planned realities are the main objects in engineering" [31,p.1].

The principal aim of CAST is thus to facilitate the utilization of methodological resources emerging from research in systems theory by developing appropriate computer software tools for practical use in systems modeling, encompassing both scientific and engineering modeling. To pursue this aim requires a cooperation of systems theorists, computer software specialists, and practical systems modelers.

Since the CAST movement is now well established, it seems useful to identify some of its characteristic branches. For example, we may distinguish CAST tools oriented to scientific modeling from those oriented to engineering modeling. Similarly, we may recognize quantitative modeling and qualitative modeling as two branches of CAST. The purpose of this paper is to characterize an emerging branch of CAST, for which I propose the name *Soft Computer Aided System Theory and Technology* (SCAST).

2 Characteristics of SCAST

The name *Soft Computer Aided System Theory and Technology* (or SCAST) is introduced here to identify a branch of CAST whose focus is on soft computing, soft systems theory, and the supporting technology. What are soft computing and soft system theory?

The term *soft computing* has recently emerged as a label for a spectrum of novel modes of computation whose principal aim is to exploit the tolerance for uncertainty to achieve tractability, robustness, and low cost. In soft computing, approximate rather than precise solution are sought to precisely formulated problems or, more typically, imprecisely formulated problems. The challenge of soft computing is to develop methods of computation that lead to acceptable solutions at low cost.

At this time, several components of soft computing are recognized in the literature. They include:

- fuzzy set theory and fuzzy logic [22,24,42];
- fuzzy measure theory [40];
- various special branches of fuzzy measure theory, such as possibility theory [6,44], evidence theory [12,13,34], or theories of imprecise probabilities [25,27,39];
- rough set theory [29];
- theory of neural computation [17];
- genetic algorithms [11,18]

These components are overviewed in Sec. 5.

The term "soft system theory" is used here to refer to any theory that deals with systems involving some kind of uncertainty (predictive, prescriptive, etc.). The uncertainty may be expressed in terms of fuzzy sets, rough sets, precise and imprecise probabilities, or other types of fuzzy measures; it may also be expressed by combining these various types of uncertainty. Again, the novel theories of uncertainty are overviewed in Sec. 5.

3 Changing Attitudes Towards Uncertainty

When dealing with real-world problems, we can rarely avoid uncertainty. At the experiential level, uncertainty is an inseparable companion of any measurement, resulting from a combination of resolution limits of measuring instruments and inevitable measurement errors. At the cognitive level, it emerges from the vagueness and ambiguity inherent in natural languages. At the social level, uncertainty results not only from the inevitable incompleteness of shared meanings obtained by people through social interaction, but it is also created and maintained by people for different purposes (privacy, secrecy, propriety, social norms, etc.).

Uncertainty is thus fundamental to human beings at all levels of their interaction with the real world. And yet, as thoroughly analyzed by Smithson [36], Western intellectual culture is notorious for its neglect of uncertainty. It is increasingly recognized, however, that attitudes towards uncertainty has been undergoing a significant change in this century. In science, this change has been manifested by a transition from the *traditional attitude*, according to which uncertainty is a plague that should be avoided by all means, to an *alternative attitude*, according to which uncertainty is fundamental to science and its avoidance is often counterproductive. Two phases of this transition can clearly be recognized, each having the characteristics of a paradigm shift (or a scientific revolution) in the sense introduced by Thomas Kuhn [26].

The *first paradigm shift* regarding uncertainty began in the late nineteenth century, when some physicists became concerned with processes at the molecular level. Although the precise laws of *Newtonian mechanics* were applicable to the study of these processes in principle, they were not applicable in practice, due to prohibitive computational demands resulting from the enormous number of entities involved. The need for a fundamentally different approach to the study of these processes motivated the development of relevant statistical methods, which eventually led to *statistical mechanics*. Precise numbers were replaced with meaningful statistical averages, and the role played in Newtonian mechanics by analytic methods based upon the calculus were in statistical mechanics replaced by statistical methods based upon probability theory.

When statistical mechanics was accepted, by and large, by the scientific community as a legitimate area of science early in this century [10], the traditional attitude towards uncertainty was for the first time revised. Uncertainty became recognized as useful, or even essential, in certain scientific inquiries. However, this recognition was strongly qualified: uncertainty was conceived solely in terms of probability theory.

While analytic methods are applicable only to problems that involve a small number of entities (e.g., variables) that are related in a predictable way, the applicability of statistical methods has exactly opposite characteristics: they require a large number of entities and a high degree of randomness. These two types of methods are thus complementary. When one type excels, the other totally fails. Despite their complementarity, these types of methods cover, unfortunately, only problems that are clustered around the two extremes of complexity and randomness scales. Warren Weaver [41] refers to them in his insightful paper as problems of *organized simplicity* and problems of *disorganized complexity*. He argues that these types of problems represent only a tiny fraction of all problems that are of interest to science. Most problems are somewhere between these two extremes. These are problems that involve nonlinear systems with large number of components and rich interactions among the components; they are often nondeterministic, but not as a result of randomness that could yield meaningful statistical averages. Weaver calls them problems of *organized complexity* and argues that they are typical in life, cognitive, social, and environmental sciences, as well as in applied fields such as modern technology, medicine, or management.

The emergence of computer technology in World War II and its rapidly growing power in the second half of this century made it possible to deal with increasingly complex problems, some of which began to resemble the notion of organized complexity. However, this gradual penetration into the domain of organized complexity revealed that the high computer power, while necessary, is not sufficient for making substantial progress in this problem domain. Analytic and statistical methods were increasingly found inadequate for dealing with the emerging problems. As a consequence, some scholars began to feel that radically new methods were needed: methods based on fundamentally new concepts and mathematical theories. An important new concept (and some mathematical theories based upon its various facets) that emerged from this cognitive tension was a broad concept of uncertainty, liberated from its narrow confines of probability theory. This broad concept of uncertainty is our primary interest in this paper.

4 Classical Types of Uncertainty

The classical mathematical theories by which certain types of uncertainty can be characterized have been set theory and probability theory. In terms of *set theory*, uncertainty is expressed by sets of alternatives in situations where one alternative is desired. For example, when medical diagnosis of a patient results in a set of possible diseases rather than a single disease, the set represents a *diagnostic uncertainty*; when an interval of values of a variable is predicted rather than a single value, the set of values in the interval represents a *predictive uncertainty*; when an unsettled historical question allows a set possible answers rather than a unique one, the set represents a *retrodictive uncertainty*.

Uncertainty expressed in terms of sets of alternatives results from the *nonspecificity* inherent in each set. Large sets result in less specific predictions, retrodictions, etc., than their smaller counterparts. Full specificity is obtained when only one alternative is possible. This kind of uncertainty can be handled, for example, by *interval analysis* [28].

The second classical theory for characterizing uncertainty, *probability theory,* expresses uncertainty in terms of a measure on subsets of a given universal set of alternatives. The measure is a function that, according to the situation, assigns a number in [0,1] to each subset of the universal set. This number, called *probability* of the subset, expresses the *likelihood* that the desired unique alternative (the actual disease of the patient, the true answer to a historical question, the correct prediction, etc.) is in this subset.

Uncertainty expressed in terms of probabilities results from the *conflict* among the likelihood claims associated with the smallest nonempty subsets of the universal set, each consisting of exactly one alternative. Since these alternatives are mutually exclusive, nonzero probabilities assigned to two or more of them conflict with one another since only one of them can be the desired one.

5 Novel Uncertainty Theories

The first paradigm shift regarding attitudes towards uncertainty took place around the turn of the century, as mentioned in Sec. 3. The emergence of new theories of uncertainty, which initiated the second paradigm shift, can be traced to the 1960s or, perhaps, to the 1950s. This paradigm shift, which is still ongoing, is undoubtedly more radical and more profound than the first one.

It is generally agreed that an important turning point in the evolution of the modern concept of uncertainty was the publication of a seminal paper by Lotfi A. Zadeh [43], even though some ideas presented in this paper were envisioned by the philosopher Max Black [1]. In his paper, Zadeh introduced a theory whose objects–*fuzzy sets*–are sets with imprecise boundaries. Membership in a fuzzy set is not a matter of affirmation or denial, but rather a matter of degree.

The importance of Zadeh's paper was that it challenged the adequacy of classical set theory and probability theory as frameworks for expressing uncertainty. By allowing imprecise boundaries, fuzzy sets acquire the capability to express concepts of natural language that are inherently vague. They also acquire the capability to bridge, in whatever crude way, mathematics and empirical reality. However, these important capabilities are gained at the cost of rejecting the classical two-valued logic. To reject the two-valued logic, upon which all classical mathematics is based, was a bold move for Zadeh to make. It is now increasingly clear, some three decades after the paper was published, that the move was worth making.

The second challenge to classical types of uncertainty came from *fuzzy measure theory* founded by Michio Sugeno [37,38], even though some basic ideas of fuzzy measures were already present in Choquet capacities two decades earlier [2]. Fuzzy measure theory is an outgrowth of classical measure theory. It is obtained by replacing the additivity axiom satisfied by classical measures with weaker axioms of *monotonicity* (with respect to set inclusion) and *continuity* or, at least, *semi-continuity* [40]. By weakening the additivity axiom, fuzzy measure theory provides us with a considerably broader framework than probability theory for formalizing uncertainty. As a consequence, it allows us to capture types of uncertainty that are beyond the scope of probability theory.

The general notion of a fuzzy measure encompasses a number of special classes of measures, each of which is characterized by a special property, in addition to the required monotonicity and continuity (or semi-continuity). Some of the best-known and most utilized special measures are plausibility, belief, possibility, and necessity measures and, of course, the classical probability measures [40]. I assume that the reader is familiar with these types of fuzzy measures and the theories of uncertainty that are based upon them: the Dempster-Shafer theory of evidence [12,13,34] and possibility theory [6,44].

Sources of uncertainty in fuzzy set theory and fuzzy measure theory are fundamentally different, even though both deal with propositions of the general type "x is P," where x is an individual from a universal set of concern and P is a relevant property. In fuzzy set theory, the qualifications for being P are vague. In fuzzy measure theory, the qualifications are precise, but our information about x is insufficient to determine whether or not x satisfies them. That is, uncertainty associated with a fuzzy set results from the *lack of precise definition* of property P, while uncertainty expressed by a fuzzy measure results from the *lack of information* regarding the given individual with respect to property P. When fuzzy measures are defined on fuzzy sets (i.e., fuzzified), both of these types of uncertainty are involved.

Another uncertainty theory, complementary to fuzzy set theory, is the *theory of rough sets* [29]. A rough set is basically an imprecise representation of a given crisp set in terms of two subsets of a partition defined on the universal set involved. The two subsets are called a lower approximation and an upper approximation. The lower approximation consists of all blocks of the partition that are included in the represented set; the upper approximation consists of all blocks whose intersection with the set is not empty.

Fuzzy sets and rough sets model different types of uncertainty. Since both types are relevant in some applications, it is useful to combine the two concepts. Rough sets based on fuzzy partitions are usually called *fuzzy rough sets*, while rough set approximations of fuzzy sets in terms of given crisp partitions are called *rough fuzzy sets* [7,8]. Again, these two combinations model different aspects of uncertainty and, consequently, have different domains of applicability.

As obvious from this brief overview, the inventory of currently available theories of uncertainty is respectable. The novel theories, subsumed under the categories of fuzzy sets, rough sets, and fuzzy measures, as well as their various combinations, have emerged over a relatively short period of the last three decades or so. While the theories themselves are now developed rather well, our understanding of the relationship among them is still not adequate. However, to fully utilize the theories, one needs a *common framework* underlying the theories. For classical set theory and probability theory, this common framework is the *classical two-valued logic* and the associated *Boolean algebra*. For the various novel uncertainty theories, this framework is too narrow. To enlarge it, our only recourse are the various *non-classical logics*.

Fuzzy set theory has usually been conceived in connection with multivalued logics, in particular infinite-valued logics whose truth values cover the whole interval [0,1]. This connection has the following three major implications for fuzzy set theory:

(1) *fuzzy set operations are truth-functional*;
(2) *fuzzy set operations are not unique;*
(3) *any choice of fuzzy set operations violates some properties of the Boolean lattice underlying classical two-valued logic and classical set theory.*

While multivalued logics play a key role in foundations of fuzzy set theory, they are not relevant to other theories of uncertainty, unless these theories are fuzzified. It turns out, as a result of recent investigations [15,21,23,32,33], that modal logics are applicable to virtually all uncertainty theories and have thus a potential to become their unifying logic framework. When fuzzy set theory is conceived in terms of modal logics, each of the three above mentioned characteristics is exactly inverted. That is, all operations are locally unique (i.e., unique for each element of the universal set) and satisfy locally all properties of the Boolean lattice of classical two-valued logic. However, none of the operations is truth-functional, which means that different operations may be employed for different elements of the universal set.

Applications of fuzzy set theory are often combined with neural networks and genetic algorithms. Neural networks [17] have effectively been used for learning membership functions, fuzzy inference rules, and other context dependent patterns [24]. On the other hand, fuzzification of neural networks extends their capabilities and applicability.

Genetic algorithms are unorthodox search or optimization algorithms, which were first proposed by John Holland [18]. As the name suggests, genetic algorithms were inspired by the processes observed in natural evolution. They attempt to mimic these processes and utilize them for solving a wide range optimization problems.

Genetic algorithms [11] have successfully been utilized to deal with various optimization problems involving fuzzy systems, e.g., to optimize fuzzy inference rules in fuzzy controllers. On the other hand, fuzzified genetic algorithms tend to be more efficient and more suitable in some applications [24].

6 Measures of Uncertainty and Information

An important aspect of every uncertainty theory is the capability to quantify the uncertainty involved. This requires that we can measure, in a unique and adequately justified way, the amount of uncertainty involved in each possible characterization of uncertainty within the theory.

Well justified measures of uncertainty of relevant types are now available not only in the classical set theory and probability theory, but also in the theory of fuzzy sets, possibility theory, and the Dempster-Shafer theory. Their coverage, which is beyond the scope of this paper, is available in Refs. [20] and [14]; the latter reference contains some important new results that are not covered in the former, more extensive reference.

Uncertainty in a problem situation, which is expressed either by a set of alternatives in terms of nonspecificity or by a probability measure in terms of conflict, is always a result of *information deficiency* pertaining to the situation. Information may be incomplete, imprecise, fragmentary, not fully reliable, vague, contradictory, or deficient in some other way. Uncertainty can be reduced only by obtaining relevant information.

To obtain desired information, we have to take some *action*. The amount of information obtained by an action can be measured by the reduction of uncertainty that results from the action. In this sense, the *amount of uncertainty* (pertaining to a problem situation) and the *amount of information* (obtained by a relevant action) are intimately connected. Since this notion of information does not capture the full richness of information in human communication, it is appropriate to refer to it as *uncertainty-based information*. To measure the amount of information of this kind, we must be able, of course, to measure the amount of associated uncertainty.

Fuzzy sets, similarly as classical sets, are capable of expressing *nonspecificity*. In addition, however, they are also capable of expressing *vagueness*. Although nonspecificity and vagueness may be viewed as special types of uncertainty, they are different in a fundamental way. Nonspecificity is a result of information deficiency, as explained in Sec. 4. Vagueness, on the other hand, emerges from imprecision of definitions, in particular definitions of linguistic terms in natural languages. Vagueness of a linguistic term in a given language results from the existence of objects for which it is intrinsically impossible to decide whether the symbol does or does not apply to them according to linguistic habits of some speech community using the language.

7 Principles of Uncertainty

Once uncertainty (and information) measures become well justified, they can very effectively be utilized for managing uncertainty and the associated information. For example, they can be utilized for extrapolating evidence, assessing the strength of relationship between given groups of variables, assessing the influence of given input variables on given output variables, measuring the loss of information when a system is simplified, and the like. In many problem situations, the relevant measures of uncertainty are applicable only in their conditional or relative terms.

Although the utility of relevant uncertainty measures is as broad as the utility of any relevant measuring instrument, their role is particularly significance in two fundamental principles for managing uncertainty, the principles of minimum and maximum uncertainty. Since types and measures of uncertainty substantially differ in different uncertainty theories, the principles result in considerably different mathematical problems when we move from one theory to another.

The *principle of minimum uncertainty* is an arbitration principle that helps us to select solutions in certain problems involving uncertainty. The principle requires that we accept only those solutions, from among all otherwise acceptable solutions, whose uncertainty (pertaining to the purpose involved) is minimal.

A major class of problems for which the principle of minimum uncertainty is applicable are *simplification problems*. When a system is simplified, it is usually unavoidable to lose some information contained in the system. The amount of information that is lost in this process results in the increase of an equal amount of relevant uncertainty. Examples of relevant uncertainties are predictive, retrodictive, or prescriptive uncertainty. A sound simplification of a given system should minimize the loss of relevant information (or the increase in relevant uncertainty) while achieving the required reduction of complexity. That is, we should accept only such simplifications of a given system at any desirable level of complexity for which the loss of relevant information (or the increase in relevant uncertainty) is minimal. When properly applied, the principle of minimum uncertainty guarantees that no information is wasted in the process of simplification. There are many simplification strategies, which can perhaps be classified into three main classes:

- simplifications made by eliminating some entities from the system (variables, subsystems, etc.);
- simplifications made by aggregating some entities of the system (variables, states, etc.);
- simplifications made by breaking overall systems into appropriate subsystems.

Regardless of the strategy employed, the principle of minimum uncertainty is utilized in the same way. It is an arbiter which decides which simplifications to choose at any given level of complexity.

Another application of the principle of minimum uncertainty is the area of *conflict-resolution problems*. For example, when we integrate several overlapping models into one larger model, the models may be locally inconsistent. It is reasonable then to require that each of the models be appropriately adjusted in such a way that the overall model becomes consistent. It is obvious that some information contained in the given models is inevitably lost by these adjustments. This is not desirable. Hence, we should minimize this loss of information. That is, we should accept only those adjustments for which the total loss of information (or total increase of uncertainty) is minimal. The total loss of information may be expressed, for example, by the sum of all individual losses or by a weighted sum, if the given models are valued differently. Similar problems arise when we have to deal with several experts opinions that are mutually inconsistent.

The second principle, the *principle of maximum uncertainty*, is essential for any problem that involves *ampliative reasoning*. This is reasoning in which conclusions are not entailed in the given premises. Using common sense, the principle may be expressed by the following requirement: in any ampliative inference, use all information available but make sure that no additional information is unwittingly added. That is, the principle requires that conclusions resulting from any ampliative inference maximize the relevant uncertainty within the constraints representing the premises. This principle guarantees that our ignorance be fully recognized when we try to enlarge our claims beyond the given premises and, at the same time, that all information contained in the premises be fully utilized. In other words, the principle guarantees that our conclusions are maximally noncommittal with regard to information not contained in the premises.

Ampliative reasoning is indispensable to science in a variety of ways. For example, whenever we utilize a scientific model for predictions, we employ ampliative reasoning. Similarly, when we want to estimate microstates from the knowledge of relevant macrostates and partial information regarding the microstates (as in image processing and many other problems), we must resort to ampliative reasoning. The problem of the identification of an overall system from some of its subsystems is another example that involves ampliative reasoning.

The principles of minimum and maximum uncertainty are well developed and tested within the classical information theory, where they are called the *principles of minimum and maximum entropy.* Perhaps the greatest skills in using these complementary principles in a broad spectrum of applications has been demonstrated by Christensen [3,4.5].

Let me now turn to the third principle to be discussed here, the *principle of uncertainty invariance.* This principle, which was introduced recently by Klir [19], attempts to establish connections among representations of uncertainty and information in alternative mathematical theories. The principle requires that the amount of uncertainty (and information) be preserved when we transfer uncertainty formalized in one mathematical theory into an equivalent formalization in another theory. The most obvious examples are the transformation from a probabilistic formalization into an equivalent possibilistic formalization and the inverse transformation [9]. The principle of uncertainty invariance guarantees that no information is unwittingly added or eliminated solely by changing the mathematical framework by which a particular phenomenon is formalized.

The principle of uncertainty invariance, when properly developed, will allow us to use simultaneously several complementary theories for representing uncertainty and information. It will enable us to convert results obtained in one theory, most suitable for the purpose, into equivalent representations in the other theories. This will make all representations equally updated and ready for further use in an opportunistic fashion. That is, given a particular problem, we will use the theory most natural for dealing with it, and then, employ the uncertainty invariance principle for up-dating representations in the other theories.

8 SCAST: A Growing Branch of CAST

It is interesting to observe the main stages in the currently ongoing paradigm shift regarding uncertainty. They conform suprisingly well to the general characterization of paradigm shifts described by Kuhn [26]. According to Kuhn, each paradigm shift is initiated by emerging problems that are difficult or impossible to be dealt with under the current paradigm. When a new paradigm is proposed, it is initially rejected in various forms (it is ignored, ridiculed, attacked, etc.) by most established scientists in the given field. Among those who support the new paradigm, most are usually very young or very new to the field and, consequently, not very influential. Since the paradigm is initially not adequately developed, the position of its proponents is weak. The progress in developing the paradigm is slow since there is little or no financial support for required research. The paradigm eventually gains its status on pragmatic grounds by demonstrating that it is more successful than the existing paradigm in dealing with problems that are generally recognized as acute. In general, the greater the scope of a paradigm shift, the longer it takes for the new paradigm to be generally accepted.

All these stages can clearly be recognized in the paradigm shift initiated by the emergence of the novel uncertainty theories overviewed in Sec. 5, primarily fuzzy set theory and fuzzy measure theory. These theories emerged from the need to bridge the gap between mathematical models and their empirical interpretations, particularly in problems that have the characteristics of organized complexity [41].

The proposal of fuzzy set theory by Zadeh [43], which has a natural outcome of the need expressed at that time, offered a radically new paradigm for mathematics and initiated the usual process of paradigm shift. Contrary to some previous publications concerned with the same need (e.g., by Black, [1], or Choquet, [2], which passed virtually unnoticed, Zadeh's paper produced a reaction in the scientific community that is typical at the beginning of each paradigm shift. The concept of a fuzzy set was ridiculed and attacked with great hostility by many, while it was supported only by a few, mostly young and not influential. A more specific characterization of this initial period, based on personal recollections, was prepared by Zadeh [45]. In spite of the initial lack of interest, skepticism, and even open hostility, the new paradigm persevered with virtually no support in the 1960s, matured significantly in the 1970s, and began to gain support in the 1980s after its pragmatic utility had been demonstrated.

The paradigm shift is still on-going and it will likely take much longer than usual to complete it. This is not suprising since the scope of the paradigm shift is enormous. The new paradigm does not affect any particular field of science, but the very foundations of science. In fact, it challenges the most sacred element of the foundations -- the Aristotelian two-valued logic, which for millennia has been taken for granted and viewed as inviolable. The acceptance of such a radical challenge is surely difficult for most scientists; it requires an open mind, enough time and considerable effort to properly comprehend the meaning and significance of the paradigm shift involved.

At this time, there is already enough evidence to demonstrate epistemological and pragmatic superiority of the broad conception of uncertainty, which represents the new paradigm, over its narrow counterpart, based solely on classical set theory and probability theory. The following are some features that make the new paradigm superior:

1.   Fuzzy sets allow us to express irreducible measurement uncertainties and make these uncertainties intrinsic to empirical data. When fuzzy data are processed, their intrinsic uncertainties are processed as well and the results obtained are more meaningful than their crisp counterparts.

2.   As universal approximators, fuzzy sets offer far greater resources for managing complexity and controlling computational cost than their crisp counterparts. This is well demonstrated by the success of fuzzy controllers.

3.   Fuzzy sets and fuzzy measures have considerably greater expressive power than classical sets and measures and, consequently, they can

effectively deal with a considerably broader class of problems. For example, they allow us to deal with information expressed in natural language.

4.   Fuzzy sets and measures, as well as their combinations, have a greater capability to capture human common-sense reasoning, decision making, and other aspects of human cognition. When employed in machine design, the resulting machines are human friendlier.

There is no doubt that our attitudes towards uncertainty have changed in a fundamental way by emergence of the novel uncertainty in Sec. 5. Uncertainty has lost its negative connotation and we begin to see its great utility. It will undoubtedly play an important role in our continuing efforts to explore the world of organized complexity.

The CAST movement, whose purpose is to bridge system theories with computing will be certainly influenced by the growing significance of soft computing and soft system theories that utilize the various types of uncertainty. In fact, this influence has already been exercised since EUROCAST '93 [30] by the appearance of contributions dealing with systems based on fuzzy set theory or possibility theory, and it is my contention that it will steadily grow in the coming years until SCAST will become a dominant branch of CAST.

REFERENCES

1. **Black, M.** ,"Vagueness: an exercise in logical analysis." *Philosophy of Science*, **4**, 1937 , pp. 427-455 (reprinted in *Intern. J. of General Systems*,**17** (2-3), 1990 , pp. 107-128).

2. **Choquet, G.** ,"Theory of capacities." *Annales de L'Institut Fourier*, **5**, 1953-54 , pp. 131-295.

3. **Christensen, R.**,*Entropy Minimax Sourcebook, Vol. IV: Applications.* Entropy, Lincoln, MA, 1981.

4. **Christensen, R.** ,"Entropy minimax multivariate statistical modeling - I: Theory." *Intern. J. of General Systems*, **11**(3), 1985 , pp. 231-277.

5. **Christensen, R.** ,"Entropy minimax multivariate statistical modeling - II: Applications." *Intern. J. of General Systems*, **12**(3), 1986 , pp. 227-305.

6. **Dubois, D. and H. Prade**,*Possibility Theory.* Plenum Press, New York, 1988.

7. **Dubois, D. and H. Prade** ,"Rough fuzzy sets and fuzzy rough sets." *Intern. J. of General Systems*, **17**(2-3), 1990 , pp. 191-209.

8. **Dubois, D. and H. Prade** ,"Putting rough sets and fuzzy sets together." In: Slowinski, R., ed., *Intelligent Decision Support.* Kluwer, Boston, 1992 , pp. 203-232.

9. **Geer, J. F. and G. J. Klir** ,"A mathematical analysis of information-processing transformation between probabilistic and possibilistic formulations of uncertainty." *Intern. J. of General Systems*, **20**(2), 1992 , pp. 143-176.

10. **Gibbs, J. W.**,*Elementary Principles in Statistical Mechanics*. Yale University Press, New Haven (reprinted by Ox Bow Press , Woodbridge, Connecticut in 1981), 1902.

11. **Goldberg, D. E.**,*Genetic Algorithms*. Addison-Wesley, Reading, Mass., 1989.

12. **Guan, J. W. and D. A. Bell**,*Evidence Theory and Its Applications, Vol. 1*. North-Holland, New York, 1991.

13. **Guan, J. W. and D. A. Bell**,*Evidence Theory and Its Applications, Vol. 2*. North-Holland, New York, 1992.

14. **Harmanec , D. and G. J. Klir** ,"Measuring total uncertainty in Dempster-Shafer theory: A novel approach." *Intern. J. of General Systems*, 22(4), 1994 , pp. 405-419.

15. **Harmanec, D., G. J. Klir and G. Resconi** ,"On a modal logic interpretation of Dempster-Shafer theory of evidence." *Intern. J. of Intelligent Systems*, (in production), 1994

16. **Hartley, R. V. L.** ,"Transmission of information." *The Bell Systems Technical Journal*, 7, 1928 , pp. 535-563.

17. **Hertz, J., A. Krogh and R. G. Palmer**,*Introduction to the Theory of Neural Computation*. Addison-Wesley, Reading, Mass., 1991.

18. **Holland, J.**,*Adaptation in Natural and Artificial Systems*. Univ. of Michigan Press, Ann Arbor, 1975.

19. **Klir, G. J.** ,"A principle of uncertainty and information invariance." *Int. J. of General Systems*, 17(2-3), 1990 , pp. 249-275.

20. **Klir, G. J.** ,"Developments in uncertainty-based information." In: Yovits, M. C., ed., *Advances in Computers, Vol. 36*. Academic Press, San Diego, 1993 , pp. 255-332.

21. **Klir, G. J.** ,"Multivalued logics versus modal logics: Alternative frameworks for uncertainty modelling." In: Wang, P. P., ed., *Advances in Fuzzy Theory and Technology, Vol. 2*. Bookwrights Press, Durham, NC., 1994

22. **Klir, G. J. and T. Folger**,*Fuzzy Sets, Uncertainty, and Information*. Prentice-Hall, Englewood Cliffs, NJ, 1988.

23. **Klir, G. J. and D. Harmanec** ,"On modal logic interpretation of possibility theory." *Intern. J. of Uncertainty, Fuzziness, and Knowledge-Based Systems*, 2(2), 1994

24. **Klir, G. J. and B. Yuan**,*Fuzzy Sets and Fuzzy Logic: Theory and Applications*. Prentice Hall, Englewood Cliffs, NJ, 1995.

25. **Kruse, R. and K. D. Meyer**,*Statistics with Vague Data*. D. Reidel, Boston, 1987.

26. **Kuhn, T. S.**,*The Structure of Scientific Revolutions*. Univ. of Chicago Press, Chicago, 1962.

27. **Kyburg, H. E.** ,"Bayesian and non-Bayesian evidential updating." *Artifical Intelligence*, 31, 1987 , pp. pp.271-293.

28. **Moore, R. E.**,*Methods and Applications of Interval Analysis*. SIAM, Philadelphia, 1979.

29. **Pawlak, Z.**,*Rough Sets: Theoretical Aspects of Reasoning About Data*. Kluwer, Boston, 1991.

30. **Pichler, F. and R. Moreno Diaz,**, (eds.),*Computer Aided Systems Theory - EUROCAST'93*. Springer-Verlag, New-York, 1994.

31. **Pichler, F. and H. Schwärtzel,**, (eds.),*CAST Methods in Modeling*. Springer-Verlag, New York, 1992.

32. **Resconi, G., G. J. Klir and U. St. Clair** ,"Hierarchical uncertainty metatheory based upon modal logic." *Intern. J. of General Systems*, **21**(1), 1992 , pp. 23-50.

33. **Resconi, G., G. J. Klir, U. St. Clair and D. Harmanec** ,"On the integration of uncertainty theories." *Intern. J. of Uncertainty, Fuzziness, and Knowledge-Based Systems*, **1**(1), 1993 , pp. 1-18.

34. **Shafer, G.**,*A Mathematical Theory of Evidence*. Princeton Univ. Press, Princeton, N.J, 1976.

35. **Simon, H. A.**,*The Sciences of the Artificial*. M.I.T. Press, Cambridge, Mass., 1969.

36. **Smithson, M.**,*Ignorance and Uncertainty: Emerging Paradigms*. Springer-Verlag, New York, 1989.

37. **Sugeno, M.**,*Theory of Fuzzy Integrals and its Applications*. (Ph. D. dissertation). Tokyo Institute of Technology, Tokyo, 1974.

38. **Sugeno, M.** ,"Fuzzy measures and fuzzy integrals : A survey." In: Gupta, M. M., G. N. Saridis and B. R. Gaines, eds., *Fuzzy Automata and Decision Processes*. North-Holland, Amsterdam and New York, 1977 , pp. 89-102.

39. **Walley, P.**,*Statistical Reasoning With Imprecise Probabilities*. Chapman and Hall, London, 1991.

40. **Wang, Z. and G. J. Klir**,*Fuzzy Measure Theory*. Plenum Press, New York, 1992.

41. **Weaver, W.** ,"Science and complexity." *American Scientist*, **36**, 1948 , pp. 536-544.

42. **Yager, R. R., S. Ovchinnikov, R. M. Tong and H. T. Nguyen,**, eds.,*Fuzzy Sets and Applications - Selected Papers by L.A.Zadeh*. John Wiley, New York, 1987.

43. **Zadeh, L. A.** ,"Fuzzy sets." *Information and Control*, **8**(3), 1965 , pp. 338-353.

44. **Zadeh, L. A.** ,"Fuzzy sets as a basis for a theory of possibility." *Fuzzy Sets and Systems*, **1**(1), 1978 , pp. 3-28.

45. **Zadeh, L. A.** ,"The birth and evolution of fuzzy logic." *Intern J. of General Systems*, **17**(2-3), 1990 , pp. 95-105.

Fundamental Systems Concepts: "The Right Stuff" for 21st Century Technology*

Bernard P. Zeigler

Electrical & Computer Eng.
The University of Arizona
Tucson, AZ 85721

Abstract

The next century will be characterized by ambitious attempts to design, construct, or manage ultra-large systems such as high bandwidth global communication networks, flexible manufacturing systems with high autonomy, and ecosystems distributed over large geographical regions. Systems concepts and principles are needed to deal with such overwhelming complexity. Fads continue to flash on the scene and fade just as fast. The enduring advances have a fundamental robustness consistent with systems-based methodology.

The goal of Computer Aided Systems Technology (CAST)[16] is to "bundle" system theoretical problem solving techniques into user-friendly, easy-to-handle and easy-to-learn packages to meet the challenges of the future. This goal would be straightforward to achieve were it not for the still immature state of systems theory in relation to Grand and National Challenge problems. In this paper, we discuss two main areas where the deficiencies in systems concepts and methodologies are apparent: 1) proliferation of modelling formalisms, and 2) incremental model-based systems engineering. We close with a discussion of how two decendents of early cybernetics, computer science and systems research, can recombine to advance CAST.

1   Proliferation of Modelling Formalisms

Systems Theory owes its utility to the fact that real systems can obey the same "system" laws and show similar patterns of behavior although they are physically very different. This potential isomorpy makes it is possible to employ common representations to treat different real systems in a uniform manner [15]. Various systems theories have been developed [16, 8, 11, 25] to provide such integrative frameworks. Although not monolithic, the system theories are generally concordant. The problem is that a great variety of independently developed modelling and knowledge representation formalisms are entering into practical

---

*This research was partially supported by NSF HPCC Grand Challenge Application Group Grant ASC-9318169 and employed the CM-5 at NCSA under grant MCA94P02

use. To name a few: discrete event dynamic systems [7], fuzzy logic[10] and neural nets [9] are being heavily investigated in the control field, as are genetic algorithms[12], qualitative simulation[3], case-based planning[6, 23], reasoning under uncertainty[19] and non-monotonic logic[4] in artificial intelligence. Computer simulation offers an attractive alternative to develop and test systems when compared with real test bed environments[29]. To support such design, an ideal simulation environment would enable designers to select and experiment with a variety of formalisms and models[14]. This requires first, a simulation environment that can accommodate new formalisms, and second a means of embedding and integrating formalisms into it. Principles were proposed by Zeigler et al.[31] for unifying the various formalisms within a systems theoretic framework whose implementation is supported by the object-oriented paradigm. The goal of this approach is to enable models to be developed as instances of formalism-based classes of dynamic systems and to facilitate combining formalisms so that multi-formalism models can be constructed. An example is the STIMS environment[18] which combines the DEVS (Discrete Event System Specification) and DESS (Differential Equation System Specification) formalisms to support combined discrete/continuous modelling in modular, hierarchical fashion. The universality of system specification formalisms, such as the combined DEV&DESS, is their implementability in a variety of simulation environments, based on diverse underlying Object-Oriented languages such as CLOS and C++. This gives them not only the power of formal rigor but also the practical capability of application to real world complex systems.

2 Incremental Simulation-Based System Engineering

The ambitious systems designs of the future will require modelling and simulation support to achieve a multiplicity of objectives. Unfortunately, when approached with differing perspectives a real system may yield various, seemingly disparate, models, obscuring the underlying unity that binds these models together. As a result, cummulative development and reuse of models is discouraged at tremendous costs of wasted resources and hindered advance in system intervention capability. Specific shortcomings include: lack of coverage in modelling objectives and consequently of essential processes, multiplicity of alternative formulations for processes that are covered, incompatibility of micromodel interfaces and resolution levels, lack of validation and independent evaluation, and absence of documentation of underlying assumptions and theoretical bases. It is remarkable to the extent to which these complaints are echoed in various domains such as human performance modelling[27], battlefield combat[1], communications systems[24] and ecosystems[20]. There is ample evidence that most of the issues are generic in nature and endemic to current practice in complex system modelling domains.

Rather than consider each model as a distinct entity, multifacetted modelling methodology[28] attempts to organize models so that a coherent whole emerges. With such support, the construction of models to meet new objectives may be accelerated, since components already existing in the model base may be exploited. To gain full advantage of the knowledge in the model base, there must be provided a strong capability for representing the components of models, their variations and their interconnection. To facilitate synthesis, models must be readily dissassembled into components, and these must be able to be easily assembled into new combinations. The System Entity Structure/Model Base (SES/MB) framework[29, 30] within multifacetted methodology provides a basis for such model development and reusability. Multifacetted methodology can be considered as a well-founded systems-theory-based proposal for long term simulation model engineering and evolution[2]. The SES/MB framework offers a concrete architecture based on the fundamental requirements for standardized simulation environments [22]. It should be noted that the ability of object-oriented programming techniques to create and maintain libraries of model components is necessary, but not sufficient, to support multifacetted modelling methodology.

Although attention has been given to the essential role of model development in system design[26], there still is relatively little adherence by practitioners to such methodologies. The remoteness of general system formalism from the immediate problem context is one major factor in this lack of adherence. Computer-assistance, whether in the form of CASE (computer-aided software engineering, computer-aided system engineering) or CAST, could do much to bridge this gap, especially if based on multiformalism principles mentioned before. However, another problem is the lack of recognition in formal systems engineering methodology of the iterative nature of all human endeavor. This has been recognized in software engineering as the downfall of the "waterfall methodology" and the rise of more flexible object-oriented analysis and design methodologies[33]. Zeigler et al.[32] extended both the system engineering and multifacetted modelling methodologies by integrating into them an approach of successive approximation. In this approach, the incremental development methodology of systems engineering is adapted to the development of domain models. During each development cycle, a complete model is developed, tested, and validated. The method begins with the development of an initial domain model that is fairly comprehensive in scope, albeit limited in the range of behaviors it can accurately reproduce. Moreover, many processes in initial models may be depicted at low levels of resolution or may be absent altogether. Through successive increments of development, the behavioral range of the model is extended by increasing the resolution of the model. This incremental approach is more responsive to challenges offered by real world multifacetted systems development and therefore should be more amenable to adoption by practitioners.

3   Relation to Computer Science and Engineering

A recent prestigious U.S. national committee report[13] called for broadening the agenda for computer science and engineering (CS&E) to include collaborative research in multidisciplinary application domains. The report advocated deriving inspiration for identifying and solving problems in these domains in order to expand the inward looking disciplinary focus of CS&E research so dominant to-day. Despite an otherwise in-depth examination of the issues, the report **fails** to observe that historically, computer science found fertile ground in the interdis-ciplinary excitement of early cybernetics. Thus, rather than radically changing the direction of the mainstream, the broader agenda would actually tend to di-vert it back along its original course. This is important since another offshoot, systems science and engineering (SS&E), continues to champion the interdisci-plinary ideal to this day. To bridge the gap with other disciplines, CS&E need not start from scratch – it can learn from SS&E. Some rapprochement of this kind has always being going on, but the pace may be picking up. Automaton pioneer, Anil Nerode has identified hybrid systems as a fertile area for the com-puter science and control theory to combine forces[5]. Systems theory founder, Yasuhiko Takahara and collaborators are formulating logical underpinnings of systems theory to facilitate their application to information systems design[21]. Much remains to be done to create a theory that is both powerful and applicable to technology problems.

4 Concluding Remarks

Continued development of the systems concepts and methodologies in response to difficult, perhaps intractable, real world "problematique" is necessary for CAST to become an accepted reality in the practictioner communities. Com-bining the practical "can do" attitude of CS&E with the conceptual orientation and experience of SS&E should prove to be the "right stuff" for building tech-nology that is truly beneficial in the 21st Century .

References

1. Davis (1992), " Introduction to Variable Resolution Concepts", Proc. Variable Resolution Modelling Symposium, Rand Tech. Report."
2. Fishwick, P. A. (1990) "Towards and Integrated Approach to Simulation Model Engineering, Vol. 17(1)", Int. J. Gen. Sys. Research, pps. 1-20.
3. P.A Fishwick and B.P. Zeigler (1991)., Qualitative Physics: Towards Automated Systems Problem Solving, J. Exp. and Theo. AI, Vol. e, pp. 219- 246.
4. Genesereth M.R., and Nilsson N.L., (1987), Logical Foundation of Artificial Intel-ligence, Morgan Kaufmann Publishers, Inc.
5. R.L. Grossman, A. Nerode, A.P. Raun and H. RIschel (1993), (eds.) Hybrid Sys-tems, Lecture Notes in Computer Science. 736, Springler-Verlag, Berlin.
6. Hammond, K.J, (1989), "Case-Based Planning", Academic Press.
7. Ho, Y-C (1989), "Special Issue on Discrete Event Dynamic Systems" Proceedings of IEEE, Vol. 77, No. 1,
8. G. J. Klir (1985), Architecture of System Problem Solving. Plenum Press, New York.
9. Kosko, B.(1992), "Neural Networks and Fuzzy Systems: A Dynamical Systems Approach to Machine Intelligence. Prentice-Hall.

10. Lee, C.C. (1990), "Fuzzy Logic in Control Systems: Fuzzy Logic Controller -Part I," IEEE Trans. Sys. Man & Cyber., Vol. 22, No. 2.
11. M.D. Mesarovic, D. Macko, and Y. Takahara (1970), Theory of Hierarchical, Multilevel, Systems, Academic Press, New York, 1970.
12. Miachalewicz, Z (1992)."Genetic Algorithm + Data Structure = Evolution Programming", Academic Press, New York, NY.
13. Computing the Future: A Broader Agenda for Computer Science and Engineering, National Research Council, National Academy Press, Washington, DC, 1992.
14. Oren, T. and Zeigler, B.P. (1979). "Concepts for advanced simulation methodologies. Simulation, Vol. 32, No. 3, pp. 69-82, 1979.
15. F. Pichler, Mathematische Systemtheorie. Walter de Gruyter, Berlin, 1975.
16. Pichler F., and H. Schwartzel (1992). CAST (Computer Aided System Theory) Methods in Modelling, Springer Verlag, New York.
17. Praehofer, H.(1991),. Systems Theoretic Foundations for Combined Discrete Continuous System Simulation. PhD Thesis, Department of Systems Theory, University Linz, Austria.
18. Praehofer, H.(1993), "An Environment for DEVS-Based Multiformalism Simulation in Common Lisp/CLOS", J. Discrete Event Dynamic Systems. Vol. 3, 1993.
19. Shafer, G., and Judea Pearl (1990), "Uncertain Reasoning," Morgan Kaufman Publishers San Mateo, CA.
20. Sklar, F.H. and R. Costanza (1991) "The development of dynamic spatial models for landscape ecology: a review and prognosis," In: M.G. Turner and R. Gardner (eds). Quantitative methods in landscape ecology, Springer-Verlag Ecological Studies 82, New York, pp.239-288.
21. S. Takahashi and Y. Takahara (1995), Logical Approach to Systems Theory, Lecture Notes in Control and Information Sciences 204, Springer-Verlag, London.
22. Tanir, O and S. Sevinc (1993), "Defining the Requirements for a Standard Simulation Environment", IEEE Computer, December.
23. Tsatsoulis, C. and R.L. Kashyap (1988), "A Case-based System for Process Planning," Robotics and Computer Integrated Manufacturing, vol 4, no. 3/4, 557-570.
24. Wheeler, T.J. and Richardson (1993), "Object Oriented Methodology for Systems of Systems", Proc. AI, Simulation and Modelling in High Autonomy Systems , IEEE Press, San Diego.
25. A.W. Wymore (1976), Systems Engineering Methodology for Interdisciplinary Teams, John Wiley, New York.
26. A.W. Wymore (1993), Model-based systems engineering : an introduction to the CRC Press, Boca Raton.
27. Young, M.J. (1993), Successively Approximating Human Performance, (HR-TP-1993-0026). Wright-Patterson AFB, OH: Armstrong Laboratory, Logistics Research Division.
28. Zeigler, B.P. (1984) Multifaceted modeling and discrete event simulation, Academic Press, London.
29. Zeigler, B.P. (1990) Object-oriented simulation with hierarchical, modular models, Academic Press, Boston, MA.
30. Zeigler, B.P. (1992) A Systems Methodology for Structuring Framilies of Models at Multiple Levels of Resolution, in: Proc. Variable Resolution Modelling Symposium, Rand Technical Report.
31. Zeigler, B.P., H. Prahofer (1993), J. Rozenblit," Integrating Systems Formalisms: How OO Supports CAST for Intelligent Systems Design", J. Systems Engnrg.

32. Zeigler, B.P. M. Young, and S. Vahie (1994), "Successive Approximation in Multifacetted Modelling Methodology: Human Performance , Int. J. Simulation (to appear).

33. G. Wilkie(1995), Object-Oriented Software Engineering, Addison-Wesley, Reading, MA..

SYNERGY:
The Design of a Systems Engineering System, I

A. Wayne Wymore

Principal Systems Engineer, *SANDS*: Systems Analysis and Design Systems
Professor of Systems Engineering Emeritus, The University of Arizona
4301 North Camino Kino, Tucson AZ 85718 USA
Voice/FAX 520 299 6663, email: wayne@sie.arizona.edu

Abstract. *SYNERGY* is an acronym for *Systems Engineering System*. The subject of this first in a series of papers is the design of a generic *SYNERGY*, deployable within any private or public organization, whose implementation will consist of people, computers and software. The inputs to *SYNERGY* shall be demands for systems to be designed, built, tested, deployed, operated and retired - and evolved. The inputted demands for systems come from any segment of society for any kind of system, in any functional or technological context, from a widget to be produced for mass consumption to the national health care delivery system. The systems engineering methodology to be employed in the design of *SYNERGY* is the same as the systems engineering methodology that *SYNERGY* will be designed to implement: model-based systems engineering which, in turn, is based on mathematical system theory. *SYNERGY* is an example of computer-aided system theory (CAST).

1   Introduction

SYNERGY is an acronym for *Systems Engineering System*. The discipline of systems engineering requires the design of a generic *SYNERGY* for the following reasons, among others possibly:

*   to codify in operational form the responsibilities of systems engineering,
*   to define operationally the metrics necessary to assess quantitatively how well systems engineering is carrying out those responsibilities,
*   to provide specifications for the implementation of *SYNERGY* in diverse organizations in order to enable and foster the practice of systems engineering in a wider spectrum of contexts,
*   to provide specifications for the design and implementation of computer tools for more efficient practice of systems engineering,

- to clarify the concepts of management in systems engineering, management of systems engineering and the technical work of systems engineering.

SYNERGY, when implemented, will be an organization of hardware, software and bioware (which last term means users and operators as well as nonhuman biological materials needed, for example, in agricultural systems, and which, henceforth, will come first in the litany: bioware, hardware, software). *SYNERGY* will contain a subsystem, herein called *SANITY*, which undoubtedly will turn out to be characterizable as Computer Aided System Theory (CAST).

The methodology to be employed for the design of *SYNERGY* is the same as the methodology that *SYNERGY* will be designed to implement: model-based systems engineering as developed in [2], and, at a more elementary level, in [1]. See the bibliography at the end of the paper; [2] will henceforth be referred to as MBSE.

2 Model-based systems engineering

The design of *SYNERGY* is based on a fundamental definition: systems engineering is the intellectual, academic and professional discipline the principal concern of which is the responsibility to ensure that all requirements for a bioware/hardware/software are satisfied throughout the life cycles of the system.

The following paragraphs will clarify this definition and explore its implications for the practice of systems engineering.

2.1 System requirements

The requirements for a system to be defined are six:

- an input/output requirement,
- a technology requirement,
- a performance requirement,
- a cost requirement,
- a tradeoff requirement and
- a system test requirement.

MBSE provides a mathematical definition for each of these requirements. The crucial observation underlying model-based systems engineering is the following: To design a system is to create a model on the basis of which a real system can be built that will satisfy all its requirements. In order to construct a theory of systems engineering it is thus necessary to choose an appropriate set of system models.

Among the criteria for an appropriate set of system models are the following:

• *Closure under coupling.* The set of system models must be closed under the operation of taking resultants of system coupling recipes. A system coupling recipe is simply a specification of the system models to be coupled as components and a specification of which output ports of which components are to be connected to which input ports of which components. Coupling recipes and resultants allow the definition of hierarchical models principal among which are models of systems buildable from components and subsystems.

The set of system models must also support additional concepts:

• *Functionality preserving simplification and elaboration.* One system model is a homomorphic image of another system model if and only if they have the same functionality (but the homomorphic image system model might be simpler and/or "smaller" and the other may be "larger" or more elaborated).

• *Model isomorphism.* One system model is isomorphic to another system model if and only if the two system models are essentially the same except for notation and have the same functionality.

• *System modes of behavior.* One system model is a system mode, or a mode of behavior of a second system model provided the sets of states, inputs and outputs of the system mode are subsets, respectively, of the sets of states, inputs and outputs of the second system model and the dynamic behavior of the second system is consistent with the system mode in the following sense: For every state x and input p of the system mode, there exists an input sequence f of the second system model, beginning with p, and a time value t, such that the second system, started in x and operating on the input sequence f will arrive at time t at the next state of the system mode from state x with input p. This concept enables the precise definition of the implementation relation: A buildable system design (model) implements a functional system design (model) if and only if the buildable system design has a system mode of which the functional system design is a homomorphic image.

• *Modelling power.* It must be possible to represent the design of the system in all life cycle phases by members of the set of system models.

The structure of the system requirements is defined in terms of the chosen set of system models. In MBSE, for pedagogical as well as practical reasons, the set of system models chosen was the set of all discrete time state machines or automata. See also [4], Wymore (1967), wherein development of a more general class of system models is reported.

2.1.1 The I/O requirement: The I/O requirement consists of six parts:

• a specification of the length of the operational life and the time scale of the system to be designed (This sets a standard time scale for all input and output trajectories.),

• definition of the set of inputs that the system to be designed shall process, manage, survive , etc.,

• definition of the set of possible (or permissible or interesting) input trajectories (scenarios or sequences or histories) which the system to be designed might experience during its operational life,

• definition of the set of outputs which the system to be designed shall produce,

• definition of the set of possible (or permissible or interesting) output trajectories (scenarios or sequences or histories) for the system to be designed and

• definition of an eligibility function that specifies which output trajectories are possible (or permissible, mandatory or desirable, - later in the design cycle) of being produced for each input trajectory by the system to be designed.

2.1.1.1 Satisfaction of an I/O requirement. A system model satisfies the I/O requirement if the sets of inputs and outputs of the system model are the same, respectively, as the sets of inputs and outputs of the I/O requirement and there is a designated state such that, when the system model is started in the designated state and fed any input trajectory from the set of input trajectories specified by the I/O requirement, the system model will produce an output trajectory that is eligible to be produced from that input trajectory according to the eligibility function of the I/O requirement.

2.1.1.2 The functionality space. The subset of all system models that satisfy the I/O requirement is called the space of functional system designs or the functionality space.

2.1.2 The technology requirement: The technology requirement or, simply, the technology, consists of a set of system models each of which represents a physical (bioware, hardware, software) component available to be used to build the real system. In practice, the technology can be unconstrained by the customer or limited by specification of certain classes of components that may not be used or by specification of certain classes of components that must be used to build the system. Here also might be included certain requirements on the architecture of the system.

2.1.2.1 Satisfaction of the technology requirement: A system model is said to be buildable in a technology, that is, satisfies the technology requirement, if the system model is the resultant of coupling together any finite number of system models in the technology.

2.1.2.2 The buildability space: The subset of all systems models that are buildable in the technology is called the space of buildable system designs or the buildability space.

2.1.2.3 The implementability space: A functional system design together with a buildable system design that implements the functional system design is called an implementable system design. The set of all implementable designs is called the implementability space.

System functional analysis produces functional system designs; physical synthesis produces implementable system designs by defining buildable system designs that implement functional system designs (MBSE and [5]).

2.1.3   The performance requirement: Operationally, the performance requirement is an algorithm for comparing any two functional system designs. Mathematically, the performance requirement is a partial order, reflexive and transitive, over the functionality space. The performance requirement is defined in terms of a comprehensive set of performance figures of merit, functions over the functionality space, precisely, quantitatively defined.

2.1.4   The cost requirement: Operationally, the cost requirement is an algorithm for comparing any two buildable system designs. Mathematically, the cost requirement is a partial order, reflexive and transitive, over the buildability space. The cost requirement is defined in terms of a comprehensive set of cost figures of merit, functions over the buildability space, precisely, quantitatively defined.

2.1.5   The tradeoff requirement: Operationally, the tradeoff requirement is an algorithm for comparing any two implementable system designs. Mathematically, the tradeoff requirement is a partial order, reflexive and transitive, over the implementability space. The tradeoff requirement is defined in terms of a comprehensive set of tradeoff figures of merit, functions over the implementability space, precisely, quantitatively defined.

The tradeoff requirement must, in addition, have a relationship with the performance and cost requirements. Since every implementable system design consists of a functional system design and a buildable system design, any two implementable system designs can be compared with both the performance requirement and the cost requirement. If, in the comparison of two implementable system designs, the first is strictly more desirable than the second with respect to the *combined* comparison by the performance requirement and the cost requirement, then the comparison of the two implementable system designs by the tradeoff requirement must result in the first being more desirable than the second. This is the essence of the notion of tradeoff. It is when the result of the *combined* comparison by the performance and cost requirements is ambiguous that an independent decision can be made by means of the tradeoff requirement.

2.1.6   The system test requirement: The system test requirement specifies what experiments shall be run on the real system, what data shall be collected, how collected, stored and processed. The objective is usually, at the very least, to obtain, for the real system, estimates of the values of all performance, cost and tradeoff figures of merit. The system test requirement must also specify the criteria, in terms of the data collected from the real system, for decisions concerning:

- compliance of the real system with the system requirements,
- conformance of the real system to the implementable system design from which it was built and
- acceptability of the real system to the customer.

2.1.6.1 The testability space. Mathematically, the system test requirement is a function defined over the set of system test possibilities, called the testability space. Intuitively, each element in the testability space consists of an implementable system design together with a system model that represents a real system that might have been built on the basis of the implementable system design. The system model representing the real system must be exactly the same as the buildable system design with the exception of the state and output behaviors of the components.

The statement of the problem of the design of a system consists of explicit definitions of each of these six requirements. The statements of the problems of the design of subsystems are derived from the statement of the problem of the design of the system by the processes of system functional analysis and physical synthesis (MBSE and [5]).

These concepts provide the foundations for the production of the systems engineering documents which lead inevitably to a design of the system that satisfies all the requirements for the system and specifies the subsystem requirements necessary for full scale engineering development, phase 3 of a system life cycle.

2.2   System life cycles

Each system life cycle to which the above definition of systems engineering refers consists of seven phases:

- Requirements development,
- Concept development,
- Full scale engineering development,
- System development,
- System test and integration,
- System operation,
- System retirement.

2.2.1   System life cycle phase 1, requirements development: In system life cycle phase 1, requirements development, systems engineering develops a statement of the problem of the design of a system whose solution will meet the customer's need, comprehensively in human terms for communication with the customer, then, in more precise (formal, mathematical) terms, consistently, for communication with "downstream" engineering and management.   Systems engineering shows that the problem has a real world solution that will meet the customer's need. Systems engineering works with the customer and an interdisciplinary team [3] so constituted that no system requirement will be overlooked.

There are four work products that systems engineering must generate in phase 1:

- Systems engineering document 1: Problem situation,
- Systems engineering document 2: Operational need,
- Systems engineering document 3: System requirements,
- Systems engineering document 4: System requirements validation.

These documents may or may not be deliverables for a given contract, but they must always be part of the systems engineering data base from which the design of the system will emerge.

The contents of Systems engineering document 1, Problem situation for *SYNERGY*, will be discussed below.

2.2.2   System life cycle phase 2, concept development: In system life cycle phase 2, concept development, systems engineering chooses the system design concept which specifies, at the least, the basic subsystem architecture of the system to be designed and the basic technology within which the system will be implemented. Systems engineering develops "design to" specifications of these subsystems to be designed in detail by "subcontractors" ("in house" or "out house"), so that the designs of the subsystems can be coupled to constitute a design that will satisfy the top level requirements. In order to accomplish this, systems engineering works with the customer and an interdisciplinary team [3] so constituted that no attractive system design concepts are overlooked, no technological errors are committed and the subcontractors will work from specifications that are meaningful to them. Each set of specifications for subcontractors will include I/O, technology, performance, cost, tradeoff and system test requirements for the assigned subsystem as well as the system design concept. Appropriate system models are developed, utilized and recorded in documents produced by systems engineering during phase 2:

- Systems engineering document 5: Concept exploration,
- Systems engineering document 6: System functional analysis,
- Systems engineering document 7: Physical synthesis.

SYNERGY shall be designed to output these seven documents in the first two life cycle phases for every inputted system acquisition demand. These seven documents will be written *for SYNERGY* in the first two life cycle phases of the design of *SYNERGY*.

Among the subsystems of *SYNERGY* already identified in MBSE are *SANITY*, the systems analysis and trade study system and *REALITY*, the real system test system. *SANITY* is used throughout the life cycle phases subsequent to the first for the comparison of designs. *REALITY* is employed throughout the life cycle phases subsequent to the fourth for the testing and ongoing evaluation of the real system.

2.2.3 System life cycle phase 3, full scale engineering development: Bioware (human factors) engineers design user/operator/maintainer actions, develop recruitment criteria and training programs. Hardware engineers design components and parts and develop "build to" specifications for manufacturing. Software engineers design data structures and algorithms. Systems engineering coordinates on the basis of models developed in phase 2. Of primary concern is control of interfaces between subsystems. Systems engineering updates the data base of system models consistent with detailed designs. In the first three phases, the system itself exists only in terms of models or designs, yet it is still the responsibility of systems engineering to ensure that a system that might be built on the basis of these models, would satisfy all the requirements.

2.2.4 System life cycle phase 4, system development: Real system components are produced or purchased: Users, operators and maintainers are recruited and trained. Hardware is manufactured or acquired. Software is coded and installed. The real system is deployed. Systems engineering coordinates on the basis of models developed in phases 1-3. Systems engineering updates models consistent with "as built" bioware , hardware and software components.

2.2.5 System life cycle phase 5, system test and integration: Systems engineering has the prime responsibility again in phase 5. Systems engineering will have performed the systems engineering functions for the design of the subsystem, *REALITY*, which will be used to test the real system or prototype. Systems engineering supervises and coordinates integration of the physical subsystems into a system which will satisfy the customer's operational need. Systems engineering documents the test results and updates models for the "as integrated" real system.

2.2.6 System life cycle phase 6, system operation: Systems engineering monitors the performance of the real system, provides periodic reevaluation to ensure that the system continues to satisfy all its requirements. Systems engineering reviews proposed engineering changes with respect to overall system impact through archived system models, supervises installation of approved changes and updates models to be consistent with engineering changes.

2.2.7 System life cycle phase 7, system retirement and replacement: Systems engineering recommends retirement and replacement of the real system and writes the proposal for the replacement system on the basis of models of the real system. The next system life cycle begins. One of the requirements for the design of *SYNERGY* shall be to demonstrate a continuing decrease in the acquisition times for the subsequent incarnations of systems.

SYNERGY must be designed to provide all these life cycle services for every inputted system acquisition demand. All these life cycle systems engineering services will be provided *for SYNERGY*.

3   Systems engineering document 1: Problem situation of
**   *SYNERGY***

3.1   The top level system function of *SYNERGY*

To provide life cycle systems engineering services for inputted demands for system acquisitions and to encourage the use of *SYNERGY* by the systems engineering community.

3.2   History of the problem and the project to acquire *SYNERGY*

The National Council on systems engineering (NCOSE) has a special interest group on systems engineering tools. There is a great demand for systems engineering tools based on MBSE but progress was limited until the publication of MBSE. The sequel to MBSE is also needed: System functional analysis, Phase 2 of model-based systems engineering, due to be published by CRC Press in 1996.

3.3   Present system(s)

The NCOSE tools committee promises to publish a survey of extant systems engineering tools which will useful here.

There are many software packages on the market claiming to be systems engineering tools for various aspects of systems engineering. It is amazing how many organizations are designing and building systems engineering tools who do not seem to know enough about systems engineering to write a useful requirements document. It is also amazing how many organizations are shopping for systems engineering tools who do not seem to know enough about systems engineering to write a useful requirements document. Perhaps, forlornly, they hope to learn systems engineering from a tool developed by people who also no little about systems engineering.

3.4   The Customer groups of *SYNERGY*, sources of requirements

3.4.1   Owners of *SYNERGY*: So far, *SANDS*: Systems Analysis and Design Systems, will own *SYNERGY*.

3.4.2   Bill payers for *SYNERGY*: So far, the design of *SYNERGY* has been strictly a bootstrap operation without a client.

3.4.3   Users of *SYNERGY*: The number of people world-wide involved in system design and systems analysis is estimated to be on the order of millions and increasing.

3.4.4   Operators of *SYNERGY*: The operators of *SYNERGY* will be primarily student and professional systems engineers and systems analysts.

3.4.5   Beneficiaries of *SYNERGY*: All stakeholders in systems designed and acquired by *SYNERGY* will benefit.

3.4.6 Victims of *SYNERGY:* All "shoot from the hip" system designers however well meaning, and the many charlatans in the systems business.

3.4.7 Technical representatives to systems engineering of *SYNERGY:* Since *SYNERGY* has, as yet, no client, there can be no technical representative to systems engineering of *SYNERGY.*

3.4.8 Other stakeholders in *SYNERGY:* The market for systems acquisition, the full-scale engineering design industry, the industry for the production of system components and subsystems, the investment resources industry all have stakes in *SYNERGY.* See Interoperability below.

It has yet to be decided just how many representatives in each group enumerated will be interviewed, how they will be chosen and how contacted.

3.5   System environment of *SYNERGY*

3.5.1 Socioeconomic impact of *SYNERGY: SYNERGY* will provide jobs for systems engineers and will make enterprises more competitive through efficient and effective systems engineering.

3.5.2 Environmental impact of *SYNERGY:* Tremendous impact can be expected through the design of better systems for conservation and protection of the environment and by including complete requirements for the design of systems of every kind.

3.5.3 Interoperability of *SYNERGY: SYNERGY* will have the following interoperability conditions:

3.5.3.1 The market for system acquisitions: SYNERGY will generate outputs as inputs to the market for system acquisitions in the form of marketing probes, designs for systems, real systems, customer satisfaction probes and will receive input from the market for system acquisitions in the form of demands for system acquisitions, payment for designs and real systems and data concerning customer satisfaction.

3.5.3.2 The full-scale engineering design industry: SYNERGY will generate outputs as inputs to the full-scale engineering design industry in the form of system design problems for the detailed design of subsystems, interface coordination, payment for detailed designs/models of subsystems and will receive input from the full-scale engineering design industry in the form of detailed designs/models of subsystems, need for interface coordination.

3.5.3.3 The production industry (bioware, hardware, software): SYNERGY will generate outputs as inputs to the production industry in the form of detailed designs/models of subsystems, interface coordination, payment for real subsystems, components and deployment and will receive input from the production industry in the form of real subsystems, real components, deployed systems and need for interface coordination.

3.5.3.4 The investment resources industry: SYNERGY will generate outputs as inputs to the investment resources industry in the form of return on investment and operating reports and will receive input from the investment resources industry in the form of investment capital, perhaps other resources.

All these relations may generate requirements for SYNERGY.

3.6   The design of the systems engineering system for SYNERGY

We do not intend to get into the infinite regression: to design the system to design the system to design the system.... SANDS will supply all systems engineering services for the life cycle of the generic SYNERGY.

4   Conclusion

This is the first in a series of papers in which details of the design, acquisition, deployment, testing, operation and retirement of the generic SYNERGY will be presented. The next few papers will summarize systems engineering documents 2-7, then papers will be published devoted to life cycle phases 3-7 of SYNERGY.

5   Bibliography

[1]   Chapman, W. L., A. T. Bahill and A. W. Wymore (1992), *Engineering Modeling and Design*, CRC Press, Inc. 2000 Corporate Blvd., N. W., Boca Raton, FL 33431.

[2]   Wymore, A. Wayne (1993), *Model-Based Systems Engineering*, CRC Press, 2000 Corporate Blvd., NW, Boca Raton FL 33431 (1 800 272 7737)

[3]   Wymore, A. Wayne (1976), *Systems Engineering Methodology for Interdisciplinary Teams*, John Wiley, NY.

[4]   Wymore, A. Wayne (1967), *A Mathematical Theory of Systems Engineering: The Elements*, John Wiley, NY.

[5]   Wymore, A. Wayne (to appear in 1996), *System functional analysis, Phase 2 of model-based systems engineering*, CRC Press, 2000 Corporate Blvd., NW, Boca Raton FL 33431 (1 800 272 7737)

The Configuration of Complex Systems

Gillian Hill

Department of Computer Science, City University
and Department of Computing, Imperial College
of Science, Technology and Medicine,
London, Great Britain

Abstract. Foundational work from logic and category theory is applied
to the practical problem of constructing complex engineering systems
from their component parts. System configuration involves keeping the
history of system construction and is carried out by applying combinators
to recursively defined system components. The results of configuration
are represented precisely by constructing diagrams in a three-dimensional
development space. A new and simple module concept is defined to manage the complexity of large engineering systems by enabling a reusable
system component to be represented and reasoned about at any level of
development.

1 Introduction

A precise foundation to the early ideas of General Systems Theory was given
by Burstall and Goguen, in [2, 8], by the provision of a categorical semantics
for the specification language Clear with the meaning of a specification denoted
by its underlying theory. The activity of specification in software engineering
was viewed as theory-building by Turski, Maibaum and Veloso in [16, 17]; and
interpretation between theories was formalized in a categorial framework in [12].
These ideas are now extended to provide a mathematical workspace within which
system components can be both structured and implemented to configure the
architecture of a final executable system. We have presented a logical framework
for systems construction in [7] and defined the mathematical workspace over
first-order logic in [10].

Our configuration language, designed in [11], is at a meta-level to a specification language and describes the operations of combinators on the objects
that represent the parts of a system. We believe that the problem of defining
a concept for a system part, at all levels of development, is crucial to the work
of constructing a system; this problem is, in our view, a *real* problem in system
development. Although modularity has been described as an essential property
for the process of structuring complex systems, in both formal and informal approaches to software development, no clear and simple definition of a module
has emerged at this general level. A precisely defined module concept is required
for the top-down decomposition of a system either by functional decomposition
or by data-abstraction; it is also required for the bottom-up configuration of
a system by the object-oriented paradigm. We offer an original definition of a

module as a uniquely named instance of a system component that does not keep the history of its construction.

This paper is arranged as follows. In Section 2, we present an intuitive and original view of system development and contrast static and dynamic system structures. Based on this intuitive approach we identify relationships between the components of a system in Section 3. Section 4 represents the 'combining' relationships as combinators in the configuration language and presents our new module concept. The mathematical workspace for configuration is introduced within the framework of category theory in Section 5. Finally conclusions are drawn in Section 6.

2 The Development of a System

Our concern has been to design a language for system configuration that is both simple and natural for software engineers to use. In this section we consider system development at an intuitive level and contrast the static system structure, based on possible dependencies between components, with the architecture of a system configuration that identifies the parts actually used to build a system as well as the 'combining' relationships between them.

2.1 The Structure of a System

The work of identifying the relationships that can exist between the component parts of a system begins with an initial description of the structure of a system. When the system is viewed at the specification level the component parts are precisely described by specifications. The meaning of the terms 'specification' and 'specification language' are in turn precisely described by some specification theory. The interfaces between specifications define the relationships that can hold between specifications.

If these interfaces make references only to objects which are defined as 'belonging' to the specification that owns the interface, context independence of the component parts of the system is achieved. Component parts can then be joined together providing that their interfaces specify that the particular method of joining is possible. This is particularly important for building a complex system from existing reusable parts, since the actual building of the component parts can be separated from the specification of those parts. The system structure of the interconnected components can then be expressed in a 'configuration' language which is distinct from the many possible specification languages. At this early stage, therefore, we distinguish between an informal notion of system configuration and the activity of specifying the system parts.

The interfaces between specifications describe the *static* structure of a system, identifying the dependencies between the components, by a hierarchical structure. In contrast a hierarchical structure of the configuration of a system describes the system's *dynamic* structure and identifies the number of component parts that are actually used to build the system as well as the relationships between the parts. In either case, a hierarchical structure is appropriate for supporting the encapsulation of subcomponents and the information hiding that enables complexity to be contained.

System configuration frequently involves the use of several components of the same type. For example, a single specification for a queue might be needed to produce several queues as identical instances of the queue specification. These instances of a queue are to be used by different parts of the system: they are 'queue modules'.

A *module* is created from a specification which expresses the properties of some system component; the module is at the same level of abstraction as its specification and may be implemented later to form an executable software component. The module is available for use and is simpler than its specification because it does not carry information about the actual construction of the component part that it represents. Both specifications and modules are manipulated by the configuration language, but the module is actually used to build the system.

In summary we view the process of system development as consisting of the following main activities:

- the precise specification of component parts
- the combination of specifications to form structured specifications
- the creation of modules and the possible further combination of modules to form the complete system

With the exception of the work on Clear, [1], and the languages based on the syntax and semantics of Clear, most of the previous work on specification has concentrated on the decomposition of specifications into smaller units in order to make them more manageable. Emphasis should now be on the combining of specifications *and* of modules and the proof of properties about the complex specifications and the complex modules which are based on those of their component specifications and modules.

We gain simplicity in our work on system configuration by emphasising, as in [13], that '*any* statement (including an entire program) is a specification'. The distinction between formal languages for programming and specification is a purely pragmatic one.

2.2 Stages in System Development

In our view the building of a system from its component parts is at a meta-level to, and so independent of, the specification of the individual component parts. The specification of the components in a chosen specification language is, in turn, at a meta-level to the implementation of components by 'concrete' modules described in some executable programming language. It follows that the languages that express:

the building of the component parts to provide system configuration
the specification of the component parts

the implementation of the component parts as programs

belong to possibly different logical systems. The complete process of system development may therefore involve the construction of a sequence of different representations and the combination of those representations at different linguistic levels and within different underlying logics.

We separate out the levels of:

- describing a system initially in terms of theories within one of several possible logics
- specifying a system in one of several possible specification theories (or specification approaches), each of which may differ in their underlying logic and each of which may offer several possible specification languages
- configuring a system by using a set of combinators which are identified as belonging to one of several possible configuration languages and which operate on both specifications and modules.

Our aim to separate the theory-building operations (where theories are specifications) from any particular underlying logical system is shared by the work of Burstall and Goguen, using Clear, in [2], of building specifications in arbitrary institutions. The large number of logical deduction systems that have been used for reasoning about specifications, as well as the large number of specification languages used to write specifications, make a level of generalization essential when we consider the building of complex specifications.

Based on the structure of a many-sorted algebra, the work on institutions unified the algebraic approach to specification by formalizing the concept of a logical system underlying a specification. Whereas our work separates configuration from specification, the theory-building operations within the framework of institutions are built into specification languages, such as Clear, which then represent a family of specification languages, each over a particular logic. Institutions have not provided the tools that are appropriate for software development, however. In our view the algebraic framework is not appropriate for programming intuitions. We suggest that a proof-theoretic approach to specification, as in [16, 17], that is based on deduction rather than interpretation in a model is closer to the intuitions required for programming.

By separating configuration from specification and specification from its underlying logic, our aim is to provide a framework which is technically more useful than institutions. We are concerned with helping software engineers at a practical level to build structured specifications. The framework of institutions seems to us to offer no more than theoretical results about the completeness and soundness of specifications. Whereas a Clear specification may express the refinement of a complex specification, our own configuration language will also express the separate refinement of simpler specifications and the final implementation by a module that represents the combination of those refined specifications. Whereas in Clear a theory is an algebra, we work with theory presentations that have a semantics provided simply by the underlying theory; composability of refinements does not then depend on the additional work of composing algebras.

2.3 The Basic Objects in a System

We have not yet made clear our idea about the nature of the objects in our model of system development, but we do know that we require both of the following sorts of object:

– specifications as named and finite representations of theories
– modules as the basic building blocks for the actual system

Since specifications belong to some formalism, the module concept cannot be independent of that formalism. It is the *creation* of an instance of a module from a specification that is the action of 'using' a specification and may be independent of the formalism. We assume the existence of an operation to *create* a module which is uniquely named after its specification:

create(queue specification)= queue module

The repeated application of the above create operation will result in the same module. In a real system, however, our create operation is complicated by the need to create more than one module from each specification. We therefore define the create operation as a binary operation on the template specification, and on the string of names of modules previously created from that specification. The first module will be created by

create(specification, λ)= module1

where λ represents the empty string of previously created module names. Our create operation, defined on the set of queue specifications is now,

create: queuespecification $\times$ queuemodules* $\rightarrow$ queuemodule

Our operation to create a module must be defined for all specifications from the most abstract description of a system component to the concrete description of that component in an executable programming language.

3 Relationships between System Components

In this section we identify the relationships that exist between components, as the system objects, in order to define the high-level combinators for the configuration language.

3.1 Specifications and Modules

The creation of a module from a specification belongs to the dynamic structure of a system, and will produce a relationship between one specification and those module instances that are 'created from that specification'. In contrast the definitional dependency between modules that are created by the same specification is part of the static structure of a system. We now increase our intuitions concerning further relationships between the objects in a system by using simple hierarchical diagrams to express the relationships that we identify.

The definitional dependency in a system between each uniquely named module and the named specification that is its class or structural template can be represented diagrammatically by a dependency-hierarchy for module creation.

Example 1. In the following dependency-hierarchy the box drawn round the queue module indicates that it is a module and therefore a usable object in the system:

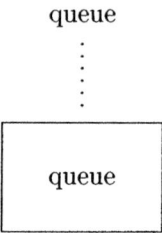

In contrast the relationship of actually creating several uniquely named modules from a specification belongs to the dynamic structure of a system.

Example 2. The creation of several instances of a queue module from its specification can be illustrated by solid lines in the use-hierarchy:

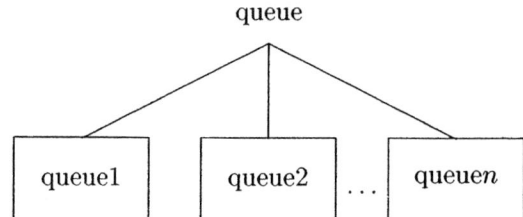

3.2 Horizontal Structuring

An important technique in the building of complex systems is the use of components *already* built to provide the basic properties required by more complex components. This greater complexity is achieved by the addition of extra properties 'horizontally', without changing the level of abstraction of the component within the stages of system development; also in the initial development of a system the client may wish to extend the information in the original requirement by adding extra properties. In contrast 'vertical development' refers to the addition of detail as an abstract component is implemented to a more concrete component. The structural complexity of components, represented by system objects, can be increased by:

- adding a group of extra properties to an object,
- adding, or including, another object to give the extra properties,
- using another object
- fitting objects together to form a composite object
- genericity.

The addition of a set of properties, or an entire object, to another object is identified as the *extension* of that former object. The use of another particular object to give extra properties is by the *parameterization* of one object by another object. However, parameterization is seen to mean the 'fitting together' of objects which are of equal importance as well as the master-slave hierarchical notion of one object 'using' another by 'leaving a hole' in its definition that is to be filled by another object. Abstraction over the sort of object that can 'fill a hole' in a definition allows parameterization with genericity.

Extension Adding extra properties enhances the basic properties of an object but does not fundamentally change the object.

Example 3. As the size of a family increases it may be useful to extend the family home by building on a bungalow for mother-in-law. This horizontal structuring is illustrated first between the specifications and then between modules.

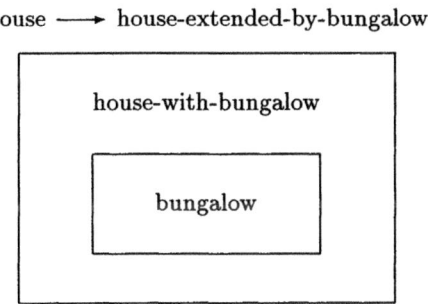

If a specification is included by name in order to extend another specification, not only are the extra properties of the included specification made available to the specification that is extended, but there is also the potential for further abstraction. This is because there is the possibility of extending by more than one module instance of the included specification. This notion of extension therefore provides data abstraction by the inclusion of abstract data type specifications by name and the creation of several module instances of the abstract data type.

If a module is first created from some specification *and then* included in another specification, or in a module created from that specification, only the operations of the single included module are available for use. By importing a module we lose the potential for data abstraction and merely gain data encapsulation for the imported abstract data type.

Parameterization The notion of using an object by parameterization involves a minimal requirement that an object is suitably matched to use another object. Equally the object to be used must fit the requirement of the object using it. Some objects may fit the requirement exactly, others may offer other properties not required by the particular using object. Checks must be made concerning the fitting of the requirement for a parameterized object; rules must take into account whether the properties possessed by an object, but not actually required by the using object, will in some way corrupt, or contradict, the properties of the using object. These checks, the same as those made when an object is extended, represent the fine tuning of system construction.

The object parameter formally represents a hole in the parameterized object's definition; in genericity this is filled by a suitable 'actual' object. The two objects are at the same level as far as the distance between an object and its implementation is concerned. Parameterization is another 'horizontal operation' that can be expressed in terms of either specifications or modules.

Example 4. In the following dependency-hierarchy the static structure expressing the use of a queue identifies use-queue as that part of the parameterized specification that is not the queue, but contains the rules for using the queue. The queue, as a parameter, defines the hole (that must be filled) in the parameterized specification.

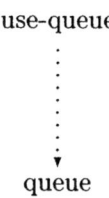

use-queue

queue

The dotted downward arrow indicates the *intended* relationship of use, or parameterization, between specifications, that is part of the static structure of the system.

The actual relationship of 'use' between specifications and between modules is part of the dynamic structure of a system and can be illustrated by a solid downward arrow in a use-hierarchy.

Example 5. The structuring of 'use-queue[queue]' is expressed in the following use-hierarchy:

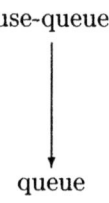

use-queue

queue

Our simple module concept allows the sharing or non-sharing of component parts to be illustrated explicitly by use-hierarchies. The interfaces that must be checked are identified in the diagrams as strips in the boxes that represent the modules.

Example 6. The configuration of two clients that share a mailbox between them is given by the use-hierarchy:

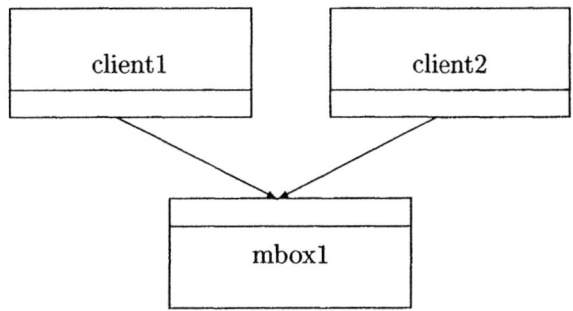

The creation of an additional module from the specification of a mailbox is needed for the configuration of two clients that use their own mailboxes.

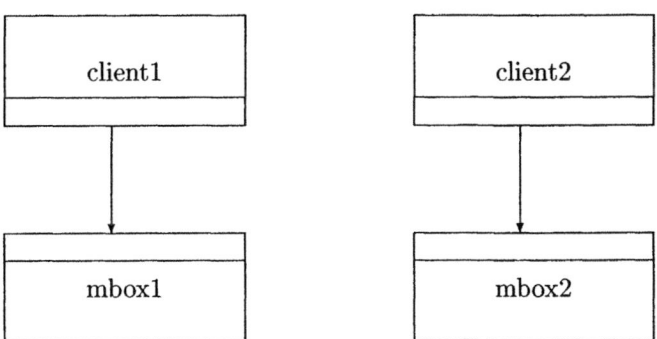

Parameterization by use relates a 'master' object to a 'slave' object. The matching of interfaces is then solely concerned with the minimal amount of information that is related to this use. Parameterization has the broader meaning, however, of *composing* objects, or fitting objects together to form a more complex composite object; the matching of interfaces may then require the checking of much larger amounts of information. A composite object can be pictured intuitively as containing holes into which the objects that it is composed of are fitted. For example, a queue 'is-made-of' elements (of the same sort) that are fitted together by being added or removed according to a particular discipline. When we configure a system by the 'bottom-up' combination of its component parts, we give a new name to each composite object as it is configured.

Example 7. A room as a composite object is a structure parameterized by the objects: ceiling, walls and floor, to form 'room[ceiling, wall, floor]'. The configuration of a room involves the fitting together of the components of the room according to the rules in its specification; the parameters identify the objects *required* for the structuring. A floor, four walls and a ceiling when fitted together 'are-a' room.

Parameterization has added power, however, if the structures formed are made more general by abstracting over the sorts of parameter that can fill the defined holes. Instead of specifically defining a queue of people we define a queue as a structure for storing any sort of object.

Example 8. A house can be viewed as an object parameterized by at least one room with the structure 'house[room]'. The actual configuration of the specification for a two-roomed bungalow could be given by a use-hierarchy which expresses the instantiation (represented by a solid vertical line) of two types of rooms as the generic parameters to the house.

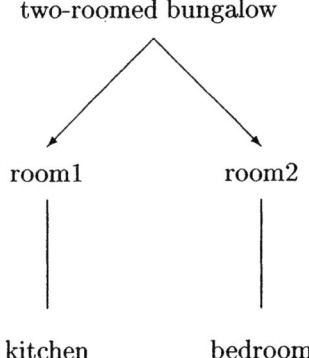

The vertical step of instantiating the generic parameters adds concrete details about the actual rooms in the bungalow.

3.3 Vertical Development

We now consider a way of structuring which adds detail to objects and results in a decrease in the level of abstraction. The process of vertical development finally results in a concrete executable object.

The implementation of an object by another object is a further relationship, that of realization, which can be described hierarchically. The implementing object adds more detail and makes the description in the specification of the implemented object less permissive. The purpose of adding detail may be to introduce those implementations which translate the language of the specification into a language more suited for execution by a machine. The realization of a specification by a program may take many implementation steps: each implementing specification *simulates* the properties of some more abstract specification.

Example 9. A parameterized queue instantiated for storing characters may be implemented by a linked list, an array or a binary tree. In turn, these implementing objects may be implemented by pointers in Pascal or arrays in Modula–2. We express some of these possible relationships by the dependency-hierarchy:

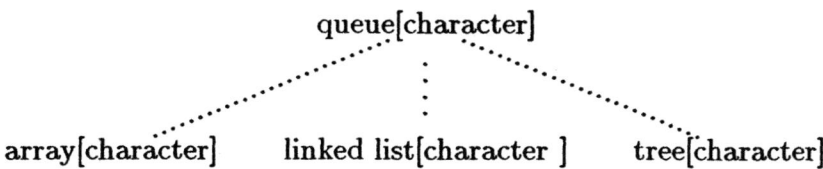

The dependency-hierarchy in Example 9 describes both a relationship of genericity and a relationship which is conceptually close to the relationship of inheritance in the object-oriented approach (expressed by Meyer in [14]). Both genericity (parameterization over types) and inheritance (allowing successive extension and specialization) are identified by Meyer as relationships that make software components more extendible and reusable. Our own aim is to add greater flexibility to our objects for configuration by providing both: generic objects which are open to variation *and* can be used as modules; and instantiated objects which are usable *and* open to refinement. We achieve this flexibility, over many possible levels of development, by distinguishing between parameterization, implementation and the creation of module instances.

Wirsing, in [20], has identified the dependency-hierarchy in Example 9 as a *reusable component*, defined as a 'top-down program development tree, where a specification is a child of another specification if it is an implementation'. We generalize on Wirsing's approach by using hierarchical structures to represent objects formed by successive applications of any of the combinators in our configuration language; the modules are reusable.

Example 10 (from Example 9). A use-hierarchy illustrates the actual implementation of one instantiated specification by another instantiated specification followed by the creation of a module instance.

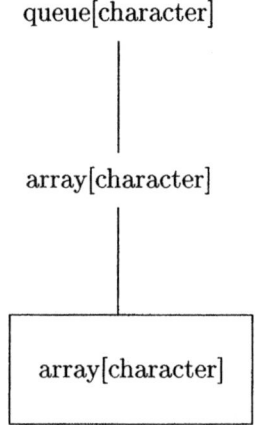

The configuration of the module binary tree[element] could be similarly described by a separate use-hierarchy. Alternatively the creation of two module instances of queue[character] could be followed by the implementation of those modules in different ways as expressed by the use-hierarchy:

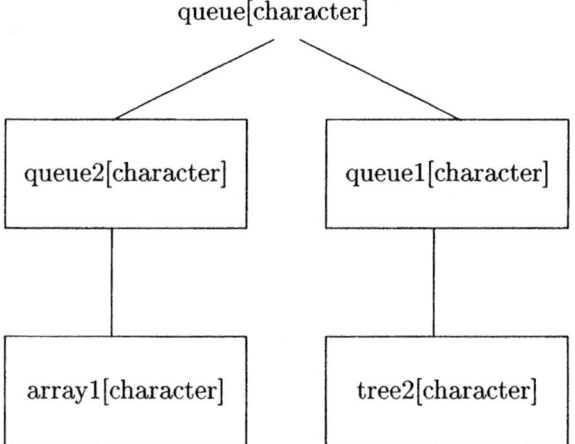

4 The Configuration of System Components

From the 'combining' relationships between system components we now identify the combinators that progressively configure the components, represented by objects in a uniform development space, into the final system.

4.1 The Create Combinator

We identify the dynamic relationship of modules created from their textual specifications as a combinator in the configuration language. Our distinction between dependency-hierarchies (for description) and use-hierarchies (for configuration) is more general than that made by Meyer, in [14], for his language Eiffel. Meyer distinguishes between description and execution:

- classes are static entities, or types, in the program text
- objects, as instances of a type, are purely dynamic in the memory of the computer.

We blend the uniformity of objects with the distinction between specifications (descriptive sorts of objects) and modules (as instances which are usable at *any* level of abstraction). Our modules are not dynamically created at run time by the instantiation of parameterization. Instead any number of module instances can be created from a (possibly configured) specification at any stage in system configuration. Modules may be created from primitive specifications before they are configured, or alternatively from complex specifications at the end of the configuring process.

The simplicity of our module concept contrasts favourably with the view of a module which is currently held in the algebraic approach after fifteen years of research into algebraic techniques for specification, In this view the module is a 'realization' of an abstract data type that may be implemented by further development steps until it is a concrete module. A precise definition is given by Ehrig in [3, 4, 5, 6] of a module as a realization of a module specification: the module specification consists of four algebraic specifications and four specification morphisms. We are concerned that modules with such a complex algebraic structure may not be an appropriate choice for software engineers to manipulate.

4.2   The Horizontal Combinators

We identify the relationships of extension and parameterization as the horizontal combinators for our configuration language; they operate on objects with sorts from the set {*specification, module*}.

Extension Intuitively it seems that for this combinator we want to be able:

- *either* to include a specification by name and then to create a module instance of the extended specification
- *or* to create a module instance of a specification and then to include that module instance.

This requirement that both specifications and modules can be included will bring a 'smoothness' into our language and will enable the user of the language to choose the order in which the combinators for the extension and the creation of modules are used. A result of this requirement is that both specifications and modules must be available in every object, which forms an inner block of scope, as the building of the system proceeds. We cannot restrict the inclusion of objects to be the inclusion of the module part only, since this would leave all specifications at the outermost level of scope; modules could then only be created at this outermost level *before* being imported to inner blocks of scope.

It is important to note that during the initial stages of system configuration the extension of a system object by a client may result in the deduction of new properties about the original object in the extended language. During the later stages of structuring objects by extension, however, it is vital to ensure that the extended object is not spoilt or tampered with by the combinator for extension. An extension which preserves the properties of the original object without allowing any new properties to be deduced in the extended language about the original object is called a *conservative extension* in [16]. Our 'planning requirement' for structuring systems is to build conservatively.

Example 11. We would surely agree that a desirable property of any suburban house extension is that it should not introduce changes in the house itself. An extension which adds a garage, an extra bedroom or even a granny-flat would not be expected to change the house into a factory or a shop. Such an extension would bring an unwelcome inconsistency into suburbia! If extensions are allowed in a suburban estate they must be conservative and not disturb the other houses.

Like Goguen, in [9], we have separated the importing (including) of modules (defined by our combinator for extension) from parameterization. Unlike Goguen we keep the notion of 'using' an object to a very general and intuitive level. In the literature 'use' has been given many meanings ranging from parameterization to a specific form of importing a module, as in [9], that is without any guarantees against junk or confusion. It seems that in the algebraic approach to specification 'use' is closely linked with what is identified by Wegner as the communication paradigm for modular programming languages, [19]. This involves only a partial interface for communication since it is restricted to the parameter part of the interface. Algebraic modules are composed horizontally by being connected at these partial interfaces; they are developed vertically by realization and refinement.

We broaden our own notion of 'use' to one that is intuitive and appropriate over all levels of software development. Extension of objects by inclusion provides complete access to the included properties. In terms of the access provided our extension relation is closer to inheritance rather than to clientship in the object-oriented approach. Clientship involves a restricted interface through which operations and data types are imported by name. Similarly in the horizontal composition of algebraic modules, described by Weber and Ehrig, in [18], modules related by 'use' are interconnected by importing names through a restricted interface. For configuration we choose inclusion as the basic mechanism for the extension of objects believing it to be both more powerful than importing through a partial interface and more appropriate for abstract specifications and modules.

Parameterization This combinator adds horizontal structure by using another object; by fitting objects together to form a composite object; or by genericity. The parameterized object is defined as an *extension* to the parameter, which only defines the hole in the object; the checks that must be made to ensure that the parameter does not corrupt the containing parameterized object are the same as those made by the combinator for extension to ensure conservativeness.

Parameterization by generic parameters is broken down into two more primitive relationships:

- the generic parameter is *conservatively extended* to the parameterized specification
- the generic parameter is instantiated by the actual parameter which also fits into the parameterized specification; instantiation as a vertical step, is made by the primitive operation of interpretation.

4.3 The Vertical Combinator for Implementation

We choose implementation as a high-level combinator for the vertical development of objects.

Traditionally there has been an unfortunate tendency to 'oversimplify' in software development by giving the specific implementation of an object at an early stage. We avoid this tendency by constructing objects which are abstract and not yet implemented, even though they have been structured horizontally by extension and parameterization. Each development step will be small, and easily comprehended, as recommended in [16]; each step will also be recorded in the textual specification part of the object.

It is important to emphasise the distinction between the meaning of 'concrete' in the Smalltalk system, [15], as instantiated by actual parameters and, in this sense, ready for use, and our own meaning of a concrete object as the implementation of some more abstract object.

The vertical step made by the implementation combinator is based on the underlying operation of interpretation to translate from the language of the object that is to be implemented to the language of the more concrete object. An extension must also be made to the language of the implementing object so that all the properties of the more abstract object can be translated by the interpretation. Implementation therefore depends on a pair of underlying operations: interpretation and extension. Whereas interpretation results in the deduction of new properties about an object, however, the extension to the language of the implementing object must not result in the deduction of any new properties about that implementing object. It is vital, as for parameterization, that this underlying extension is 'conservative' in order to avoid the corruption of the extended implementing object.

4.4 The Primitive Combinators

We have noticed that parameterization and implementation are themselves composed of more basic operations that we now identify as primitive, or low-level combinators:

- $c \, . \, e$ denoting a conservative extension between objects that preserves the properties of the extended object and does not allow the deduction of new properties about the extended object in the extension

- *e* denoting extension between objects that may be non-conservative
- *i* denoting interpretation between objects that adds details of primitive operations by translating to a new and more concrete language.

Since the extension of an object involves the *inclusion* of the language of the extending object in the language of the extended object, it is a special case of translating to a new language. Within our mathematical workspace for configuration, therefore, the primitive operation of interpretation underlies the semantics of all our combinators when we include the operation of extension of an object as a special case of interpretation. As a primitive combinator interpretation gives a precise definition to the higher-level and more user-friendly combinators.

A result in [16] concerning the *stability* of a modular system puts a requirement on the underlying logic for our configuration theory. Any logic that satisfies the Craig interpolation property (based on the Craig interpolation lemma) is shown to preserve the stability of the structuring mechanisms used for modularity. The notion of a conservative extension as the *safe* way of enlarging an object is fundamental, therefore, to the achievement of modularity in system configuration.

The important implication of such stability in a modular system is that data in the interfaces between component parts of a system is preserved under interpretations. The Craig interpolation property guarantees that if an interpretation is made between specifications, a conservative extension of the original specification is translated, under the interpretation, to an extension of the interpreted specification which is also conservative.

In summary, the high-level combinator for extension is defined in terms of e or $c \cdot e$, whereas parameterization is defined only in terms of $c \cdot e$. The instantiation of generic parameters is defined in terms of i. Implementation is defined by the pair $(i, c \cdot e)$. Configuration to form a concrete object involves choosing appropriate $(i, c \cdot e)$ pairs and composing implementations that have already been configured; composition is dependent on the property of Craig interpolation in the underlying logic.

5 Configuration within a Mathematical Workspace

Objects are configured precisely in a mathematical workspace by operations on categorical diagrams and and by constructing the colimit of the resulting diagram. A graphical tool in the computer aided environment for system configuration will provide a user interface for building diagrams. Each new diagram represents the environment of the object being configured and is labelled with 'simpler' objects at the nodes and primitive combinators as the arrows. Underlying each diagram is a simple shaped graph, of nodes and arrows, that expresses, by its shape, the result of the high-level combinator that is being used to build the specification. Configuration is completed by first constructing the colimit object at the vertex of the cocone over the diagram and then flattening the cocone to form the extended diagram that represents the specification of the new configured object. The shape of this extended diagram is added to the environment and is available for further configuration.

In our examples of configuration we label the nodes of diagrams by particular objects with sorts in the set $\{s, m\}$, and the arrows by the particular primitive combinators operating on the objects. This enables us to distinguish between configured objects which have the same shape but have different actual specifications at the nodes.

Example 12 (from Example 8).

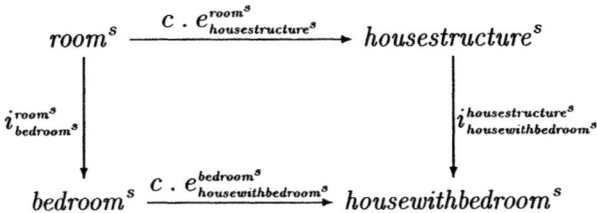

is the pushout diagram representing the instantiation of a room by a bedroom in the specification of a house. The colimit object, housewithbedroom^s, at the vertex of the cocone is made into a single node diagram and the whole diagram is then flattened to the new extended diagram that represents the specification for the configured object 'house-with-bedroom'.

The advantage of applying categorical constructions in the system development space is that connections between objects are made precisely. The amalgamated sum joins objects by identifying different representations of common subobjects; it forms a 'best' cocone. The construction of the colimit of the diagram representing a configured specification creates a unique module instance that is structurally identical to the specification. We fix as our configuration theory the category of all first-order theories and theory morphisms between them, paired with its sub-category of all conservative theory morphisms. Our definition of the theory for configuration ensures that the structure of objects that are conservatively extended is preserved under subsequent interpretation. Building systems precisely means building systems safely by preserving the structure and properties of the system components.

6 Conclusions

We have suggested a solution to the practical problem of constructing systems-in-the-large. Our solution provides an architecture for the configuration of systems that is independent of the variety of approaches that exist for the specification, design and coding of systems. The key idea is to focus on configuring systems from reusable modules that can be created and reasoned about at any stage in system development. Each module is precisely defined as an instance of its textual specification.

By identifying explicit combinators between system objects rather than one general notion of compositionality based on 'use' between modules, we have simplified the module construct itself to be an instance of a specification. The module does not contain structural information; this belongs to the combinators that are applied to the recursively defined system objects. Our notion of a module is not, therefore, an implementation of an abstract data type; it can itself be implemented by the combinator for implementation. In this paper we have indicated only briefly how system configuration can be defined precisely within a mathematical workspace.

Acknowledgments

The many discussions with Tom Maibaum have been very valuable. Anthony Finkelstein, Chris Hankin and Steve Vickers have also made helpful suggestions. Paul Taylor's macros were used for the diagrams.

References

1. R. A. Burstall and J. A. Goguen. An informal introduction to specifications using Clear. In Boyer and Moore, editors, *The Correctness Problem in Computer Science*. Academic Press, 1981.
2. R. M. Burstall and J. A. Goguen. The semantics of Clear, a specification language. In *Abstract Software Specifications, LNCS 86*. Springer-Verlag, 1979.
3. H. Ehrig, W. Fey, H. Hansen, M. Lowe, and D. Jacobs. Algebraic concepts for software configuration development. Technical Report 88–19, Technical University, Berlin, August 1988.
4. H. Ehrig, W. Fey, H. Hansen, M. Lowe, D. Jacobs, and A. Langen. Algebraic specification of modules and configuration families. Technical Report 88–17, Technical University, Berlin, August 1988.
5. H. Ehrig, W. Fey, H. Hansen, M. Lowe, and F. Papisi-Presicce. Algebraic theory of modular specification development. Technical report, Technical University, Berlin, 1987.
6. H. Ehrig, F. Papisi-Presicce, W. Fey, and E. K. Blum. Algebraic theory of module specifications with constraints. In *Proceedings MFCS LNCS 233*, pages 59–77, 1986.
7. N. Fenton and G. Hill. *Systems Construction and Analysis: A Mathematical and Logical Framework*. McGraw-Hill International (UK), 1993.
8. J. Goguen. Categorical foundations for general systems theory. In *Advances in Cybernetics and Systems Research*, pages 121–130. Transcripta books, 1973.
9. J. Goguen. Principles of parameterized programming. Technical report, Stanford University, August 1987. Draft Version.
10. G. Hill. Category theory for the configuration of complex systems. In T. Rus M. Nivat, C Rattray and G. Scollo, editors, *Algebraic Methodology and Software Technology, Entschede, 1993*, pages 193–200. Proceedings of the Third International Conference on Algebraic Methodology and Software Technology, University of Twente, The Netherlands, 21–25 June 1993, Springer-Verlag, 1994. Workshops in Computing series.

11. G. Hill. *A Language for System Configuration*. PhD thesis, Department of Computing, Imperial College, University of London, 1994. draft.

12. T. Maibaum, J. Fiadeiro, and M. Sadler. Stepwise program development in Π institutions. Technical report, Imperial College, November 1990.

13. T. S. E. Maibaum and W. M. Turski. On what exactly is going on when software is developed step-by-step. In *Proceedings of 7th International Conference on Software Engineering*, pages 528–533, 1984.

14. B. Meyer. Reusability: The case for object-oriented design. *IEEE Software*, pages 50–63, March 1987.

15. M. Stefik and D. Bobrow. Object-oriented programming: Themes and variations. *The AI Magazine*, 6(4), 1986. Winter.

16. W. M. Turski and T. S. E. Maibaum. *The Specification of Computer Programs*. International Computer Science Series. Addison Wesley, 1987.

17. P. A. S. Veloso. Program development as theory manipulations. Technical report, PUC/RJ Departamento de Informatica, Rio de Janeiro, Brazil, May 1985. Series: Monografias em Ciencia da Computacao No4/85.

18. H. Weber and H. Ehrig. Specification of modular systems. *IEEE Transaction on Software Engineering*, SE–12:786–798, June 1986.

19. P. Wegner. The object-oriented classification paradigm. In *Research Directions in Object-Oriented Programming*, pages 479–560. MIT Press, 1987.

20. M. Wirsing. Algebraic description of reusable software components. Technical report, University of Passau, 1987. Draft Version.

On the Expressibility of Discrete Event Specified Systems

Herbert Praehofer

Systems Science
Johannes Kepler University
A-4040 Linz, Austria

Bernard P. Zeigler

Electrical & Computer Eng.
The University of Arizona
Tucson, AZ 85721

Abstract

In this paper we address the question of the expressibility of discrete event specified systems (DEVS), i.e., we characterize the subclass of dynamical systems which can be homomorphically represented by DEVS models. We show that causal dynamical systems with piecewise constant input and output segment spaces are DEVS-representable. Moreover, DEVS-representable dynamical systems are closed under coupling, i.e., that a valid coupling of DEVS-representable dynamical systems is a DEVS-representable dynamical system. This justifies hierarchical, modular construction of both DEVS models and real-world (continuous or discrete) counterpart systems. Furthermore, we investigate a subclass of dynamical systems in detail, viz., discretely interacting continuous systems with internal dynamics specified by differential equations. This class is important in the rapidly developing field of hybrid autonomous systems control. As an application to CAST (computer-aided system technology), we show that this class is amenable to DEVS representation and therefore, also new forms of high performance parallel/distributed discrete event simulation.

1 Introduction

Discrete event modeling and simulation is finding evermore applications to the analysis and design of complex technological and natural manufacturing, computer, ecological and many other systems. Discrete event simulations were made possible by, and evolved with, the growing computational power of the digital computer. The prime requirement for conducting simulation runs was to be able to program the computer appropriately. However, general understanding of the nature of discrete event dynamical systems *per se* is still in its relative infancy compared to continuous systems. The form of system description intrinsic to the computer-based discrete event simulation languages was not mathematically well characterized. Code implementation details obscured the formal core of such models and it was often tacitly assumed that no useful code-independent model description could be given. However, recently formal treatment of discrete event dynamical systems is receiving increasing attention[7].

Work on a mathematical foundation of discrete event dynamic modeling and simulation began in the 70s [20, 21, 23]. DEVS (discrete event system specification) was introduced as an abstract formalism for discrete event modeling in a manner parallel to the differential equation specified system formalism (DESS) for continuous modeling. Because of its system theoretic basis, DEVS stakes a claim to be the most inclusive formalism for discrete event dynamical systems[7] (DEDS). Indeed, DEVS claims to be exactly the short-hand needed to specify systems whose input, state and output trajectories are piecewise constant[25]. The step-like transitions in the trajectories are identified as discrete events. This claim raises the general question of the *expressive power* or *expressibility* of the DEVS formalism. By *expressibility* of a formalism we mean roughly the range of systems that it can represent. This is one (mathematical) aspect of the adequacy of a formalism to describe real world phenomena[21].

Ashai and Zeigler[3] characterize the input/output behavior which can be represented by DEVS and provide a method to create a DEVS from a functional input/output observation frame (IOFO)[20]. Here we address the question of DEVS expressibility by characterizing the class of dynamical systems which can be represented by DEVS models. We prove that any dynamical system, which has piecewise constant input and output segments and which is causal, can be represented by a DEVS in a well-defined homomorphic manner. We call this class of systems *DEVS-Representable*[25].

As a very important special class of dynamical systems, we will discuss DESS systems with piecewise constant input and output. Such systems are prominent in such areas as intelligent control[2, 12] and reactive system design[5]. In these areas continuous processes are controlled by high-level, symbolic, event-driven control schemes. Hence, sensing and actuation in such systems is mainly event-like. The DEVS-extension for multi-formalism combined discrete/continuous modeling and simulation introduced recently[15, 14] offers a powerful modeling formalism capable of expressing both discretely and continuously interacting continuous systems. Here we will show how to characterize the DEVS abstractions of discretely interacting systems.

Closure under coupling [21, 15] is a desirable property for subclasses of dynamical systems since it guarantees that coupling class instances forms a system in the same class. This facilitates construction of hierarchical models. We will show that the class of DEVS-representable dynamical systems is closed under coupling.

In the following, the second section defines the concept of DEVS-representable dynamical system and presents a proof that any dynamical system with piecewise constant input and output can be represented by DEVS. The third section then investigates continuous systems as an important subclass of dynamical systems. We review DEV&DESS formalism as a system theoretic formalism for combined discrete/continuous modeling and define a subclass of such systems with piecewise constant input and output. Then we apply the DEVS-representation concepts introduced earlier for this special type of system. In the fourth section we prove closure under coupling for DEVS-representable systems. Finally, we discuss the application of the theoretical results to high performance simulation of discretely interacting continous systems.

2 The Class of DEVS-Representable Dynamical Systems

2.1 General Dynamical Systems

Based on [13, 20, 19] we define a general dynamical system as follows:

$$DS = (T, X, Y, \Omega, Q, \Delta, \Lambda)$$

with T is the time base, X is the set of input-values, Y is the set of output-values, Ω is the set of admissible input segments $\omega :< t_1, t_2 > \rightarrow X$ over T and X and Ω is closed under concatenation as well as under left segmentation [20], Q is the set of states, $\Delta : Q \times \Omega \rightarrow Q$ is the global state transition function, and $\Lambda : Q \times X \rightarrow Y$ is the output function.

The global state transition function of the general dynamical system DS has to fulfill the consistency property, i.e.,

$$\Delta(q, \omega_{(t,t>)}) = q, \tag{1}$$

the semigroup property, i.e., $\forall \omega :< t_1, t_2 > \rightarrow X \in \Omega, t \in < t_1, t_2 > \Rightarrow$

$$\Delta(q, \omega_{<t_1,t_2>}) = \Delta(\Delta(q, \omega_{<t_1,t>}), \omega_{<t,t_2>}), \tag{2}$$

and the causality property, i.e., $\forall \omega, \bar{\omega} \in \Omega$ if $\forall t \in < t_1, t_2 >: \omega(t) = \bar{\omega}(t)$ then

$$\Delta(q, \omega_{<t_1,t_2>}) = \Delta(q, \bar{\omega}_{<t_1,t_2>}). \tag{3}$$

Causality, the semigroup property and closure of admissible segments under left segmentation justifies defining the state trajectory resulting from every initial state $q \in Q$ and input segment $\omega :< t_1, t_2 > \rightarrow X \in \Omega$ by $STRAJ_{q,\omega} :< t_1, t_2 > \rightarrow Q$ with $\forall t \in < t_1, t_2 >$

$$STRAJ_{q,\omega}(t) = \Delta(q, \omega_{<t_1,t>}). \tag{4}$$

Similarly we define the output trajectory $OTRAJ_{q,\omega} :< t_1, t_2 > \rightarrow Y$ by: for every initial state $q \in Q$, input segment $\omega :< t_1, t_2 > \rightarrow X \in \Omega$ and $t \in < t_1, t_2 >$

$$OTRAJ_{q,\omega}(t) = \Lambda(STRAJ_{q,\omega}(t), \omega(t)). \tag{5}$$

Now the input/output behavior R_{DS} of the dynamical system is given by

$$R_{DS} = \{(\omega, OTRAJ_{q,\omega}) : \omega \in \Omega, q \in Q\}. \tag{6}$$

2.2 The Discrete Event Specified System (DEVS) Formalism

An atomic discrete event specified system is a structure [20, 21, 23]

$$DEVS = (X_M, Y_M, S, \delta_{ext}, \delta_{int}, \lambda, ta)$$

where X_M is the set of inputs, Y_M is the set of outputs, S is the set of sequential states, $\delta_{ext} : Q \times X \rightarrow S$ is the external state transition function, $\delta_{int} : S \rightarrow S$ is the internal state transition function, $\lambda : S \rightarrow Y$ is the output function, and $ta : S \rightarrow \mathcal{R}_0^+ \cup \{\infty\}$ is the time advance function.

A DEVS specifies a dynamical system DS in the following way:

- the time base T is the real numbers $\mathcal{R}$;
- $X = X_M \cup \{\Phi\}$, i.e., the input set of the dynamical system is the input set of the DEVS together with a symbol $\Phi \notin X_M$ specifying the non-event,
- $Y = Y_M \cup \{\Phi\}$, i.e., the output set of the dynamical system is the output set of the DEVS together with Φ,
- $Q = \{(s,e) : s \in S, 0 \le e \le ta(s)\}$ the set of states of the dynamical system consists of the sequential states of the DEVS paired with a real number e giving the elapsed time since the last event,
- the admissible input segments is the set of all DEVS-segments over X and T (DEVS-segments [20] are characterized by the fact that for any $\omega :< t_1, t_2 >\to X \in \Omega$, there is only a finite number of event times $\{\tau_1, \tau_2, \ldots, \tau_n\}, \tau_i \in< t_1, t_2 >$ with $\omega(\tau_i) \ne \Phi$. As a special case, a so-called Φ segment has no events.
- for any DEVS input segment $\omega :< t_1, t_2 >\to X \in \Omega$ and state $q = (s,e)$ at time t_1 the global state transition function Δ is defined as follows:
$\Delta(q, \omega_{<t_1,t_2>}) =$

$$(s, e + t_2 - t_1) \tag{7}$$

if $e + t_2 - t_1 < ta(s)$ and ω is the Φ segment,

$$\Delta((\delta_{int}(s), 0), \omega_{<t_1+ta(s)-e,t_2>}) \tag{8}$$

if $e + t_2 - t_1 \ge ta(s)$ and $\neg\exists t \in< t_1, t_1 + ta(s) - e >: \omega(t) \ne \Phi$,

$$\Delta((\delta_{ext}(s, e + t - t_1, \omega(t)), 0), \omega_{(t,t_2>}) \tag{9}$$

if $\omega(t) \ne \Phi$ and $t \le t_1 + ta(s) - e$ and ω restricted to $< t_1, t >$ is a Φ segment.
- the output function Λ of the dynamical system is given by

$$\Lambda((s,e), x) = \lambda(s). \tag{10}$$

From the above definition we see that for a given input segment ω, three different cases are identified:

- First, if ω is a non-event segment and an internal event does not occur in interval $< t_1, t_2 >$, then the sequential state is left unchanged but the elapsed time component e of the total state has to be updated (7).
- Second, if a time event scheduled by the time advance function does occur in the time interval $< t_1, t_2 >$ and there is no input event prior to the internal event time, then the internal transition function defines a new sequential state, the elapsed time component is set to zero, and the rest of the input segment is applied to this new state (8).
- Third, if the input segment ω defines an input event prior to the occurrence of an internal event, then the external state transition function is applied with the external input value and elapsed time and the elapsed time is set to zero. The rest of the input segment is applied to this new state (9).

2.3 DEVS-Representation of Constant Input/Output Dynamical Systems

We define a *piecewise constant trajectory* $\omega :< t_1, t_2 > \rightarrow X$ in the following way: there is a finite (possibly empty) subset $\{\tau_1, \tau_2, \ldots, \tau_n\}$, $\tau_i \in < t_1, t_2 >$ for which $\omega(\tau_i) \neq \omega(\tau_i - \epsilon)$ and $\epsilon \in \mathcal{R}^+$ is arbitrarily small. Piecewise constant trajectories are equivalent to DEVS-segments in the sense that they can be transformed into each other. The transformations are as follows:

- *Piecewise constant trajectory ω to DEVS-segment $\bar{\omega}$:* For every event time $\tau_i \in \{\tau_1, \tau_2, \ldots, \tau_n\}$ of ω there is an value $\bar{\omega}(\tau_i) = \omega(\tau_i)$ and for all $\tau \neq \tau_i \Rightarrow \bar{\omega}(\tau) = \Phi$.
- *DEVS-segment $\bar{\omega}$ to piecewise constant trajectory ω:* Let $\{\tau_1, \tau_2, \ldots, \tau_n\}$, $\tau_i \in < t_1, t_2 >$ be the event times of the DEVS-segment $\bar{\omega}$ with $\bar{\omega}(\tau_i) \neq \Phi$. Then for every $\tau \in < t_1, t_2 >$ we define the respective value of the piecewise constant trajectory $\omega(\tau)$ by the value $\bar{\omega}(\tau_x)$ with τ_x being the largest time in $\{\tau_1, \tau_2, \ldots, \tau_n\}$ with $\tau_x \leq \tau$. If such a number τ_x does not exist then $\omega(\tau) = \Phi$.

For $x \in X$ let $x_{<t_1,t_2>}$ denote the segment $\omega :< t_1, t_2 > \rightarrow X$ with constant value x, i.e., $\forall t \in < t_1, t_2 > \omega(t) = x$. If it is clear from the context, we write x for $x_{<t_1,t_2>}$.

We now define a *constant input/output dynamical system* as a dynamical system whose input trajectories $\omega \in \Omega$ and associated output trajectories $OTRAJ_{q,\omega}$ are piecewise constant only. Our main interest is to show how such a system can be represented in the DEVS formalism.

Given a constant input/output dynamical system $DS = (T, X, Y, \Omega, Q, \Delta, \Lambda)$, we define a $DEVS_{DS} = (X_M, Y_M, S, \delta_{ext}, \delta_{int}, \lambda, ta)$ in the following way: $X_M = X$, $Y_M = Y$, $S = Q \times X$, and for every total state $((q, x), 0)$ and input segment $\omega :< t_1, t_2 > \rightarrow X$ we define

$$ta((q, x)) = min\{e | OTRAJ_{q,x}(t_1) \neq OTRAJ_{q,x}(t_1 + e), \tag{11}$$

$$\delta_{int}((q, x)) = (STRAJ_{q,x}(t_1 + ta((q, x))), x), \tag{12}$$

$$\delta_{ext}((q, x), e, x') = (STRAJ_{q,x}(t_1 + e), x'), \tag{13}$$

$$\lambda((q, x)) = \Lambda(q, x). \tag{14}$$

Theorem 1: The $DEVS_{DS}$ as constructed above and the original DS are behaviorally equivalent, i.e., they always have the same input/output behavior when started in the same state.

Proof: The main work is to show that for all input trajectories, ω, and initial states, q, the states in the constructed discrete event system $DEVS_{DS}$ are equal to the states of the original dynamical system DS at both the event times $\tau_e \in \{\tau_{e1}, \tau_{e2}, \ldots, \tau_{en}\}$ of ω and the event times $\tau_i \in \{\tau_{i1}, \tau_{i2}, \ldots, \tau_{in}\}$ of the associated output trajectory $OTRAJ_{q,\omega}$. This is done by examining the state and output trajectories of the dynamical system $\bar{DS} = (\bar{T}, \bar{X}, \bar{Y}, \bar{\Omega}, \bar{Q}, \bar{\Delta}, \bar{\Lambda})$

specified by $DEVS_{DS}$. From the definition of the output in (14), and from the fact that outputs do not change between event times it follows that the output behavior of the discrete event system $DEVS_{DS}$ is equal to that of the original dynamical system DS. The formal proof showing state equality at event times in the original dynamical system DS and the dynamical system $\tilde{DS}$ specified by $DEVS_{DS}$ is presented in the Appendix.

3 Discrete Event Abstraction of Continuous Processes

In this section we will investigate a special subclass of dynamical systems – those whose state dynamic is defined by a set of first order differential equations. Discrete event abstractions of continuous systems are of special interest because in many areas of discrete event simulation – transport systems, traffic systems, ecological systems[4], and others – the models employed are actually abstractions of continuous processes. Another area receiving increasing attention is hybrid system control [1, 17, 8, 18, 9, 6]. In hybrid control the plant itself is modeled by a set of differential equations but its interface to the world imposes piecewise constant input/output behavior: it receives high level command-like control inputs and produces threshold sensor outputs.

For combined discrete/continuous system modeling, the DEV&DESS formalism [15, 14] affords a systems theory-based integration of the DEVS and DESS formalism. The output-partitioned DEV&DESS is a subclass of DEV&DESS [16] intended for modeling hybrid control systems. An output-partitioned DEV&DESS is a DEV&DESS which only accepts discrete inputs and whose outputs are quantized. The next section will review these formalisms in greater detail. Then we will show how a DEVS-representation of an output-partitioned DEV&DESS can be derived.

3.1 The DEV&DESS Formalism Reviewed

In this section we review the DEV&DESS formalism for combined discrete/continuous simulation modeling and introduce the output-partitioned DEV&DESS as a subclass of DEV&DESS with piecewise constant input and output.

An DEV&DESS is a structure [15, 14]:

$$DEV\&DESS = (X, Y, S, \delta_{ext}, \delta_{state}, \delta_{int}, \lambda_d, ta, f, \lambda_c)$$

where $X = X_c \times X_d$ is the set of inputs, the Cartesian product of continuous inputs X_c and discrete inputs X_d, $Y = Y_c \times Y_d$ is the set of outputs, the Cartesian product of continuous outputs Y_c and discrete outputs Y_d, $S = S_c \times S_d$ is the set of states, the Cartesian product of continuous states S_c and discrete states S_d, $\delta_{ext} : Q \times X \to S$ is the external state transition function, with $Q = \{(s, e) | s \in S, 0 \le e \le ta(s_d)\}$ is the set of total states, $\delta_{state} : Q \times X_c \to S$ is the state event state transition function, $\delta_{int} : S \to S$ is the internal state transition function, $\lambda_d : S \times X_c \to Y_d$ is the discrete output function, $ta : S_d \to \mathcal{R}_0^+ \cup \infty$ is the time advance function, $\phi : S \times X_c \to S_c$ is the rate of change function, and $\lambda_c : S \times X_c \to Y_c$ is the continuous output function.

The DEV&DESS formalism inherits the discrete parts and the event behavior from the DEVS formalism, likewise, the continuous parts and the continuous behavior derive from the DESS (differential equation specified system) formalism. Hence, a DEV&DESS has an input and output interface split up into discrete and continuous input and output ports through which all the interactions with the environment occur. It has both continuous and discrete state variables. The external transition and the internal transition together with the discrete output and the time advance functions are employed to model input events and time scheduled events, respectively. The DEV&DESS additionally introduces a new type of event – *state events*Q are generated from threshold crossings of continuous states and inputs. The state event transition function models these events. The derivative function and the continuous output function inherited from DESS specify the continuous behavior.

The output-partitioned DEV&DESS is intended to support modeling systems found in hybrid system control applications.

Definition: An output-partitioned DEV&DESS (op-DEV&DESS) is a DEV&DESS with the following constraints:

- $X_c = \{\}$, $Y_c = \{\}$, i.e., the system only has discrete inputs and outputs and therefore the input/output behavior is piecewise constant,
- $\forall s \in S : ta(s) = \infty$, i.e., the system does not have time events,
- $S_d = \{\}$, i.e., the system only has continuous states,
- $S_c = S_{c1} \times S_{c2} \times \ldots \times S_{cn}$, i.e., the continuous state space is n-dimensional (it has n state variables),
- the state dynamic is defined by a set of first order differential equations ϕ, i.e., the global state transition function is defined by

$$\Delta(s_c, \omega_{<t_1, t_2>}) = s_c + \int_{t_1}^{t_2} \phi(STRAJ_{s_c, \omega}(\tau), \omega(\tau)) d\tau, \tag{15}$$

- for every state component S_{ci} we identify a set of *threshold values*

$$\{f_{i,1}, f_{i,2}, \ldots, f_{i,b_{m_i}}\}, f_{i,b_i} \in S_{ci} \cup \{-\infty, +\infty\}$$

which define a partition of the state set into a set of mutual exclusive blocks

$$\{< f_{i,1}, f_{i,2} >, < f_{i,2}, f_{i,3} >, \ldots, < f_{i,b_{i,m}-1}, f_{i,b_{m_i}} >\}$$

which we call *output partition* in the sequel,

- Y_d is a finite output set and each element $y_d \in Y_d$ corresponds to a block in the output partition. For the discrete output function λ_d we require that states in the same block of the output partition have the same output, i.e.,

$$\lambda_d(s_c) = \lambda_d(\bar{s}_c) \Leftrightarrow \forall i \exists f_{i,b_i} : s_{ci} \in (f_{i,b_i-1}, f_{i,b_i}] \wedge \bar{s}_{ci} \in (f_{i,b_i-1}, f_{i,b_i}] \tag{16}$$

— state events arise from state trajectory crossings of output partition boundaries only. The discrete output set typically is the Cartesian product of threshold sensor output sets. Therefore, the threshold sensor values define the output partition of the state set. Threshold sensors only react at the threshold values, what is modeled by state events.

As an example of output-partitioned DEV&DESS let us discuss systems of *discretely interacting continuous component systems (DICS)*[10]. When a set of continuous components is coupled in such a way that outputs are sent to inputs only at discrete moments, we call the resultant object a system of discretely interacting continuous systems. Such systems can be described by output-partitioned DEVS&DESS in which each component is described as a DESS and outputs are generated at quantizer thresholds by state events. As an example, consider Figure 1. Each component is an integrator

$$s(t) = \int x(t) \, dt \qquad (17)$$

where $s(t)$ is the state variable and $x(t)$ is the piecewise constant input. The output is the state after having been quantized i.e.

$$y(t) = D \times \left[Integer Round \left(\frac{s(t)}{D} \right) + \frac{1}{2} \right] \qquad (18)$$

where $y(t)$ is output, and D is the quantization size. Note that the state space is partitioned into equal sized blocks $\{ \ldots, (-D, 0], (0, D], (D, 2*D], (2*D, 3*D], \ldots \}$ and that the output $y(t)$ changes at discrete points in time where the state events occur (crossing of the quantizer thresholds).

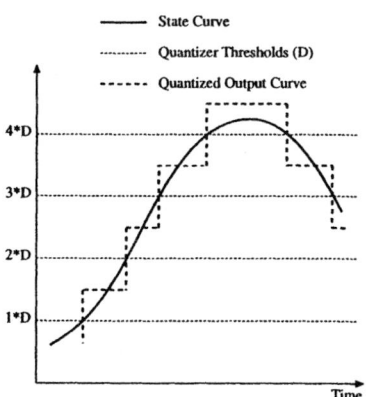

Fig. 1. The integrator models.

The DICS is an example of a so-called Sparse Output DEVS [10]. Let H be the time taken for each step in computation of the next state in an integrator.

The smaller the parameters, D and H are, the more the output trajectories approach the correct curves. The *sparseness* of a quantized component can be described as:

$$Sparseness = \frac{D}{H}. \tag{19}$$

The greater the sparseness, the larger the time between output events, the greater the opportunity to exploit parallel/distributed discrete event simulation.

3.2 DEVS Abstraction of Output-Partitioned DEV&DESS

We apply the process of DEVS-representation of constant input/output dynamical systems to define the DEVS-representation $DEVS = (X_M, Y_M, S, \delta_{ext}, \delta_{int}, \lambda, ta)$ of an output-partitioned DEV&DESS. The latter is given by: $opDEV\&DESS = (X_d, Y_d, S_c, \delta_{ext}, \delta_{state}, \lambda_d, f)$:

- $X_M = X_d$, $Y_M = Y_d$, $S = S_c \times X_d$,
- from (11), (15), and (16) it follows that $ta((s_c, x)) =$

$$min\{e|s_c + \int_{t_1}^{t_1+e} \phi(STRAJ_{s_c,x}(\tau), \omega(\tau))d\tau = (s'_{c1}, \ldots, f_{i,b_i}, \ldots, s'_{cn})\}, \tag{20}$$

i.e., the time advance is given by the time it takes to reach the next threshold f_{i,b_i} given the current input x,
- from (12), (15) and (20) it follows that $\delta_{int}((s_c, x)) =$

$$(s_c + \int_{t_1}^{t_1+ta(s)} \phi(STRAJ_{s_c,x}(\tau), x)d\tau, x) = ((s'_{c1}, \ldots, f_{i,b_i}, \ldots, s'_{cn}), x) \tag{21}$$

i.e., the next state at an internal transition is the continuous state at the next threshold crossing and the same input as before,
- from (13) and (15) it follows that $\delta_{ext}((s_c, x), e, x') =$

$$(s_c + \int_{t_1}^{t_1+e} \phi(STRAJ_{s_c,x}(\tau), x)d\tau, x') \tag{22}$$

i.e., the state at an external event is given by the continuous state value which is reached when the system in state s_c has received a constant input x for e time units and the new input value x',
- from (14) and (16) it follows that

$$\lambda((s_c, x)) = \lambda_d(s_c, x). \tag{23}$$

Based on the definition above, a DEVS model of the quantized integrator in Figure 1 can be depicted as shown in Figure 2. The continuous behavior is abstracted to events when output changes occur. The time to the next output event is modeled by the time advance function ta.

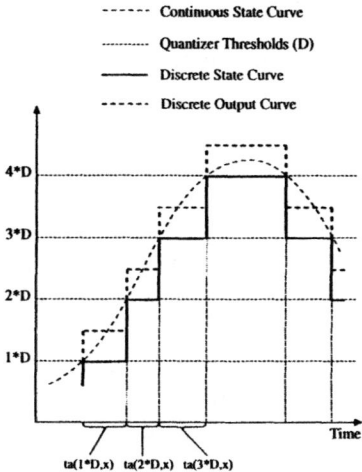

Fig. 2. DEVS representation of integrator model.

3.3 Computation of Event Times and States

From the definition of DEVS-representations of continuous processes as outlined above, we see that the discrete event model of the system affords a more efficient simulation than a differential equation model of the same system. The differential equation model requires a stepwise generation of successive model states while the discrete event form computes state changes only at event times. However to make discrete event simulation feasible, one has to have a means to compute the time advance values and the continuous states at event times. Zeigler [22, 23] identifies several approaches to this which depend on the knowledge about the original system. A situation met in practice occurs where input changes do not affect the occurrence of the next state crossing. Discrete event modeling as done in many classical application areas satisfies this requirement. In this case, updating the state between internal events is straight forward[11].

4 Closure Under Coupling of DEVS-Representable Systems

In this section we investigate the question of closure under coupling for DEVS-representable systems. We show that a proper coupling of DEVS-representable components is a DEVS-representable. We do that for the most general modular coupled system form as introduced in [20] and presented in [15].

4.1 Modular Coupled System Formalism

A most general formalism for modular coupled systems is a structure [20, 15]

$$N = (X_N, Y_N, D, M, I, Z)$$

where X_N is the set of inputs of the network, Y_N is the set of outputs of the network, $D = \{1, 2, \ldots, n\}\}$ is index set for components, $M = \{M_d | d \in D\}$ is the set of causal component systems, $I = \{I_d | d \in D \cup N\}$ and $I_d \subset D \cup N$ is the set of influencers of component d or of the external network output, $Z = \{Z_d | d \in D \cup N\}$ and $Z_d : \underset{e \in I_d}{\times} Y X_e \rightarrow X Y_d$ is the interface mapping for component d or for external network output, with $Y X_e = Y_e$ if $e \neq N$, $Y X_e = X_N$ if $e = N$, $X Y_d = Y_N$ if $d = N$, and $X Y_d = X_d$ if $d \neq N$.

The most general multi-component system specification described above consists of the specification of the input / output interface of the network (X_N and Y_N), the specification of the component systems $\{M_d\}$ and the coupling scheme ($\{I_d\}$ and $\{Z_d\}$). The set of influencers I_d of the components d also may contain the network N. Through the general form of the coupling specification scheme employing interface maps, arbitrary couplings can be realized. The interface map Z_d specifies how the input values of component d are derived from the outputs or external inputs of its influencers $e \in I_d$. The interface map Z_N for network N specifies the output based on outputs of the component systems and eventually its own inputs.

4.2 The Dynamical System Specified by a Modular Coupled System Specification

Based on the interface mappings $Z_d : \underset{e \in I_d}{\times} Y X_e \rightarrow X Y_d$ we define the functions $\bar{Z}_d : \underset{e \in I_d}{\times} \Omega_e \rightarrow \Omega_d$ by $\bar{Z}_d(\ldots, \omega_e, \ldots) = \omega_d$ with $\omega_d(t) = Z_d(\ldots, \omega_e(t), \ldots)$.

For a properly specified dynamical system the following constraints have to be fulfilled:

1. the components have to be causal dynamical systems,
2. the component systems must have a common time base T,
3. the interface segment maps $\bar{Z}_d$ have to define admissible input segments for the component systems d,
4. no algebraic loops are allowed [20, 15], i.e., in every coupling loop there is at least one component whose output is defined on the state only and is not dependent on its input value.

A modular coupled system specification N defines a dynamical system $DS = (T, X, Y, \Omega, Q, \Delta, \Lambda)$ in the following way:

- the time base T is the common time base of the component systems,
- $X = X_N, Y = Y_N$,
- $Q = \underset{d \in \{1, 2, \ldots, n\}}{\times} Q_d$ the set of states of the dynamical system is defined as the Cartesian product of states of the components,

- the admissible input segments $\omega :< t_1, t_2 > \rightarrow X \in \Omega$ are constrained in such a way that the input segments $\omega_d = \bar{Z}_d(\ldots, \omega, \ldots)$ of components d with $N \in I_d$ are admissible input segments for d,
- the global state transition function Δ is defined by the Cartesian product of the state transitions of the components:

$$\Delta(q, \omega) = (\Delta_1(q_1, \omega_1), \ldots, \Delta_d(q_d, \omega_d), \ldots, \Delta_n(q_n, \omega_n)) \tag{24}$$

with

$$\omega_d = \bar{Z}_d(\ldots, OTRAJ_{q_e, \omega_e}, \ldots) \tag{25}$$

with $e \in I_d$ and $OTRAJ_{q_e, \omega_e} = \omega$ if $e = N$,
- the output function Λ of the dynamical system is given by

$$\Lambda(q, x) = Z_N(\underset{e \,\in\, I_N}{\times} OTRAJ_{q_e, \omega_e}) \text{ with } e \in I_N. \tag{26}$$

Proposition 1: The dynamical system as constructed as outlined above is well defined, i.e., it fulfills the consistency, semigroup, and causality properties.

Proof: Through constraint 3 and 4 and equation (25) it is guaranteed that the input segments of the components are well defined. As the component systems are causal and the global transition function of the coupled system is composed of the transition functions of its components, it follows that the coupled system is well defined. q.e.d.

4.3 DEVS-Representable Coupled Systems

From the definition of DEVS-representability and proposition 1 it easily follows that DEVS-representable systems are closed under coupling. As a consequence, a coupling of components which are DEVS-representable and which does not contain algebraic loops is a DEVS-representable dynamical system. This is stated in:

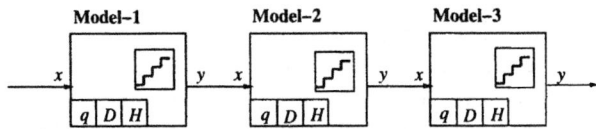

Fig. 3. Coupled DICS model.

Theorem 2: A modular coupled system whose components are DEVS-representable and which does not contain algebraic loops is a DEVS-representable dynamical system.

Proof: We show the correctness of theorem 2 by establishing that the coupling of DEVS-representable systems fulfills the constraints 1 to 3 for properly specified coupled systems (constraint 4 is required in the theorem). (1) DEVS-representable systems are causal dynamical systems, (2) the real numbers $\mathcal{R}$ is the time base of all components, (3) as the output segments of all components are piecewise constant and the input segments of the network have to be piecewise constant, also the input segments $\bar{Z}_d(\ldots, OTRAJ_{q_e, \omega_e}, \ldots)$ are piecewise constant and hence are admissible input segments.

All the constraints are fulfilled and from proposition 1 it follows that the coupled system is causal. The output trajectory of the coupled system is defined by $\bar{Z}_N(\ldots, OTRAJ_{q_e, \omega_e}, \ldots)$. Since $OTRAJ_{q_e, \omega_e}, e \in I_d$ are piecewise constant, also the output segment of the coupled system is piecewise constant. But by theorem 1 any causal system with piecewise constant input and output is DEVS-representable. q.e.d.

Figure 3 depicts a DICS which is a coupling of quantized integrators. As integrators do not feed forward the input directly to the output, no such DICS can have an algebraic loop (regardless of whether it has a feedback loops). Hence DICS of quantized integrators are DEVS-representable.

5 Summary

In this paper we showed that any causal dynamical system which has piecewise constant input and output segments can be represented by a DEVS model in a direct homomorphic mapping. We have named this class of systems *DEVS- representable*. Furthermore, we showed that DEVS-representable systems are closed under coupling, i.e., that a coupling of DEVS-representable components itself is a DEVS-representable. This justifies hierarchical, modular construction of both DEVS models and real-world (continuous or discrete) counterpart systems. Hierarchical, modular construction enables us to build, verify and validate, complex system models in a stagewise incremental fashion that is increasingly recognized as the preferred approach to engineering reliable systems.

These two results have practical relevance in many more respects. The formulation of the abstraction of discrete event models from more detailed descriptions throws light on the intrinsic nature of discrete event modeling. It can also be a basis for automatic abstraction processes being developed[16]. Discrete event abstractions normally afford a more efficient simulation than detailed, e.g. continuous, counterparts and they are candidates for distributed simulations. We investigated a subclass of dynamical systems in detail, viz., discretely interacting continuous systems with internal dynamics specified by differential equations. This class is important in the rapidly developing field of hybrid autonomous systems control. As an application to CAST (computer-aided system technology), we show that this class is amenable to DEVS abstraction. New forms of high performance simulation of spatially distributed systems are enabled with this approach[10]. Indeed, we have employed DEVS abstraction to enable parallel/distributed discrete event simulation of large scale, high resolution watershed hydrological models, thereby demonstrating watershed dynamics at a degree of realism greater than any previously attempted [24].

References

1. P.J. Anstaklis, M.D. Passino, and S.J. Wang. Towards intelligent autonomous control systems. *Journal of Intelligent and Robotic Systems*, 1(4):315–342, 1989.
2. P. J. Antsaklis, K. M. Passino, and S. J. Wang. Towards intelligent autonomous control systems. *J. of Intelligent and Robotic Systems*, 1(4):315–342, 1989.
3. T. Asahi and B. P. Zeigler. Behavioral characterization of discrete event systems. In *Proc. of AI, Simulation and Planning in High-Autonomy Systems*, pages 127–132, Tucson AZ, Sept 1993. IEEE/CS Press.
4. M. Perestrello de Vasconcelos and B. P. Zeigler. Discrete event simulation of forest landscape response to fire disturbances. *Ecological Modeling*, 65:177–198, 1993.
5. D. Harel et al. Statemate: A working environment for the specification of compex reactive system. *IEEE Trans. on SE*, 16(4):403–414, 1990.
6. R.L. Grossman, A. Nerode, A.P. Raun, and H. Rischel, editors. *Hybrid Systems*, volume 736 of *Lecture Notes in Computer Science*. Springer-Verlag, Berlin, 1993.
7. Y. C. Ho. Special issue on discrete event dynamic systems. *Proceedings of the IEEE*, 77(1), 1989.
8. M.D. Lemmon and P.J. Anstaklis. Hybrid systems and intelligent control. In *Proc. of the 1993 IEEE International Symposium on Intelligent Control*, pages 174–179, Chicago, IL, 1993.
9. Y.-H. Li. Hybrid synthesis of optimal control for discrete event systems. In *Proc. of the 1993 IEEE International Symposium on Intelligent Control*, pages 308–313, Chicago, IL, 1993.
10. C. Liao, A. Motaabbed, D. Kim, and B.P. Zeigler. Distributed simulation algorithms for sparce output DEVS. In *Proc. of AI, Simulation and Planning in High-Autonomy Systems*, pages 171–177, Tucson AZ, Sept 1993. IEEE/CS Press.
11. C.J. Luh and B. P. Zeigler. Abstracting event-based control models for high autonomy systems. *IEEE Trans. on Systems, Man and Cybernetics*, 23(1):42–54, 1993.
12. A. Meystel. Intelligent control: A sketch of the theory. *Journal of Intelligent and Robotic Systems*, 2(2):97–107, 1989.
13. L. Padula and M. A. Arbib. *Systems Theory*. Saunders, Philadelphia, 1974.
14. F. Pichler and H. Schwaertzel, editors. *CAST Methods in Modelling*. Springer-Verlag, 1992.
15. H. Praehofer. *System Theoretic Foundations for Combined Discrete-Continuous System Simulation*. PhD thesis, Johannes Kepler University of Linz, Linz, Austria, 1991.
16. H. Praehofer, P. Bichler, and B. P. Zeigler. Synthesis of endomorphic models for event-based control. In *Proc. of AI, Simulation and Planning in High-Autonomy Systems*, pages 120–126, Tucson AZ, Sept 1993. IEEE/CS Press.
17. P.J.G. Ramadge and W.M. Wohnham. The control of discrete event systems. *Proceedings of the IEEE*, 77(1):81–98, 1989.
18. J. A. Stiver and P.J. Anstaklis. Extracting discrete event models from hybrid control systems. In *Proc. of the 1993 IEEE International Symposium on Intelligent Control*, pages 298–301, Chicago, IL, 1993.
19. G. Wunsch, editor. *Handbuch der Systemtheorie*. Akademie-Verlag, Berlin, 1986.

20. B. P. Zeigler. *Theory of Modelling and Simulation.* John Wiley, New York, 1976.
21. B. P. Zeigler. *Multifacetted Modelling and Discrete Event Simulation.* Academic Press, London, 1984.
22. B. P. Zeigler. DEVS representation of dynamical systems: Event-based intelligent control. *Proceedings of the IEEE,* 77(1):72–80, 1989.
23. B. P. Zeigler. *Object-Oriented Simulation with Hierarchical, Modular Models.* Academic Press, London, 1990.
24. B. P. Zeigler, George Ball, and Doo Hwan Kim. Hierarchical distributed genetic algorithms control of simulation-based optimization. In *Proceedings of the Petaflops Frontier Workshop, 1995 Conference on Massively Parallel Computation,* Washington, DC, 1995. IEEE/CS Press.
25. B. P. Zeigler and W. H. Sanders. Preface to special issue on environments for discrete event dynamic systems. *Discrete Event Dynamic Systems: Theory and Application,* 3(2):110–119, 1993.

Appendix: Proof of Theorem 1

In the following we proof that states at event times are equal in the original piecewise constant input/output dynamical system DS and the dynamical system $\bar{DS}$ of the DEVS abstraction $DEVS_{DS}$ of DS.

Let $\omega :< t_1, t_2 >\to X$ be any valid piecewise constant input trajectory and $\bar{\omega} :< t_1, t_2 >\to X \cup \{\Phi\}$ the corresponding DEVS-segment for $DEVS_{DS}$, let q be the state of the dynamical system DS and $((q, x), 0)$, $x = \omega(t_1)$ the corresponding total state of $\bar{DS}$ at time t_1, and let τ_e the smallest event time of ω and τ_i be the smallest event time of $OTRAJ_{q,\omega}$, then we identify two cases:

1. $\tau_e \le \tau_i$ (external event): The state of $\bar{DS}$ at time τ_e is $ST\bar{R}AJ_{\bar{q},\bar{\omega}}(\tau_e)$ which through (4) is equal to $\bar{\Delta}(\bar{q}, \bar{\omega}_{<t_1,\tau_e>})$ which through (9) and because $\bar{\omega}(\tau_e) \ne \Phi$ is defined by $(\delta_{ext}((q,x), 0 + \tau_e - t_1, \bar{\omega}(\tau_e)), 0)$ which through (13) is equal to $((STRAJ_{q,x}(t_1 + \tau_e - t_1), \bar{\omega}(\tau_e)), 0)$ which is $((STRAJ_{q,x}(\tau_e), \bar{\omega}(\tau_e)), 0)$ and because ω is constant with value $x = \omega(t_1)$ until τ_e this is equal to $((STRAJ_{q,\omega}(\tau_e), \bar{\omega}(\tau_e)), 0)$.

2. $\tau_i < \tau_e$ (internal event): From (11) it follows that the time advance $ta(s) = \tau_i - t_1$. The state of $\bar{DS}$ at time τ_i is $ST\bar{R}AJ_{\bar{q},\bar{\omega}}(\tau_i)$ which through (4) is equal to $\bar{\Delta}(\bar{q}, \bar{\omega}_{<t_1,\tau_i>})$ which through the value of ta and the dynamic definition of DEVS in (8) is defined by $(\delta_{int}((q, x)), 0)$ which through (12) is equal to $((STRAJ_{q,x}(t_1 + ta(s)), x), 0)$ which is $((STRAJ_{q,x}(t_1 + \tau_i - t_1), x), 0)$ which is $((STRAJ_{q,\omega}(\tau_i), x), 0)$ because ω is constant with value $x = \omega(t_1)$ until τ_i.

As the state of DS and state component q of the $DEVS_{DS}$ as well as the input value $\omega(\tau)$ are equal to the state component x of the DEVS at the first event time τ, for the new state and the rest of the input trajectory $\omega_{<\tau,t_2>}$ the same procedure can be applied to show that the states at the next event time are equal.

Q.e.d.

An Object-Oriented Architecture for Possibilistic Models

Cliff Joslyn

Mail Stop 522.3, NASA Goddard Space Flight Center
Greenbelt, MD 20771, USA
joslyn@kong.gsfc.nasa.gov
http://groucho.gsfc.nasa.gov/joslyn

Abstract. An architecture for the implementation of possibilistic models in an object-oriented programming environment (C++ in particular) is described. Fundamental classes for special and general random sets, their associated fuzzy measures, special and general distributions and fuzzy sets, and possibilistic processes are specified. Supplementary methods—including the fast Möbius transform, the maximum entropy and Bayesian approximations of random sets, distribution operators, compatibility measures, consonant approximations, frequency conversions, and possibilistic normalization and measurement methods—are also introduced. Empirical results to be investigated are also described.

1 Introduction

Possibility theory [4] is an alternative information theory to that based on **probability**. Although possibility theory is logically independent of probability theory, they are related: both arise in **Dempster-Shafer evidence theory** as **fuzzy measures** defined on **random sets**; and their distributions are both **fuzzy sets**. So possibility theory is a component of a broader **Generalized Information Theory** (GIT), which includes all of these fields [18].

Possibility theory was originally developed in the context of fuzzy systems theory [28]. More recently, possibility theory is being developed independently of both fuzzy sets and probability. In particular, the author is developing the mathematics and semantics of possibility theory [9, 12, 13]. These methods include **possibilistic measurement** procedures [8, 10] and **possibilistic processes** such as **possibilistic automata** [11]—generalizations of nondeterministic processes whose non-additive weights adhere to the laws of mathematical possibility theory.

This paper describes an architecture for implementing possibilistic models in an object-oriented environment. The approach is based on the mathematics of consistent random sets as the basis for the representation of measured possibility distributions and possibilistic processes.

2 Possibility Theory

Mathematical possibility theory can only be briefly introduced here. See [4, 12, 18] for details and proofs.

2.1 Mathematical Possibility in GIT

Given a finite universe $\Omega := \{\omega_i\}, 1 \leq i \leq n$, the function $m: 2^{\Omega} \mapsto [0, 1]$ is an **evidence function** (otherwise known as a **basic probability assignment**) when $m(\emptyset) = 0$ and $\sum_{A \subseteq \Omega} m(A) = 1$. Denote a **random set** generated from an evidence function as $\mathcal{S} := \{\langle A_j, m_j \rangle : m_j > 0\}$, where $\langle \cdot \rangle$ is a vector, $A_j \subseteq \Omega, m_j := m(A_j)$, and $1 \leq j \leq N := |\mathcal{S}| \leq 2^n - 1$. A random set $\mathcal{S}$ is essentially a subset-valued random variable on 2^{Ω}, with m its "probability distribution". Denote the **focal set** of $\mathcal{S}$ as $\mathcal{F} := \{A_j : m_j > 0\}$ with **core** and **support** respectively $\mathbf{C}(\mathcal{F}) := \bigcap_{A_j \in \mathcal{F}} A_j, \mathbf{U}(\mathcal{F}) := \bigcup_{A_j \in \mathcal{F}} A_j$.

The **plausibility** and **belief** measures on $\forall A \subseteq \Omega$ are $\mathrm{Pl}(A) := \sum_{A_j \cap A \neq \emptyset} m_j$ and $\mathrm{Bel}(A) := \sum_{A_j \subseteq A} m_j$. These are non-additive **fuzzy measures** [27], and are dual, in that $\forall A \subseteq \Omega, \mathrm{Bel}(A) = 1 - \mathrm{Pl}(\overline{A})$. In general only plausibility will be considered below. Bel (and Pl as its dual) determines the evidence function according to the **Möbius inversion**

$$m(A) = \sum_{B \subseteq A} (-1)^{|B-A|} \mathrm{Bel}(B). \tag{1}$$

The **plausibility assignment** (otherwise known as the **one-point coverage function**) of $\mathcal{S}$ is $\mathrm{Pl} = \langle \mathrm{Pl}_i \rangle := \langle \mathrm{Pl}(\{\omega_i\}) \rangle$, where

$$\mathrm{Pl}_i := \sum_{A_j \ni \omega_i} m_j. \tag{2}$$

Random set **inclusion** is defined by the formula

$$\mathcal{S}_1 \subseteq \mathcal{S}_2 \quad := \quad \forall A \subseteq \Omega, \mathrm{Pl}_1(A) \leq \mathrm{Pl}_2(A). \tag{3}$$

The best justified formula for combining two random sets $\mathcal{S} := \mathcal{S}_1 \odot \mathcal{S}_2$, yielding a combined evidence function $m = m_1 \odot m_2$, is **Dempster's rule**

$$\forall A \subseteq \Omega, \qquad m(A) := \frac{\sum_{A_1 \cap A_2 = A} m_1(A_1) m_2(A_2)}{\sum_{A_1 \cap A_2 \neq \emptyset} m_1(A_1) m_2(A_2)}. \tag{4}$$

Since $\forall \omega_i, \mathrm{Pl}_i \in [0, 1]$, therefore Pl is the membership function of a **fuzzy subset** of Ω, denoted $\widetilde{\mathrm{Pl}}$. Conversely, any fuzzy subset $\widetilde{F} \subseteq \Omega$ with membership function $\mu_{\widetilde{F}}: \Omega \mapsto [0, 1]$ can be mapped to an equivalence class of one-point equivalent random sets on Ω. If $\sum_i \mu_{\widetilde{F}}(\omega_i) \geq 1$, then $\widetilde{F}$ can be taken as a plausibility assignment of any of an equivalence class of random sets; similarly, if $\sum_i \mu_{\widetilde{F}}(\omega_i) \leq 1$, then $\widetilde{F}$ can be taken as the one-point assignment of a belief function of any of a different equivalence class of random sets. Under some conditions the evidence values m_j are determined by the plausibility assignment values Pl_i. Then $N \leq n$, and Pl is a **distribution** of $\mathcal{S}$.

When $\forall A_j \in \mathcal{F}, |A_j| = 1$, then $\mathcal{S}$ is **specific**, and $\Pr(A) := \text{Pl}(A) = \text{Bel}(A)$ is a **probability measure** which is additive in the traditional way

$$\forall A, B \subseteq \Omega, \qquad \Pr(A \cup B) = \Pr(A) + \Pr(B) - \Pr(A \cap B). \tag{5}$$

Then $\mathbf{p} = \langle p_i \rangle := \text{Pl}$ is a **probability distribution** with additive normalization and operator

$$\sum_i p_i = 1, \qquad \Pr(A) = \sum_{\omega_i \in A} p_i. \tag{6}$$

$\mathcal{S}$ is **consonant** ($\mathcal{F}$ is a nest) when (without loss of generality for ordering, and letting $A_0 := \emptyset$) $A_{j-1} \subseteq A_j$. Now $\Pi(A) := \text{Pl}(A)$ is a **possibility measure** and $\eta(A) := \text{Bel}(A)$ is a **necessity measure**. Since results for necessity are dual to those of possibility, only possibility will be discussed in the sequel.

As Pr is additive, so Π is **maximal**, in that $\forall A, B \subseteq \Omega$, $\Pi(A \cup B) = \Pi(A) \vee \Pi(B)$, where $\vee$ is the maximum operator. As long as $\mathbf{C}(\mathcal{F}) \neq \emptyset$ (this is required if $\mathcal{F}$ is a nest), then $\boldsymbol{\pi} = \langle \pi_i \rangle := \text{Pl}$ is a **possibility distribution** with maximal normalization $\bigvee_i \pi_i = 1$ and operator

$$\Pi(A) = \bigvee_{\omega_i \in A} \pi_i. \tag{7}$$

However, **consistency** $\mathbf{C}(\mathcal{S}) \neq \emptyset$, not consonance, is all that is necessary for Pl to be a maximally normalized possibility distribution π, even though the plausibility *measure* of a consistent, non-consonant random set is not a possibility measure. But given a possibility distribution π, there is a unique consonant one-point equivalent random set, denoted $\mathcal{S}^\pi$ and called the **consonant approximation**. It is determined by taking the ω_i in order of descending π_i, and letting $A_i = \{\omega_1, \omega_2, \ldots, \omega_i\}$ and $m_i = \pi_i - \pi_{i+1}$, where $\pi_{n+1} := 0$.

Nonspecificity N and **strife S** are two **uncertainty measures** which are defined on random sets, respectively

$$\mathbf{N}(\mathcal{S}) := \sum_j m_j \log_2 |A_j|, \qquad \mathbf{S}(\mathcal{S}) := -\sum_j m_j \log_2 \left[\sum_{k=1}^{n} m_k \frac{|A_j \cap A_k|}{|A_j|} \right]. \tag{8}$$

They measure respectively the possibilistic and probabilistic aspects of the uncertainty or information represented in the random set, and together form the **total uncertainty** $\mathbf{T} := \mathbf{N} + \mathbf{S}$. They have special forms for distributions, and in the probabilistic case the uncertainty collapses to stochastic entropy $\mathbf{H}(\mathbf{p})$, while in the possibilistic case the strife is bounded above by a small number.

Generalized **processes** are defined on distributions when a generalized **disjunction operator** $\oplus$ and a generalized **conjunction operator** $\otimes$ on $[0, 1]$ are available such that:

- $\langle \oplus, \otimes \rangle$ form a semiring ($\otimes$ distributes over $\oplus$),

- $\oplus = \sqcup$ and $\otimes = \sqcap$ are a triangular **conorm** and **norm**: monotonic, associative, commutative operators with identity 0 and 1 respectively [4], so that $\langle \oplus, \otimes \rangle = \langle \sqcup, \sqcap \rangle$ is a **conorm semiring**,
- and $\oplus$ is the operator of the distributions in question.

For $x, y \in [0, 1]$, the operation $x \sqcup_m y := (x + y) \wedge 1$ is a conorm, where $\wedge$ is the minimum operator. So stochastic normalization (6) forces $+$, when acting on a probability distribution and restricted to $[0, 1]$, to be a conorm $\oplus = \sqcup_m = +$. Then $\langle +, \times \rangle$ is the unique semiring for stochastic processes. For possibility, $\oplus = \vee$ is the unique conorm operator, but there are many semirings of the form $\langle \vee, \sqcap \rangle$ for a generic norm $\sqcap$. $\wedge$ and $\times$ are two of the more popular norms, as is $x \sqcap_m y := 0 \vee (x + y - 1)$.

Given an appropriate semiring, then **joint, marginal**, and **conditional** distributions are available, with novel possibilistic forms. In particular, **conditional possibility** is parameterized by the choice of norm operator $\sqcap$, and even once a norm is fixed, $\pi(y|_\sqcap x)$ is not always unique.

Given a particular **current state distribution** (probabilistic $\mathbf{p}$ or possibilistic $\boldsymbol{\pi}$), and a table of conditional probabilities or possibilities, then the **next state distribution** $\mathbf{p}'$ or $\boldsymbol{\pi}'$ is derived from a generalized linear matrix composition operation (the familiar form in the probabilistic case).

A **possibilistic process** is a system $\mathcal{Z}_\pi := \langle \Omega, \sqcap, \boldsymbol{\Pi}, \boldsymbol{\pi}^0 \rangle$ for some $\sqcap$, where:

- $\boldsymbol{\Pi} = [\boldsymbol{\Pi}_{ij}] = \left\langle \boldsymbol{\Pi}^{(j)} \right\rangle$ is a matrix of conditional possibilities, so that for $1 \leq j \leq n, \boldsymbol{\Pi}^{(j)}$ is the vector representation of a conditional possibility distribution function $\pi(\cdot|\omega_j): \Omega \mapsto [0, 1]$, $\boldsymbol{\Pi}$ is a vector of such conditional distributions, and $\boldsymbol{\Pi}_{ij} := \pi(\omega_i|\omega_j)$;
- $\boldsymbol{\pi}^t$ is a possibility distribution on Ω with $\boldsymbol{\pi}^0$ given, and $\forall t > 0$,

$$\boldsymbol{\pi}^t := \boldsymbol{\pi}^{t-1} \circ \boldsymbol{\Pi}, \qquad \pi_i^t := \bigvee_{j=1}^{N} \pi_j^{t-1} \sqcap \pi(\omega_i|\omega_j), \qquad (9)$$

where $\circ$ is $\langle \vee, \sqcap \rangle$ matrix composition.

$\pi_i^t = \pi^t(\omega_i)$ is the possibility of being in state ω_i at time t, and $\boldsymbol{\Pi}_{ij} = \pi(\omega_i|\omega_j)$ is the possibility of transiting from state ω_j to ω_i.

It has been demonstrated [12, 11] that while both stochastic and possibilistic processes generalize deterministic processes, only possibilistic processes are generalizations of *nondeterministic* processes. When possibility values are restricted to $\{0, 1\}$, then the strict nondeterministic case is recovered.

Possibilistic versions of some of the standard stochastic systems theoretical forms have been defined [12]. In particular, **possibilistic Markov processes** and **Monte Carlo methods** are available, and **possibilistic automata** are constructed by extending possibilistic processes to include possibilistically distributed input and output alphabets and functions.

Finally, it should be noted that both probabilistic and possibilistic processes are specializations of fuzzy processes using fuzzy matrix composition [18]. In each special case the conorm semiring $\langle \sqcup, \sqcap \rangle$ is restricted to special forms, resulting in the various normalization and other conditions mentioned above, which would not otherwise be required.

2.2 Possibilistic Models

A possibilistic model requires possibilistic measurement and prediction procedures. Possibilistic prediction is based on the possibilistic processes briefly outlined above. **Possibilistic measurement** methods have also been developed by the author [8, 10]. The essential requirement is the collection of the frequency of occurrence of subsets or intervals which are partially overlapping. If their global intersection is nonempty (the empirical random set is consistent), then (2) yields an empirical possibility distribution. Otherwise, possibilistic normalization procedures [9] would be required.

3   CAST Implementations of Possibilistic Models

The Computer-Aided Systems Theory (CAST) movement is predicated on the idea that computer implementations are at least useful, and sometimes even necessary, for the development and application of systems theoretical methods. This has been remarked on by Klir.

> Systems knowledge can also be obtained experimentally. Although systems (knowledge structures) are not objects of reality, they can be simulated on computers and in this sense made real. We can then experiment with the simulated systems for the purpose of discovering or validating various hypotheses in the same way as other scientists do with objects of their interest in their laboratories. In this sense, computers, may be viewed as laboratories of systems science. Experimentation with systems on computers is not merely possible, but it may give us knowledge that is otherwise unobtainable. [16, p. 102]

Horgan notes [6] that as the complexity of problems increases, this situation is becoming common generally in mathematics. The result is the growth of so-called "experimental" or "computer-aided" mathematics, where computer-based tools are used to empirically investigate the properties of mathematical systems.

CAST implementations of possibilistic systems in particular are crucial not only as platforms for the application of possibilistic qualitative modeling, but also for the empirical investigation of the properties of possibilistic processes. There are still many open questions, as described more fully elsewhere [12].

Existing CAST implementations (for example Pichler's [22] and Zeigler's [29]) are deterministic. The extension of these implementations, and the development of new environments, to include representations of indeterminism, uncertainty, and information is crucial. For example, existing systems could not implement neural networks with (stochastic) noise.

It is also clear that the fundamental categories for the representation of uncertainty and information should be included in the foundational, primitive levels of CAST systems, from which implementations of more complex and specific systems theoretical methods should then be constructed. This includes the entire repertoire of GIT, allowing, for example, the use of methods from probability, statistics, and fuzzy theory, as well as random sets and possibility theory.

GIT-based CAST implementations should allow the handling of hybrid sources and representations of uncertainty, the integration of multiple sources of information, and the transformation between representational forms of information. For example, Klir's General Systems Problem Solver (GSPS) [15] was designed specifically to accommodate both probabilistic and possibilistic representations of information, and is best implemented in a GIT-based general systems theoretical CAST environment.

There has recently been an explosion of fuzzy systems implementations for both the commercial and academic markets ([7, 23] are examples). The same is not the case for general GIT methods, however, and certainly not for possibility theory. One exception is the work of Galway [5], who has implemented a system for manipulating random subsets of $\mathbb{R}^2$.

4 Object-Oriented Environments

One of the most successful programming paradigms of recent years is the **object-oriented** approach [3]. This methodology is based on the concepts of **objects**, which are complex data elements, and **classes**, or "intelligent data types" for objects, which isolate type-specific procedures within type-specific levels. Logical relations among classes allow for the **inheritance** of procedures from more general classes to their specialized cases. Classes and objects have **attributes**, either other objects (data attributes) or **methods** (procedural attributes). A class **invariant** is a logical condition which must always be true of every object of the class in order for it to be existing in a legal state.

The target language for the proposed architecture below is C++ [24]. It was selected for its popularity and efficiency, and because of the availability of standard, inexpensive compilation environments and software support libraries.

The main results of this paper are summarized in Figs. 1–3, which show Entity-Relationsip (ER) diagrams [3] of the proposed class hierarchies. These are a slight modification of the standard form presented by Coad and Yourdon [3]. Each node denotes a class (data type). Nodes are linked by labeled arcs, each indicating one of the following relations, where X and Y are classes:

- $X \xrightarrow{\text{is-a}} Y$: Y inherits from X, so that Y is a **specification** of X and has all the properties of X. X is called a **parent** and Y a **child**. For engineering reasons, or to capture efficiencies present in the special cases, some attributes may be implemented redundantly in child classes. For example, the formula for random set nonspecificity $\mathbf{N}(\mathcal{S})$ is greatly simplified in the consonant case by calculating the nonspecificity of the possibility distribution $\mathbf{N}(\pi)$.

- $X \xrightarrow{\text{collection}} Y$: Y objects are implemented as a **collection** (for example, a list, set, bag, or vector) of X objects.
- $X \xrightarrow{\text{has-a}} Y$: Y is a **component** of X, so that each X-object contains a Y sub-object. When Y is a sub-object of X, then Y may have access to X-specific information. For engineering or efficiency reasons, Y may be implemented separately from X, or a Y may be constructed from an X, copying the appropriate X-specific information into the Y object. In fact, at the strictly logical level this relation simply requires that each X object **determines** a unique Y object, or that a procedure exists to construct a Y object from an X object. This is the sense that will frequently be used below.

Note that while arrows always move from the more general to the more specific, this is not always mirrored by the English translations. For example, $X \xrightarrow{\text{is-a}} Y$ is read that "Y is an X", while $X \xrightarrow{\text{has-a}} Y$ is read that "X has a Y".

5 Fundamental Classes

Each of the ER diagrams below describes a different portion of the overall architecture, and is accompanied by a set of descriptions of the classes included in the figure. Only the most basic methods and procedures are included here; supplementary methods are described in Sec. 6.1.

The ER diagrams and class descriptions are written in a kind of non-C++-specific class design "pseudo-code". Only the *logical* relations among the classes are described. For example, in Fig. 1 it is not specified exactly how a plausibility assignment is determined from a plausibility measure. It will be presumed that if $X \xrightarrow{\text{has-a}} Y$, then Y will have access to X-specific information. The figures share the class `Poss_Dist`, for possibility distributions, in common.

5.1 Random Sets

Fig. 1 shows the class hierarchy including random sets and their fuzzy measures and distributions. The class `Random_Set` is the most general, and therefore one of the most heavily laden, classes in the proposed architecture.

`Random_Set` — A random set S.

Data Attributes	Universe	Integer $n = |\Omega|$
	Card	Integer $N = |S|$
	Data	A list of pairs $\langle a_j, m_j \rangle$, with floats m_j and integers a_j, where a_j is an n-long bit-mask determining the characteristic function χ_{A_j} of the focal element $A_j \subseteq \Omega$
Invariants		$1 \leq n, \quad 0 \leq m_j \leq 1, \quad \sum_{j=1}^{N} m_j = 1$ $1 \leq a_j \leq 2^n - 1, \quad 1 \leq j \leq N \leq 2^n - 1$
Methods	Strife	$S(S)$
	Nonspec	$N(S)$
	Total	$T(S)$
	Core	$C(S)$

Random sets, continued.

Methods	Monte_Set	Select a focal element A_j by a possibilistic Monte Carlo method
	Monte	Select a universe element ω_i by a possibilistic Monte Carlo method
	+	The Dempster combination operator $\odot$ (4) which combines this Random_Set with another, producing a combined Random_Set
	<=	The random set inclusion relation $\subseteq$ (3), a boolean reporting whether this Random_Set is included within another
	Complete?	Boolean: is S complete?

Consistent_RS — A consistent random set.

Invariant	$C(S) \neq \emptyset$

Consonant_RS — A consonant random set.

Data Attribute	Ordering	A list $\langle a_j \rangle$ permuting the a_j according to the inclusion relation among the A_j
Invariants		$\forall 1 \leq \bar{\jmath}_1 \leq \bar{\jmath}_2 \leq N, A_{\bar{\jmath}_1} \subseteq A_{\bar{\jmath}_2}, \quad 1 \leq N \leq n$

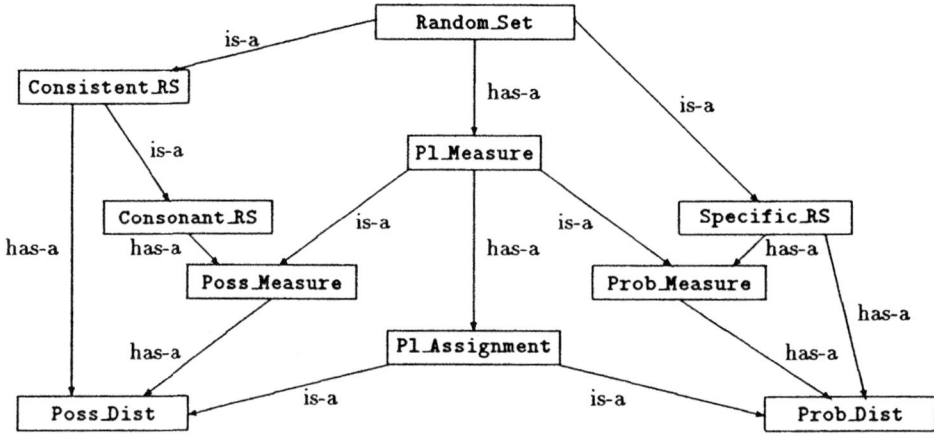

Fig. 1. Random sets, evidence measures, and distributions.

Specific_RS — A specific random set.

Invariants	$\forall 1 \leq j \leq N, |A_j| = 1, \quad 1 \leq N \leq n$

Pl_Measure — A plausibility measure Pl on a random set.

Method	Value	Given $0 \leq a_j \leq 2^n - 1$, returns a float $\mathrm{Pl}^j := \mathrm{Pl}(A_j)$.

Poss_Measure — A possibility measure Π.

Invariant$\|\forall a_1, a_2, \mathrm{Pl}(A_1 \cup A_2) = \mathrm{Pl}^1 \vee \mathrm{Pl}^2$.

Prob_Measure — A probability measure Pr.

Invariant$\|\forall a_1, a_2, \mathrm{Pl}(A_1 \cup A_2) = \mathrm{Pl}^1 + \mathrm{Pl}^2 - \mathrm{Pl}(A_1 \cap A_2)$.

Pl_Assignment — A plausibility assignment Pl from a random set.

Data Attribute	Data	A list of floats $\langle \mathrm{Pl}_i \rangle$, where $\mathrm{Pl}_i = \mathrm{Pl}(\{\omega_i\})$
Invariant		$1 \le i \le n$

Poss_Dist — A possibility distribution π.

Invariant		$\bigvee_i \mathrm{Pl}_i = 1$
Methods	Nonspec	$N(\pi)$
	Strife	$S(\pi)$
	Total	$T(\pi)$
	Core	$C(\pi)$

Prob_Dist — A probability distribution p.

Invariant		$\sum_i \mathrm{Pl}_i = 1$
Method	Entropy	$H(p)$

5.2 General Distributions

Fig. 2 shows the class hierarchy of distributions and fuzzy sets.

Element — A generic element of a distribution or fuzzy set.

Data Attribute	Value	A floating-point "fit" (fuzzy digit), f.
Invariant		$0 \le f \le 1$

Dist_Elem — An element of a distribution.

Methods	+	The generalized disjunction $\oplus$ which aggregates this Dist_Elem with another, producing an aggregated Dist_Elem.
	*	The generalized conjunction operator $\otimes$ which combines this Dist_Elem with another, producing a combined Dist_Elem.

Fuzzy_Set — A collection of elements comprising μ.

Data Attributes	Universe	Integer $n = |\Omega|$
	Data	A list of Elements $\langle f_i \rangle$.
Invariant		$1 \le i \le n$

Poss_Elem — Element of a possibility distribution π_i.

Invariant$\| f_1 \sqcup f_2 = f_1 \vee f_2$.

Poss_Dist — A possibility distribution π.

Data Attributes	Universe	Integer $n = |\Omega|$
	Data	A list of Poss_Elems $\langle \pi_i \rangle$
Invariant		$1 \le i \le n, \quad \bigvee_{i=1}^{n} \pi_i = 1$

Prob_Elem — Element of a probability distribution p_i.

Invariant$\| f_1 \sqcup f_2 = f_1 + f_2$.

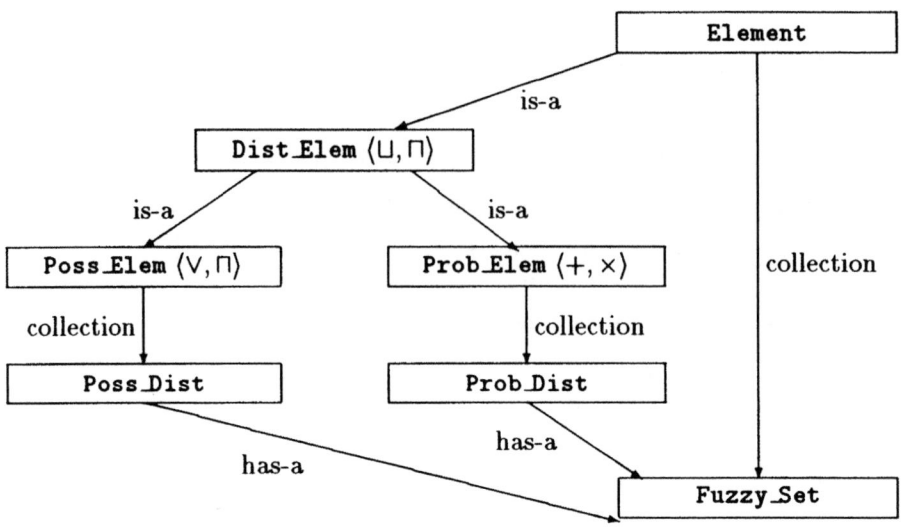

Fig. 2. Distributions and fuzzy sets.

`Prob_Dist` — A probability distribution p.

Data Attributes	Universe	Integer $n = |\Omega|$
	Data	A list of `Prob_Elems` $\langle p_i \rangle$
Invariant		$1 \leq i \leq n, \qquad \sum_{i=1}^{n} p_i = 1$

Note that `Poss_Dist` and `Prob_Dist` are repeated here as collections of their elements, inheriting from `Pl_Assignment` from Fig. 1.

5.3 Possibilistic Processes

Finally, Fig. 3 shows the class hierarchy of possibilistic processes. `Poss_Dist` has been specified above.

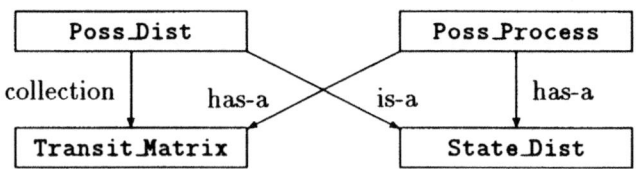

Fig. 3. Possibilistic processes.

`Transit_Matrix` — A possibilistic transition matrix Π.

Data Attributes	Universe	Integer $n = |\Omega|$
	Data	A list of `Poss_Dists` $\langle \Pi^{(j)} \rangle$.
Invariant		$1 \leq j \leq n$

State_Dist — The current state possibility distribution π^t.

Data Attribute	Time	The current time, an integer t
Invariant		$0 \le t$

Poss_Process — A possibilistic process $\mathcal{Z}_\pi$.

| Method | Advance | Determine next state function $\pi^t = \pi^{t-1} \circ \Pi$. |

6 Extensions to the Basic Architecture

Of course, the architecture described above is merely the core for a broader implementation of possibilistic models, which must also involve a variety of measurement methods and links into CAST implementations of other aspects of GIT, let alone the input/output routines necessary for any software system.

6.1 Supplementary Methods

There are a number of other procedures [12] which are special or supplementary to the basic procedures, but which are still beneficial to implement explicitly. Most of these are transformations of one of the classes to another. The following relations can be appended to the basic diagrams above as appropriate.

Möbius Transform: The **Möbius transform** [14] or **fast Möbius transform** [26] is an algorithm utilizing the Möbius inversion formula (1) to calculate among belief measures and evidence functions. The fast Möbius transformation is extremely efficient, and will be used in implementing the relation Random_Set $\xrightarrow{\text{has-a}}$ Pl_Measure from Fig. 1. Let the $\omega_i \in \Omega$ be taken in an arbitrary order, and assume a random set $\mathcal{S}$.

  1. Assume the evidence function m of $\mathcal{S}$, and let $m_0 := m$. Then $\forall A \subseteq \Omega$, and $1 \le i \le n$, determine m_i by the algorithm

$$m_i(A) := \begin{cases} m_{i-1}(A) + m_{i-1}(A - \{\omega_i\}), & \omega_i \in A \\ m_{i-1}(A), & \omega_i \notin A \end{cases}. \quad (10)$$

Then $m_n = \text{Bel}$, where Bel is the belief function of $\mathcal{S}$.

  2. Assume the belief function Bel of $\mathcal{S}$, and let $m_n := \text{Bel}$. Then $\forall A \subseteq \Omega$, and $n \ge i \ge 1$, determine m_i by the algorithm

$$m_{i-1}(A) := \begin{cases} m_i(A) - m_i(A - \{\omega_i\}), & \omega_i \in A \\ m_i(A), & \omega_i \notin A \end{cases}. \quad (11)$$

Then $m_1 = m$, where m is the evidence function of $\mathcal{S}$.

Distribution Operations:

$$\text{Poss\_Dist} \xrightarrow{\text{has-a}} \text{Poss\_Measure}, \qquad \text{Prob\_Dist} \xrightarrow{\text{has-a}} \text{Prob\_Measure}. \quad (12)$$

From (7), given a possibility distribution π, a possibility measure Π can be constructed; similarly, from (6), given a probability distribution p, a probability measure Pr can be constructed.

Probabilistic Approximations:

$$\texttt{Random\_Set} \xrightarrow{\text{has-a}} \texttt{Prob\_Dist}. \tag{13}$$

Probability distribution approximations of random sets are available [12] either as the **maximum entropy probability distribution**

$$p^{\mathcal{S}}(\omega) := \sum_{A_j \ni \omega} \frac{m_j}{|A_j|}, \tag{14}$$

or the **Bayesian approximation** $\bar{p}$ determined from

$$\bar{p}(\omega_i) := \frac{\sum_{A_j \ni \omega_i} m_j}{\sum_{j=1}^{N} m_j |A_j|} = \frac{\text{Pl}_i}{\sum_{i=1}^{n} \text{Pl}_i}. \tag{15}$$

Compatibility Measures: Given a `Prob_Dist` p and `Poss_Dist` π, a **compatibility measure** $\gamma(\pi, p)$ [12] represents the degree of compatibility of consistency between p and π. These are very useful in possibilistic normalization and transformation procedures. The most prominent is the Zadeh-compatibility $\gamma_Z(\pi, p) = \sum_i p_i \pi_i$.

Consonant Approximation:

$$\texttt{Consistent\_RS} \xrightarrow{\text{has-a}} \texttt{Consonant\_RS}. \tag{16}$$

As mentioned in Sec. 2, the consonant approximation random set $\mathcal{S}^\pi$ is uniquely determined by the maximally normal plausibility assignment of a consistent random set.

Frequency Conversions:

$$\texttt{Prob\_Dist} \xrightarrow{\text{has-a}} \texttt{Poss\_Dist}, \qquad \texttt{Poss\_Dist} \xrightarrow{\text{has-a}} \texttt{Prob\_Dist}. \tag{17}$$

Probability and possibility distributions are co-determining according to a wide variety of different methods [12].

Possibilistic Normalization:

$$\texttt{Random\_Set} \xrightarrow{\text{has-a}} \texttt{Consistent\_RS}, \qquad \texttt{Pl\_Assignment} \xrightarrow{\text{has-a}} \texttt{Poss\_Dist}. \tag{18}$$

Even given an inconsistent random set, there are a variety of **possibilistic normalization** methods [9, 12] which allow for the construction of a consistent random set, and of a normal possibility distribution from a sub-normal plausibility assignment.

Measurement: The result of the possibilistic measurement methods developed by the author [8, 10], and described in Sec. 2.2, is the construction of a random set, hopefully consistent, from measured data. Therefore, while they require explicit implementation, they fall outside of the regular class hierarchy which has one root in the class `Random_Set`.

6.2 Other Extensions

There are further extensions to link the implementation of possibilistic methods with other GIT methods and CAST implementations. These extensions can be considered either as a part of this research program, or as extensions to the research programs which have been or may be launched by others.

- The CAST program originated with the implementations of deterministic finite state machines of Pichler and his colleagues [22]. There is a clear relation to the possibilistic approach described here, and the opportunity to generalize to a variety of different GIT-based representations of finite state machines with uncertainty, including nondeterministic and stochastic machines.
- There has been some work [2, 21] on the implementation of Klir's GSPS system [15]. It would certainly be very valuable to relate these efforts directly.
- In addition to the measurement methods mentioned above, **possibilistic clustering methods**, including **possibilistic c-means** [20] and the **mountain method** [1], can be integrated.
- While we have criticized the traditional dependence of possibility theory on fuzzy theory, their relation certainly cannot be ignored. And although it is not our specific focus, there is value in relating possibilistic implementations with those of the variety of fuzzy set operations and concepts. There is by now a huge literature on these methods (see Kosko [19] and Terano, Asai, and Sugeno [25] for just two examples), and many academic and corporate efforts to develop fuzzy theoretical systems. Hopefully other researchers are building CAST-based implementations of fuzzy systems methods, which could then be integrated with this specifically possibilistic system. In particular, as mentioned in Sec. 2, the action of possibilistic and stochastic processes is just a case of fuzzy relation composition, so a generalized fuzzy CAST system implementing fuzzy relations can be used as a base for possibilistic automata.
- Finally, possibility measures, as extreme plausibility measures, exist within the more general Dempster-Shafer evidence theory. The extension of possibility theoretical implementations to necessity measures (as the extreme belief measures dual to possibility measures), and general belief/plausibility pairs, may be very useful. Beyond that, both belief and plausibility measures are special fuzzy measures [27], and so the ultimate extension is to the construction of CAST-based systems for fuzzy measures in general.

7 Empirical Investigations

As noted in Sec. 3, it is common in systems theory that computer-based implementation and simulation are necessary in order to investigate the properties of the systems under consideration, and this is the case with possibilistic systems, processes, and models. There are a number of issues which it is desirable to investigate empirically.

Nonspecificity Calculations: Determination of informational properties, and in particular nonspecificity values and the changes in these values, is of great interest. This would include, for example:

- Calculation of $N(\pi^t)$ of the state vector of a possibilistic process as a function of t;
- Calculation of $N(\pi)$ where π is a possibilistic histogram, and the dependence of $N(\pi)$ on the measurement method used;
- The change from $T(S)$ to $T(\pi)$ under possibilistic normalization (transforming an inconsistent random set S to a consistent random set with possibility distribution π), and the dependence on both the general normalization method chosen and the various sub-choices required within some of the methods.
- Determination of $N(\pi)$ under the cases where π is a special fuzzy number [10], for example a parallelogram or triangle.

Possibilistic Processes: Aside from nonspecificity calculations, there are a number of other properties of possibilistic processes which require empirical investigation, for example the dependence of the form of possibilistic processes on the choice of norm operator $\sqcap$ and the choice of conditional possibility measure.

Measurement Methods: It is natural to explore the properties of the various measurement methods empirically, comparing the results of one data source using multiple methods.

Uncertainty Invariance Transformations: The Uncertainty Invariance Principle (UIP) [18] is a method for transformation among GIT structures on the basis of conservation of total uncertainty. While Klir and Parviz [17] have begun to empirically examine some results of frequency conversion methods, including the UIP, there are still many unanswered questions about the UIP. In the context of this work, it would be interesting to compare the time evolution of similar stochastic and possibilistic processes, and then compare those against their respective UIP transformations.

References

1. Barone, Joseph M; Filev, Dimitar P; and Yager, Ronald R: (1993) "Exponential Possibility Distributions and the Mountain Method", in: *Proc. NAFIPS 1993*, ed. M. McAllister, pp. 118-122, Allentown PA
2. Cellier, Francois: (1987) "Qualitative Simulation of Technical Systems Using the GSPS Framework", *Int. J. of General Systems*, v. **13**:4, pp. 333-344
3. Coad, Peter and Yourdon, Edward: (1991) *Object-Oriented Analysis*, Prentice-Hall, Englewood Cliffs NJ
4. Dubois, Didier and Prade, Henri: (1988) *Possibility Theory*, Plenum Press, New York
5. Galway, L: (1986) "Workstation Based Environment for the Statistical Analysis of Set-Valued Data", in: *Computer Science and Statistics: Proceedings of the 18th Symposium on the Interface*, pp. 332-334
6. Horgan, John: (1993) "Death of Proof", *Scientific American*, v. **269**:4, pp. 74-89

7. Janzen, Thomas E: (1993) "C++ Classes for Fuzzy Logic, Sets, and Associative Memories", *C User's Journal*, v. **11/93**, pp. 55-71

8. Joslyn, Cliff: (1992) "Possibilistic Measurement and Set Statistics", in: *Proc. NAFIPS 1992*, v. **2**, pp. 458-467, Puerto Vallerta

9. Joslyn, Cliff: (1993) "Empirical Possibility and Minimal Information Distortion", in: *Fuzzy Logic: State of the Art*, ed. R Lowen and M Roubens, pp. 143-152, Kluwer, Dordrecht

10. Joslyn, Cliff: (1993) "Some New Results on Possibilistic Measurement", in: *Proc. NAFIPS 1993*, pp. 227-231, Allentown PA

11. Joslyn, Cliff: (1994) "On Possibilistic Automata", in: *Computer Aided Systems Theory—EUROCAST '93*, ed. F. Pichler and R. Moreno-Diáz, pp. 231-242, Springer-Verlag, Berlin

12. Joslyn, Cliff: (1994) *Possibilistic Processes for Complex Systems Modeling*, PhD dissertation, SUNY-Binghamton, UMI Disseration Services, Ann Arbor MI

13. Joslyn, Cliff: (1995) "Towards an Independent Possibility Theory with an Objective Semantics", in: *Proc. 1995 Int. Workshop on Foundations and Applications of Possibility Theory*, to appear

14. Kennes, Robert: (1992) "Computational Aspects of Möbius Transformations of Graphs", *IEEE Trans. on Systems, Man, and Cybernetics*, v. **22**:2, pp. 201-223

15. Klir, George: (1985) *Architecture of Systems Problem Solving*, Plenum, New York

16. Klir, George: (1991) *Facets of Systems Science*, Plenum, New York

17. Klir, George and Parviz, Behvad: (1992) "Probability-Possibility Transformations: A Comparison", *Int. J. of General Systems*, v. **21**:1, pp. 291-310

18. Klir, George and Yuan, Bo: (1995) *Fuzzy Sets and Fuzzy Logic*, Prentice-Hall, New York

19. Kosko, B, ed.: (1990) *Neural Networks and Fuzzy Systems*, Prentice-Hall, Englewood Cliffs NJ

20. Krishnapuram, Raghu and Keller, James: (1993) "Possibilistic Approach to Clustering", *IEEE Trans. on Fuzzy Systems*, v. **1**:2, pp. 98-110

21. Orchard, Robert: (1990) "Knowledge Processing: Semantics for the Klir Hierarchy of General Systems", in: *Computer Aided Systems Theory*, ed. Franz Pichler, pp. 33-40, Springer-Verlag, Berlin

22. Pichler, Franz and Praehofer, Herbert: (1988) "CAST:FSM – Computer Aided Systems Theory: Finite State Machine", in: *Cybernetics and Systems '88*, ed. R. Trappl, pp. 737-742, Kluwer, Boston

23. Sosnowski, Zenon A and Pedrycz, W: (1992) "FLISP: Representing and Processing of Uncertain Information", in: *Fuzzy Logic for the Management of Uncertainty*, ed. L Zadeh and J Kacpryzk, pp. 495-521, Wiley, New York

24. Stroustrup, Bjarne: (1991) *C++ Programming Language*, Addison Wesley

25. Terano, Toshiro; Asai, Kiyoji; and Sugeno, Michio: (1987) *Fuzzy Systems Theory and Its Applications*, Academic Press, New York

26. Thoma, H Mathis: (1991) "Belief Function Computations", in: *Conditional Logic in Expert Systems*, ed. IR Goodman et al., pp. 269-308, North Holland

27. Wang, Zhenyuan and Klir, George J: (1992) *Fuzzy Measure Theory*, Plenum Press, New York

28. Zadeh, Lotfi A: (1978) "Fuzzy Sets as the Basis for a Theory of Possibility", *Fuzzy Sets and Systems*, v. **1**, pp. 3-28

29. Zeigler, BP: (1990) *Object-Oriented Simulation with Hierarchical Modular Models*, Academic Press, San Diego

Fuzzy Expert System Technology

I.B. Türkşen

Principal Investigator
Intelligent Systems Laboratory
Department of Industrial Engineering
University of Toronto, Toronto
Ontario, M5S 1A4, Canada

Abstract. Fuzzy expert systems are the second generation expert systems. In real world activities, system states have various attributes that at best could be assessed as a matter of degree. If we express our knowledge of system states in terms of an all or nothing paradigm then much valuable information does not appear in system's models. However, if we express our knowledge of systems with fuzzy sets, much valuable information stays embedded in the rules of behaviour in system's models. Thus, the fuzzy set paradigm gives our knowledge of systems an effective representation.

For this purpose, four levels of knowledge representation are identified as: (i) linguistic, (ii) meta-linguistic, (iii) propositional, and (iv) computational. In particular, four classes of fuzzy rule base schema are identified within a unified framework of implication statements. That is we identify at least four classes of propositional expressions that represent linguistic expression of rules. Interval valued fuzzy sets are defined as particular Type II fuzzy sets in order to expose second order semantic uncertainty embedded in linguistic expressions.

1. Introduction

Generally, experts communicate their knowledge of a particular systems' behaviour via linguistic expressions of a natural language. Systems analysts, on the other hand, transform these linguistic expressions with their interpretive skills into certain "meta-linguistic" expressions, such as logical, mathematical, graphical, etc., in order to develop a particular analysis of a context dependent system's description. Next, such meta-linguistic expressions are transformed into certain "propositional expressions" again based on some interpretations and assumptions. Finally, propositional expressions are transformed into certain "computational expressions". At the end, after an appropriate analysis of computational expression, we usually need to return to linguistic expressions in order to interpret the "meaning" representation that is embedded in these computational expressions. This last step is needed in order to verify that an expert's knowledge is transformed properly to reflect a system's behaviour pattern. This cycle may be repeated a number of times until an expert's knowledge is properly encoded in computational expression through the transformation between the four levels of knowledge representation.

In this paper, we discuss a particular subset of these four levels of knowledge representation through fuzzy sets and logics paradigm.

1.1 Natural language

Since experts' communication commences with a natural language, let us review some relevant aspects of a natural language and its compostionality within the context of fuzzy expert systems technology.

The basic fact about a natural language that inspires a compositional view is that we humans are able to parse and produce novel sentences in finite time, based on finite knowledge. In concrete linguistic contexts, we are able to interpret sentences we have never heard before and respond to them with sentences we have never constructed before. It appears undeniable that the understanding which is the result of such novel interpretations must depend in large measure on some sort of prior understanding of the "meanings" of the component terms. And this "prior understanding" must be dependent on the new content in some way and to some "degree". This is the sense with which L.A. Zadeh, [6, 8] proposed fuzzy sets and logics for the representation of these context dependent "meanings".

On the contrary, the proponents of the classical set and logic presuppose that these general "meanings" are entirely context independent, at least so far as the particular speakers/hearers are concerned. For them, these context independent "meanings" are "objective", "literal", "idealized", etc. They are treated as "sharply delimited" well defined little objects that are able to line up next to each other with certain "constant connectives", known as classical logic connectives "and", "or", and governed by well defined rules of "grammar". The problem of novelty becomes entirely combinatorial under the presuppositions of the proponents of classical set and logic.

However, "idealized meanings" require idealized structures, methods, procedures and understandings. There is no risk and no surprises in this idealized world of classical logic.

What impresses us most about a natural language is its vivacity, robustness, flexibility. There is the extraordinary resourcefulness of speakers and listeners, experts and analysts in the face of complex and overwhelming circumstances that are imprecisely and vaguely describable.

The vast majority of our words are nothing more than gross, rough hewn stop gaps in the face of "imprecise information" realities of our everyday experiences. Despite evidently "organized" reality that finite objects have "relatively" well defined boundaries, it is still inescapable that the vast majority of detail we constantly perceive must be ignored and summarized into clusters of information now known as fuzzy sets.

1.2 Linguistic Expressions

Within the context of fuzzy expert systems, expert knowledge is expressed with the following type of rules:

(i) (a) "If interest rates increase alot, then the speculative demand for stocks would decrease quite a bit";

  (b) "If the temperature is high then the comfort level would be low";

(ii) "if inventory is low and demand is high, then production should be high," etc.

The first two expressions are simple rules with a single left hand side known as the "antecedent" and a single right hand side known as the "consequent". The second expression is a bit more complex with two antecedents and a single consequent.

In general, expert rules may contain multiple antecedents and multiple consequents. For the sake of brevity, we will discuss only the single antecedent and single consequent rules which are essentially the building blocks for more complex rules. There are now well know decomposition algorithms that reduce these complex rules to simple ones and then combine their results [3].

1.3 Meta-Linguistic Expressions

Meta-linguistic expressions are symbolic abstract representations of linguistic expressions. At times they are known as "object-language" expressions. The abstract meta-linguistic expression of the simple linguistic rule expression is often stated as

$$\text{IF A THEN B} \tag{1}$$

$$\text{or} \quad A \rightarrow B \tag{2}$$

$$\text{or} \quad \text{NOT A OR B} \tag{3}$$

$$\text{or} \quad \text{NOT [A AND NOT B]} \tag{4}$$

The expressions (1) - (4) are shorthand form alternative expressions for:

$$\text{IF X is A THEN Y is B}$$

where X and Y are linguistic variables that represent the antecedent and consequent variables, respectively, and A and B are vague, imprecise linguistic values that are assigned to X and Y, respectively. In fuzzy set paradigm, A and B are represented by Type I fuzzy sets which are defined by their membership functions $A(x)$ and $B(y)$, $x \in X$ and $y \in Y$ as:

$$A(x) : X \rightarrow [0,1] , x \in X , B(y) : Y \rightarrow [0,1] , y \in Y ,$$

[6]. These membership functions are computational expressions .

During the first thirty years of research and development on fuzzy set and logic theories, most researchers were content with the assignment of the class of connectives known as t-norms and conorms or pseudo t-norms and conorms to linguistic connectives, "and", "or", etc. As a consequence of this practice, the combination of two or more fuzzy sets are defined to be Type I fuzzy sets by such crisp operators. It has been found in experimental research that the natural language use of linguistic connectives "and", "or", etc., are interpreted by experts as a matter of degree [9]. Therefore, meta-linguistic connectives "AND", "OR", etc., need to be interpreted as connectives that are variable. Thus they ought not be assigned directly to crisp connectives such as classical logic connectives: "and" $\equiv (\bullet)$ "or" $\equiv (+)$, nor should they be assigned to t-norms and conorms or pseudo t-norms and conorms of fuzzy sets and logics.

Therefore it is suggested that the combination of two or more vague linguistic values should be represented by interval-valued fuzzy sets. A particular formation of interval-valued fuzzy sets are defined by the disjunctive and conjunctive propositional expressions that can be assigned to meta-linguistic expressions as they will be discussed next. In fact, it is shown that "compensatory 'AND'" operators [9] fall into the interval-valued fuzzy sets specified by disjunctive and conjunctive normal forms [2].

At this junction, it is worthwhile to note that interval-valued fuzzy sets are contained within the power set of [0,1] and they are known as Type II fuzzy sets [8]. Type II fuzzy sets are defined as:

$$A(x) : X \rightarrow P[0,1], x \in X$$

where A which is a combination of two Type I fuzzy sets, say A_1 and A_2, and $A(x)$ is the membership values of a combined fuzzy set A, at $x \in X$, a point on the universal set X and P[0,1] is the power set of [0,1].

2. Type II Fuzzy Sets

As noted in the last section, the interpretation that the combination of two or more vague linguistic values ought to identify a Type II fuzzy sets was discovered about a decade or so ago [1]. At that time it was proposed that a particular Type II fuzzy sets called interval valued fuzzy sets be defined by the disjunctive and conjunctive normal forms in fuzzy set paradigm.

In the following discussion of the four classes of normal forms, we will discuss both the propositional expressions and the corresponding computational expression within a framework of axiomatic restrictions and relaxations.

2.1 Normal Form Expressions

In this view of axiomatic foundations, there are four classes of fuzzy sets and logics. These four classes are next discussed from the most restricted to the most relaxed class and only for the meta-linguistic expressions of the form "IF A THEN B" type. A discussion of other meta-linguistic expressions may be found on our other writings [4].

2.1.1 Class IV Normal Forms

This class of normal forms are the unique Boolean Normal forms. They are based on two-valued set and logic paradigm where meta-linguistic connectives "AND", "OR", "NOT" are given by classical truth table definition over the lattice {0,1} in the usual manner.

It is well known that the disjunctive and conjunctive normal forms, DNF and CNF, respectively, of the sixteen basic combination of concepts (see Table 1) are equivalent in Boolean set and logic paradigm, i.e., $DNF(\bullet) \equiv CNF(\bullet)$ (see Table 2). It is worthwhile to emphasize that this equivalence is a consequence of the well known axioms of the two-valued set and logic paradigm (see Table 3).

Table 1. Meta-Linguistic Expressions of Combined Concepts for any A and B

Number	Meta-linguistic Expressions
1	UNIVERSE
2	EMPTY SET
3	A OR B
4	NOT A AND NOT B
5	NOT A OR NOT B
6	A AND B
7	A IMPLIES B
8	A AND NOT B
9	A OR NOT B
10	NOT A AND B
11	A IF AND ONLY IF B
12	A EXCLUSIVE OR B
13	A(AFFIRMATION)(AFF(A))
14	NOT A(COMPLEMENTATION)(COM(A))
15	B(AFFIRMATION)(AFF(B))
16	NOT B(COMPLEMENTATION)(COM(B))

Table 2. Classical Disjunctive and Conjunctive Normal Forms, where $\cap$ is a conjunction, $\cup$ is a disjunction and c is a complementation operator

#	Disjunctive Normal Forms	Conjunctive Normal Forms
1	$(A\cap B)\cup(A\cap c(B))\cup(c(A)\cap B)\cup(c(A)\cap c(B))$	I
2	$\varnothing$	$(A\cup B)\cap(A\cup c(B))\cap(c(A)\cup B)\cap(c(A)\cup c(B))$
3	$(A\cap B)\cup(A\cap c(B))\cup(c(A)\cap B)$	$(A\cup B)$
4	$(c(A)\cap c(B))$	$(A\cup c(B))\cap(c(A)\cup B)\cap(c(A)\cup c(B))$
5	$(A\cap c(B))\cup(c(A)\cap B)\cup(c(A)\cap c(B))$	$(c(A)\cup c(B))$
6	$(A\cap B)$	$(A\cup B)\cap(A\cup c(B))\cap(c(A)\cup B)$
7	$(A\cap B)\cup \qquad (c(A)\cap B)\cup(c(A)\cap c(B))$	$(c(A)\cup B)$
8	$(A\cap c(B))$	$(A\cup B)\cap(A\cup c(B))\cap \qquad (c(A)\cup c(B))$
9	$(A\cap B)\cup(A\cap c(B))\cup \qquad (c(A)\cap c(B))$	$(A\cup c(B))$
10	$(c(A)\cap B)$	$(A\cup B)\cap \qquad (c(A)\cup B)\cap(c(A)\cup c(B))$
11	$(A\cap B)\cup \qquad (c(A)\cap c(B))$	$(A\cup c(B))\cap(c(A)\cup B)$
12	$(A\cap c(B))\cup(c(A)\cap B)$	$(A\cup B)\cap \qquad (c(A)\cup c(B))$
13	$(A\cap B)\cup(A\cap c(B))$	$(A\cup B)\cap(A\cup c(B))$
14		
15	$(c(A)\cap B)\cup(c(A)\cap c(B))$	$(c(A)\cup B)\cap(c(A)\cup c(B))$
16	$(A\cap B)\cup \qquad (c(A)\cap B)$ $(A\cap c(B)) \qquad (c(A)\cap c(B))$	$(A\cup B)\cap \qquad (c(A)\cup B)$ $(A\cup c(B))\cap \qquad (c(A)\cup c(B))$

Table 3. Axioms of two-valued set and logic theory

Involution:	$c(c(A)) = A$
Commutativity:	$A \cup B = B \cup A$
	$A \cap B = B \cap A$
Associativity:	$(A \cup B) \cup D = A \cup (B \cup D)$
	$(A \cap B) \cap D = A \cap (B \cap D)$
Distributivity:	$A \cap (B \cup D) = (A \cup B) \cap (A \cup D)$
	$A \cup (B \cap D) = (A \cap B) \cup (A \cap D)$
Idempotence:	$A \cup A = A$
	$A \cap A = A$
Absorption:	$A \cup (A \cap B) = A$
	$A \cap (A \cup B) = A$
Absorption of complement:	$A \cup (c(A) \cap B) = A$
	$A \cap (c(A) \cup B) = A$
Absorption by I and $\varnothing$:	$A \cup I = I$
	$A \cap \varnothing = \varnothing$
Identity:	$A \cup \varnothing = A$
	$A \cap I = A$
Law of contradiction:	$A \cap c(A) = \varnothing$
Law of excluded middle:	$A \cup c(A) = I$
De Morgan's laws:	$c(A \cup B) = c(A) \cap c(B)$
	$c(A \cap B) = c(A) \cup c(B)$

This class of normal forms for a rule of the form "IF A THEN B", or $A \rightarrow B$ for short, are written as:

$$FDNF^{(4)}(A \rightarrow B) = DNF(A \rightarrow B) = (c(A) \cap c(B)) \cup (c(A) \cap B) \cup (A \cap B)$$

$$FCNF^{(4)}(A \rightarrow B) = CNF(A \rightarrow B) = c(A) \cup B$$

where $c(A)$ and $c(B)$ are complements of the crisp sets A and B.

Furthermore, the connectives $\cap, \cup$ are the classic crisp "and", "or" connectives. Due to the equivalence stated above, we have:

$$DNF(A \rightarrow B) \equiv CNF(A \rightarrow B)$$

It is clear that this is a very degenerate Type II representation. However, it is well known that this expression when evaluated with the assignments from {0,1} lattice becomes a vertex of the second order lattice of the classical logic. Also it is well known that classical implication so structure is created to model an idealized rule structure for an idealized world model.

In this sense, classical expert systems, i.e., first generation expert systems, non-fuzzy expert systems, require perfect and precise information based rules to represent a systems' behaviour. It is however known that the real world is neither so idealized nor information about its behaviour are so perfectly or precisely well known. For such reasons, first generation expert systems are either applicable to simple well defined cases or alternately they require many well defined rules in order to respond many well defined situations.

2.1.2 Class III Normal Forms

This class of normal forms are formed by original Zadeh relaxations. These normal forms are obtained with: (i) the extension of the two-valued sets to infinite valued sets by way of generalizing the valuation set of the {0,1} lattice to the [0,1] interval, and (ii) the relaxation of the law of excluded middle and its dual, the law of contradiction. In fact, they form the first non-degenerate Type II fuzzy sets. They may be constructed either (i) by an application of the valuation set of the [0,1] interval values to the class IV normal forms [1] discussed above in Section 2.1.1, or (ii) by directly deriving fuzzy disjunctive and conjunctive normal forms, $FDNF(\bullet)$ and $FCNF(\bullet)$, respectively from Fuzzy Truth Tables and with the proper application of associativity, commutativity and idempotency axioms [4]. Thus they are equivalent to class IV normal forms in propositional expressions only but they are different in the valuation content. In fact, it is shown that

$$FDNF^{(3)}(\bullet) = DNF(\bullet) \subseteq CNF(\bullet) = FCNF^{(3)}(\bullet)$$

where $FDNF^{(3)}(A \rightarrow B) = (c(A) \cap c(B)) \cup (c(A) \cap B) \cup (A \cap B)$, and $FCNF^{(3)}(A \rightarrow B) = c(A) \cup B$. For this class with Zadeh's De Morgan Triple, i.e., Max-Min $= (\vee, \wedge)$ and $N(a) = 1-a$, we get the Fuzzified Kleene-Dines implication as:

$$\text{FDNF}^{(3)}(A(x)\to B(y))=(N(A(x))\wedge N(B(y)))\vee(N(A(x))\wedge B(y))\vee(A(x)\wedge B(y))$$

and $\quad \text{FCNF}^{(3)}(A(x)\to B(y))=N(A(x))\vee B(y).$

This class of normal forms are suitable to real life systems, which have rules where the linguistic and hence meta-linguistic connectives are variables and the propositional connectives $\cap,\cup$ are idempotent, associative and commutative. However, there are no laws of excluded middle and contradiction for such rules, i.e., $A\cup c(A)\subseteq I$ and $A\cap c(A)\supseteq\varnothing$. In fact, all the rules of such an expert system are true as a matter of varying degrees specified by the interval $I^{(3)}(A\to B)=\left[\text{FDNF}^{(3)}(A\to B),\text{FCNF}^{(3)}(A\to B)\right]$ which represents the Fuzzified Kleene-Dienes Implication.

2.1.3. Class II Normal Forms

This class of normal forms are constructed by the relaxation of idempotency and hence distributivity and absorption axioms as well as keeping the evaluation set to the [0,1] interval. The linguistic and meta-linguistic connectives of this class of rules are again variable. In this class there are infinitely many meta-linguistic connectives that are definable by the t-norm - conorm class of computational operators. The well known examples of these operators are: (i) the Bold or Lukasiewicz operators, i.e., $T_L(a,b)=\max\{0,a+b-1\}$ and $S_L(a,b)=\min\{1,a+b\}$; and (ii) algebraic sum and product operators, i.e., $T_P(a,b)=ab$ and $S_S(a,b)=a+b-ab$, etc. The structure of these normal forms are derived directly from fuzzy Truth Tables with the application of associativity and commutative axioms but idempotency is not applicable. These normal forms are:

$$\text{FDNF}^{(2)}(A\to B)=(c(A)\cap c(B))\cup(c(A)\cap c(B))\cup(c(A)\cap B)$$

$$\cup(c(A)\cap B)\cup(A\cap B)\cup(A\cap B),$$

$$\text{FCNF}^{(2)}(A\to B)=(c(A)\cup B)\cap(c(A)\cup B).$$

For this class, with Lukasiewicz De Morgan triple, i.e., (S_L,T_L,N) we get the fuzzified Lukasiewicz implication:

$$\text{FDNF}^{(2)}(A(x)\to B(y))=(N(A(x))T_L N(B(y)))S_L(N(A(x))T_L N(B(y)))$$

$$S_L(N(A(x))T_L B(y))S_L(N(A(x))T_L B(y))$$

$$S_L(A(x)T_L B(y))S_L(A(x)T_L B(y))$$

$$\text{FCNF}^{(2)}\big(A(x)\to B(y)\big)=\big(N(A(x))S_L B(y)\big)T_L\big(N(A(x))S_L B(y)\big).$$

With the Algebraic Sum and Product Triple, i.e., (S_S, T_P, N), we get Fuzzified Reichenbach implication:

$$\text{FDNF}^{(2)}\big(A(x)\to B(y)\big)=\big(N(A(x))T_P N(B(y))\big)S_S\big(N(A(x))T_P N(B(y))\big)$$

$$S_S\big(N(A(x))T_P B(y)\big)S_S\big(N(A(x))T_P B(y)\big)$$

$$S_S\big(A(x)T_P B(y)\big)S_S\big(A(x)T_P B(y)\big)$$

$$\text{FCNF}^{(2)}\big(A(x)\to B(y)\big)=\big(N(A(x))T_P B(y)\big)S_S\big(N(A(x))T_P B(y)\big)$$

The meta-linguistic connectives of this class are again variable. Each fuzzy expert system rule defines a Type II fuzzy set identified by

$$\text{FDNF}^{(2)}(A\to B) \text{ and } \text{FCNF}^{(2)}(A\to B),$$

In this class, $\text{FDNF}^{(2)}(A\to B) \neq \text{FCNF}^{(2)}(A\to B)$. But it is found that at times

$$\text{FDNF}^{(2)}(A\to B) \subseteq \text{FCNF}^{(2)}(A\to B)$$

while at other times

$$\text{FDNF}^{(2)}(A\to B) \supseteq \text{FCNF}^{(2)}(A\to B)$$

depending on t-norm-conorm De Morgan triple that needs to be applied. Nevertheless, it is true that there is an interval $I^{(2)}(A\to B)$ defined by the values of $\text{FDNF}^{(2)}(A\to B)$ and $\text{FCNF}^{(2)}(A\to B)$.

At the computational level, this interval identifies a region of uncertainty defined by:

$$\big|I^{(2)}\big(A(x)\to B(y)\big)\big|=\big|\text{FDNF}^{(2)}\big(A(x)\to B(y)\big)-\text{FCNF}^{(2)}\big(A(x)\to B(y)\big)\big|,$$

where $I^2\big(A(x)\to B(y)\big)$ represents the Fuzzified Class II implications such as Fuzzified Lukasiewicz implication, etc.

Thus the degree of confidence that we have in the applicability of an expert rule, $A\to B$, can be computed as $\big|I^{(2)}\big(A(x)\to B(y)\big)\big|$. This means that we know that the membership values associated with the expert rule is in the interval $I^{(2)}\big(A(x)\to B(y)\big)$ but we are not sure where a particular membership value would be. Thus there is a second order semantic uncertainty measure defined by $I^{(2)}\big(A(x)\rightarrow B(y)\big)$.

2.1.4 Class I Normal Forms

This class of normal forms are constructed by the further and the final relaxations of the associativity and commutativity axioms. This class of meta-linguistic connectives have only the boundary and monotonicity axioms and thus they form the foundation of the most general monotonic reasoning schema. This class of meta-linguistic connectives are also a class of variable connectives and they are generated by the computational operators known as pseudo t-norms and conorms.

The normal forms of this class are

$$\text{FDNF}^{(1)}(A\rightarrow B)=\big(c(B)\cap c(A)\big)\cup\big(B\cap c(A)\big)\cup\big(B\cap A\big)$$

$$\cup\big(c(A)\cap c(B)\big)\cup\big(c(A)\cap B\big)\cup\big(A\cap B\big),$$

$$\text{FCNF}^{(1)}(A\rightarrow B)=\big(B\cup c(A)\big)\cap\big(c(A)\cup B\big)$$

An example of this class of pseudo t-norms and conorms is:

$$\hat{T}_{min}(a,b)=\begin{cases}0 & \text{if } a+b\leq 1\\ b & \text{otherwise,}\end{cases} \text{ and}$$

$$\hat{S}_{max}(a,b)=\begin{cases}1 & \text{if } a+b\geq 1\\ b & \text{otherwise,}\end{cases}$$

For this class with the pseudo min-max De Morgan triple, i.e., $\big(\hat{S}_{max},\hat{T}_{min},N\big)$ defined above, we get the fuzzified Godel implication:

$$\text{FDNF}^{(1)}\big(A(x)\rightarrow B(y)\big)=\big(N(B(y))\hat{T}_{min}N(A(x))\big)\hat{S}_{max}\big(B(y)\hat{T}_{min}N(A(x))\big)$$

$$\hat{S}_{max}\big(B(y)\hat{T}_{min}A(x)\big)\hat{S}_{max}\big(N(A(x))\hat{T}_{min}N(B(y))\big)$$

$$\hat{S}_{max}\big(N(A(x))\hat{T}_{min}B(y)\big)\hat{S}_{max}\big(A(x)\hat{T}_{min}B(y)\big)$$

$$\text{FCNF}^{(1)}\big(A(x)\rightarrow B(y)\big)=\big(B(y)\hat{S}_{max}N(A(x))\big)\hat{T}_{min}\big(N(A(x))\hat{S}_{max}B(y)\big).$$

Again each one of fuzzy expert system rules of this class defines a Type II fuzzy set specified by $\text{FDNF}^{(1)}(A\rightarrow B)$ and $\text{FCNF}^{(1)}(A\rightarrow B)$. In this class again, we have:

$$\text{FDNF}^{(1)}(A\rightarrow B)\neq \text{FCNF}^{(1)}(A\rightarrow B).$$

Again it is found that the containment changes for different $x \in X$. However the interval $I^{(1)}(A(x) \to B(y))$ is well defined. It specifies a region of uncertainty defined by:

$$\left|I^{(1)}(A(x) \to B(y))\right| = \left|FDNF^{(1)}(A(x) \to B(y)) - FCNF^{(1)}(A(x) \to B(y))\right|.$$

Once again this interval $I^{(1)}(A(x) \to B(y))$ identifies a second order semantic uncertainty. Fuzzy expert system rules that generate such intervals again has a degree of confidence specified by $\left|I^{(1)}(A(x) \to B(y))\right|$ and the uncertainty is associated with not knowing where the particular membership value is in the interval $I^{(1)}(\bullet)$.

2.2 Knowledge Representation with Fuzzy Rules

It is clear from Section 2.1 that four classes of systems behaviour knowledge can be represented by the four classes of fuzzy normal forms and the associated fuzzy rules of a given real life system.

It is however a challenging problem to find out which expert rules and their linguistic and meta-linguistic connectives are subject to which axioms in order to identify the class of fuzzy normal forms that need be applicable to model a real-life system effectively. In fact, an extensive experimental analysis of expert rules together with input-output data is an essential and inescapable part of modeling such real-life systems.

It should be pointed out here for re-emphasis that we have been discussing knowledge representation of an expert's rules with three transformations from a linguistic to meta-linguistic to propositional and to computational levels. This approach is quite different from the fuzzy control methodology discussed in many current applications of fuzzy paradigm. In particular, such fuzzy control applications are to be gathered under the general heading of fuzzy graph theory. Whereas, the particular view presented here need to be identified under the general heading of fuzzy logic theory. Where the word "logic" is used in the sense of the discipline of "Logic" and fuzzy logic as it generalization from $\{0,1\}$ lattice to infinite valued $[0,1]$ interval with four classes of crisp operators under different axiomatic restrictions.

3. Fuzzy Inference

Once a system's context dependent specific knowledge has been encoded in terms of a Type II fuzzy set representation, then the next stage is its utilization via fuzzy inference algorithms to deduce fuzzy conclusions for a new system observation either for prediction or for diagnosis of a system's behaviour and control, etc.

The fuzzy inference algorithm is proposed by Zadeh [7] as the compositional rule of Inference, CRI, better known as Generalized Modus Ponens, GMP. Again within the last thirty years much research has been devoted to this subject. However, all this research has concentrated in the main on the derivation of consequences that are restricted to Type I fuzzy set and logic paradigm as:

$$A'o(A \rightarrow B) = B'$$

where A, A', B, B' are all considered to be specified with Type I fuzzy sets as well as $A \rightarrow B$ and $A'o(A \rightarrow B) = B'$.

When we extend our horizon to Type II fuzzy sets as it was discussed in Section 2, we can as a natural consequence extend fuzzy inference to Type II fuzzy Inference as well. In this domain of inquiry, we are able to construct once again four distinct classes of inference procedure associated with each of the four classes of fuzzy knowledge representation presented in Section 2. Page limitation on this paper does not allow us to venture in this direction. However interested readers can find the details of Type II fuzzy inference methodology in Türkşen [4].

References:

1. I.B. Türkşen: Interval-Valued Fuzzy Sets Based on Normal Forms. Fuzzy Sets and Systems 20, 191-210 (1986)

2. I.B. Türkşen: Interval-Valued Fuzzy Sets and "compensatory AND". Fuzzy Sets and Systems 51, 295-307 (1992)

3. I.B. Türkşen, K. Demirli: Rule and Operation Decomposition in CRI. In: P.P. Wang (ed) Advances in Fuzzy Theory and Technology Vol I, 219-256 (1993)

4. I.B. Türkşen: Fuzzy Normal Forms. Fuzzy Sets and Systems 69, 319-346 (1995)

5. I.B. Türkşen: Type I and II Fuzzy Sets. In P.P. Wang (ed) Advances in Fuzzy Theory and Technology Vol III (to appear).

6. L.A. Zadeh: Fuzzy Sets. Information and Control 8, 338-353 (1965)

7. L.A. Zadeh: Outline of a new Approach to the Analysis of Complex Systems and Decision Processes: IEEE-Tran on Systems, Man and Cybernetics 3, 28-44 (1973)

8. L.A. Zadeh: Concept of a Linguistic Variable and its Application to Approximate Reasoning I, II, III. Information Sciences 8, 199-249, 301-357 (1975); 9 43-80 (1975)

9. H.J. Zimmerman, P. Zysno: Latent Connectives in Human Decision Making. Fuzzy Sets and Systems 4, 37-51 (1980).

Deciding Boundedness for Systems of Two Linear Communicating Finite State Machines

Abderrahim Benslimane

Laboratoire d'Informatique, Institut Polytechnique
de Sévenans
90010 Belfort cedex France
Tel. : (33).84.58.31.26
E-mail : Abder.Benslimane@utbm.fr

Abstract. Since 1981, we know that the unboundedness is an undecidable problem for systems of finite state machines that communicate exclusively by exchanging messages over unidirectional, FIFO channels. This led some authors to reduce the general model in order to find restricted classes in where the problem becomes decidable. In this paper, we present a new class of systems constituted of two linear Communicating finite state machines. A linear machine is a directed graph where each strongly connected component is reduced to an elementary circuit. We show for this class that the boundedness problem is decidable. Consequently, the decision procedures can be automatized and integrated into the frame of Computer-Aided Systems Technology. Examples will be used to illustrate our purpose.

1 Introduction

In view of the increasing use of the Computer-Aided Systems Technology (CAST), it becomes more important to characterize the programs in point of view their termination. The termination problem is well known in the domain of distributed systems. In particular, for data communication protocols it corresponds to the boundedness problem. In order, to decide this problem and others, we need to find effective, mathematical methods of specification and validation of protocols. For this, several models have been successfully used to study the communication protocols. They are the ''finite state machines'' model, the ''Petri nets'' model and ''programming language'' model. In this paper, we are interested in the model of Communicating Finite State Machines.

Systems of Communicating Finite State Machines (CFSM's, for short) are useful in modeling [6, 16], analysis [3, 4, 9] and synthesis [9, 19] of communication protocols and distributed systems. Some real protocols have been modelled, analyzed, or synthesized as systems of CFSM's. For example, we find the alternating-bit protocol [3], the call establishment/clear protocol in X.25 [9], and X.21 [14].

A principal goal of modeling communication protocols by systems of CFSM's is to detect many protocol design errors like unspecified reception, deadlock and unboundedness. To accomplish this, one tries to generate exhaustively reachable global states. Nevertheless, in the finite state model, if the communicating channels are unbounded then the reachability graph is infinite. To allow implementation of the validation procedure, some authors [17, 20] consider the channel capacity to bounded by an integer value. In this case the reachability graph is finite. This will get us to ask the following question : Given a type of design error, is it sufficient to take a value as bound for the channels in order to detect all errors of this type ? Unfortunately, the response is no. Nevertheless, if we know in advance that the reachability graph is finite, its construction allows us to deduce some interesting results such as channel capacity or deadlock.

Since 1981, we know that the finiteness of the reachability graph is an undecidable problem for a system of arbitrary number of CFSM's and so for a system of only two CFSM's [4]. This led some authors to reduce the general model in order to find restricted classes for which the problem becomes decidable [7, 8, 10, 12, 13, 15, 18].
In fact, these restrictions affected :
- the number of CFSM's (two or more),
- the capacity of the channels (bounded by a linear function or constant),
- the input languages of the channels (linear language, tally language),
- the type of states in a CFSM (mixed state),
- the structure of the CFSM (linear, cyclic).
In this paper, we are interested in the structure of the CFSM's. We consider a system of two CFSM's, where every CFSM is linear. For this class, we show that the boundedness problem is decidable.

We first study the systems of two CFSM's where each CFSM is constituted of one hanging circuit. After, we generalize the previous results to the systems of two linear CFSM's. Those results and others about particular systems of two identical CFSM's are briefly presented in [2].

The paper is organized as follows :
- in section 2, we give the preliminaries and the general overall definitions concerning the model utilized throughout the paper,
- section 3 gives a characterization of the communication between two machines,
- in section 4, we discuss the framework of our result and we show that the boundedness is a decidable problem for a system of two linear CFSM's. We then, illustrate our purpose by giving some examples,
- finally a conclusion of the paper is presented in section 5.

2 Systems of CFSM's

2.1 Preliminaries

Let X be an alphabet (i.e., finite set) whose elements are called letters. The concatenation operator "." allows us to construct words on X. A word x on X is a sequence of letters of X. The empty word is denoted by ε. A language L is a set of words. X^* is the set of finite words on X (X^* contains the empty word) and X^+ is equal to $X^*-\{\varepsilon\}$; X^ω is the set of infinite words on X. We write x^ω the concatenation of an infinite number of x and x^* the concatenation of finite number of x. We denote by $|x|$ the length of x. We have $|x.x'|=|x|+|x'|$, for every pair of words x, $x' \in X^*$, and $|\varepsilon|=0$. For a set A, we use the notation $|A|$ for the cardinality of A.

A word x is a prefix of word y if there exists a word z such that y=x.z, we note x≤y. We denote by FG(y) the set of prefix of word x. We write x[n] for the prefix of the word x whose length is equal to n. Given a language L⊆X* and a subalphabet Y⊆X, the projection of L on Y is written $proj_Y(L)$ and is defined by the following morphism : $proj_Y(t.x)$= if t∈ Y then $t.proj_Y(x)$ else $proj_Y(x)$ and $proj_Y(ε)$=ε.

2.2 Model

We consider a system S of communicating finite state machines consisting of a set of two finite state processes that send messages to, and receive messages from, each other. The communication medium from one process to another is assumed to be error-free and unbounded FIFO buffer. A send event causes a message to be enqueued in a FIFO buffer, while a receive event dequeues a message from a buffer. Consequently, send events can never block whereas receive event can.

We represent each process P_i, i=1..2, by a CFSM A_i.
A communicating machine is a directed labelled graph with two types of edges called sending and receiving edges.

Now, we define formally the model of CFSM A_i as a four-tuple $<Q_i, A_i, q_{i0}, δ_i>$ where
- Q_i denotes the finite set of states of A_i.
- A_i denotes the finite set of possible interactions of A_i. It is the set of actions written in the form '-a' that can be sent from A_i to the other CFSM and those written in the form '+a' that can be received by A_i from the other CFSM.
- q_{i0} is the initial state of A_i.
- $δ_i$ is a partial function mapping, $δ_i : Q_i × A_i → Q_i$
$δ_i(q_i,'-a')$ is the state that A_i move to from state q_i after sending a message 'a' to the other CFSM. $δ_i(q_i,'+a')$ is the state that A_i move to from state q_i after receiving a message 'a' sent by the other CFSM.

We denote by M_i the alphabet of messages that can be sent by one CFSM to the other.

Remark
In the set A_i there are two types of events : sending events and receiving events. We note E_i (respectively R_i) the set of sending events (respectively of receiving events) in CFSM A_i. Let q_i be a state in CFSM A_i, we note $E_i(q_i)$ (respectively $R_i(q_i)$) the set of sending actions (respectively receiving actions) labelling the outgoing transitions from state q_i. ◊

Definition 1
We define over A_i^*, i=1..2, an isomorphism φ that transform the emissions onto receptions and inversely by : ∀ x∈ A_i^*, ∀ a∈ M_i, φ(+a.x)= -a.φ(x), φ(-a.x)= +a.φ(x) and φ(ε)= ε.

Also, we define the morphism ψ from E_i^*, i=1..2, to M_i^* to extract the messages sent in a sequence of sending actions.
Let s = (-a).s' be a sequence of actions in E_i^*, then $\psi(s)=a.\psi(s')$ and $\psi(\varepsilon)=\varepsilon$. $\Diamond$

Definition 2
The set of global states system S is defined by :

$$EG(S)=Q_1 \times Q_2 \times M_1^* \times M_2^*$$

(q_1, q_2, w_1, w_2) will be used to denote a global state, where
- q_i is a state of CFSM A_i, i=1..2,
- w_1 is the content of the channel c_{12} from A_1 to A_2 and
- w_2 is the content of the channel c_{21} from A_2 to A_1.
$e_0=(q_{10}, q_{20}, \varepsilon, \varepsilon)$ denotes the initial global state, where ε is the empty sequence and q_{i0} is the initial state of A_i, i=1..2. $\Diamond$

Definition 3
Let $e=(q_1, q_2, w_1, w_2)$ a global state, then a transition t is executable if and only if there exist i, i in [1..2], and state q_i such that one of the following conditions is verified :

i) $t= -a \in E_i(q_i)$.
ii) $t= +a \in R_i(q_i)$ and $w_j = a.x$ (j=1+(i mod 2))

Let $e'=(q_1', q_2', w_1', w_2')$ a global state reached from e, then the execution of t is defined by :

i) In the first case, the message 'a' sent from A_i to A_j (j=1+(i mod 2)) is appended at the end of w_j, and the CFSM A_i moves from state q_i to a state q_i'.
All the elements of e' are equal to those of e except :
$$q_i'=\delta_i(q_i, -a) \text{ and } w_i'=w_i.a.$$

ii) In the second case, the message 'a' sent by A_j (j=1+(i mod 2)) is removed from $w_j=$ a.x, and the CFSM A_i moves from state q_i to a state q_i'.
All the elements of e' are equal to those of e except :
$$q_i'=\delta_i(q_i, +a) \text{ and } w_j'=x.$$

We say that e' is <u>directly reachable from e</u>, by the execution of t denoted by $e \rightarrow^t e'$.
We denote :
$e \rightarrow e'$ for $\exists t\ e \rightarrow^t e'$,
$e \rightarrow \perp$ if from e there is no reachable state. $\Diamond$

Definition 4
Let $\rightarrow^*$ the reflexive and transitive closure of $\rightarrow$. A global state e is <u>reachable</u> if and only if $e_0 \rightarrow^* e$.

We use $e\to^s e'$ to denote that the global state e' is reachable from e by the event sequence s. ◊

Definition 5
Let S a system of two CFSM's, we define a <u>set of global states reachable</u> from e_0 by
$$EA(S) = \{e\in EG(S) \mid \exists s\in A^* : e_0\to^s e \}$$
Remark that $EA(S) \subseteq EG(S)$.

The <u>reachability tree</u> of a system S of CFSM's is denoted by $AA(S)$ and it is a rooted tree defined in the following way :
1) the root r is labelled by the initial global state e_0,
2) a node n, labelled by $e\in EA(S)$, has a directed successor n', labelled by $e'\in EA(S)$, if and only if there exists an executable transition t such that $e\to^t e'$,
3) a node labelled by $e\in EA(S)$ has no successor if and only if $e\to\perp$.

The <u>reachability graph</u> $GA(S)$ is obtained from the reachability tree $AA(S)$ by identifying nodes which have the same label.

The <u>language</u> $L(S)$ of the system S is constituted of words labelling all branches in $AA(S)$. ◊

Definition 6
1) A global state $e=(q_1, q_2, w_1, w_2)$ is said to be a <u>deadlock state</u> if and only if $e\to\perp$ and $w_1=w_2=\epsilon$.
2) If in each global state the content of the channel c_{ij} consists of no more than h messages, then the channel is said to be <u>h-bounded</u>. The channel is said to be <u>bounded</u> if it is h-bounded for some h.

3) If in each execution, the sequence of messages that $\mathcal{A}_i$ send to $\mathcal{A}_j$ forms a sentence in a given language L, then it is said that the input language of the channel c_{ij} is L. We note $L_I(c_{ij})$ the input language of c_{ij}. We define formally, $L_I(c_{ij})=FG(\psi(proj_{E_i}(L(S))))$.

4) We distinguish some input languages of channels which have been studied for certain classes of systems [7, 10, 11] : monogeneous, linear and word-linear.

Let $L\subseteq X^*$ be a language where X is an alphabet.
L is said to be <u>strictly monogeneous</u> if and only if there exist $x,y\in X^*$ such that $L\subseteq FG(x.y^*)$.
L is said to be <u>monogeneous</u> if and only if it is in finite union of languages strictly monogeneous :

$$L\subseteq \bigcup_{i=1}^{N} FG(x_i y_i^*).$$

L is said to be k-linear (or k-bounded) if and only if there exist k letters $a_1, ..., a_k \in$ X such that

$$L \subseteq a_1^* ... a_k^*.$$

L is said to be semi-linear if and only if it is in a finite union of k-linear languages.

$$L \subseteq \bigcup_{i=1}^{N} a_{i1}^* a_{i2}^* ... a_{ik}^*.$$

L is said to be k-word-linear if and only if there exist k words $z_1, ..., z_k \in X^*$ such that
$$L \subseteq z_1^* ... z_k^*.$$

L is said to be semi-word-linear if and only if it is in a finite union of prefix of k-word-linear languages.

A system S of CFSM's is said to be monogeneous, linear or word-linear if the input languages of all channels of S are respectively monogeneous, linear or word-linear.◊

Definition 7
An elementary circuit is constituted by distinct edges only. A hanging circuit is an elementary and simple circuit which has one non-empty path from the initial state. A linear CFSM is a directed graph where the strongly connected components are reduced to elementary circuits. Remark that the reduced graph of such graph is without circuit. In such CFSM, we consider that, if there is a circuit then it is a hanging circuit. ◊

Figure 2, shows a system of two CFSM's. In CFSM $\mathcal{A}_1$, the non-empty path $P_1=(-1)(+1)(-2)$ leads to the hanging circuit $C_1=(+1)(-2)(+1)(-2)(-2)$. In CFSM $\mathcal{A}_2$, the non-empty path $P_2=(-1)(+1)$ leads to the hanging circuit $C_2=(-1)(+2)$.

In a linear CFSM, each circuit is reachable by a finite path. When we leave one circuit we don't came back to it another time.

3 Communication characterization

We will consider, now, a system S constituted with two CFSM's $\mathcal{A}_1$ and $\mathcal{A}_2$ where every sequence of actions labelling one circuit contains at least one receiving action. We have removed the case where a given circuit is labelled by a sequence of only sending actions because if this circuit is reachable then the reachability tree is necessarily infinite.

Reduction function

The following function, allows the reduction of the first letters of two sequences of actions.
In the first rule, the message 'a' sent from $\mathcal{A}_1$ to $\mathcal{A}_2$ is appended at the end of w_1.

In the second rule, the message 'a' received by $\mathcal{A}_1$ is removed from the head of $a.w_2$. The rules iii) and iv) are symmetrical to i) and ii).

The rest of the rules means a blockage of one of the two CFSM's.

Definition 8

Let $x_1 \in A_1^*$ and $x_2 \in A_2^*$ be two sequences of actions (in CFSM $\mathcal{A}_1$ and CFSM $\mathcal{A}_2$ respectively).

Let $w_1 \in M_1^*$ and $w_2 \in M_2^*$ be two words representing the content of the channel c_{12} and c_{21} respectively.

Reduction R (morphism anti-Dyck) of two words is defined as a function

R from $(\times_{i=1..2} A_i^*) \times (\times_{i=1..2} M_i^*)$ to $(\times_{i=1..2} A_i^*) \times (\times_{i=1..2} M_i^*) \cup \{\perp\}$

by the following rules :

i) $R((-a)x_1, x_2, w_1, w_2) = (x_1, x_2, w_1.a, w_2)$

ii) $R((+a)x_1, x_2, w_1, a.w_2) = (x_1, x_2, w_1, w_2)$

iii) $R(x_1, (-a)x_2, w_1, w_2) = (x_1, x_2, w_1, w_2.a)$

iv) $R(x_1, (+a)x_2, a.w_1, w_2) = (x_1, x_2, w_1, w_2)$

v) $R((+a)x_1, x_2, w_1, b.w_2) = \perp$

vi) $R(x_1, (+a)x_2, b.w_1, w_2) = \perp$

vii) $R((+a)x_1, (+b)x_2, \varepsilon, \varepsilon) = \perp$

where $\perp$ denotes non-progress of at least one CFSM, when executing the two sequences. ◊

Notation

We denote by $R^P(x_1, x_2, w_1, w_2)$ the application of the function R p times. In the following, we apply R at sequences x_1 and x_2 until one of them becomes empty or a non-progress arise. We use R instead of R^P.

To simplify the expressions, we will use the notations :

$^-x = proj_{E_i}(x)$ the projection of sequence x, where $x \in A_i^*$, on emissions,

$^+x = \phi(proj_{R_i}(x))$ the transformation of the receptions in x onto emissions. ◊

Definition 9

Given two infinite sequences x and y.

x and y communicate infinitely if :

$\forall n \exists p, q \geq n R(x[p], y[q], \varepsilon, \varepsilon) = (\varepsilon, \varepsilon, w_{1n}, w_{2n})$, where $w_{1n} \in M_1^*$ and $w_{1n} \in M_2^*$. ◊

4 Decidability about reachability tree and boundedness

4.1 Framework of the result

In this section, we resume the results concerning the finiteness of the reachability graph problem for the systems of CFSM's where the input languages of the channels are :

- monogeneous : $x.y^*$ with x, $y \in A^*$ [7, 8],
- linear : $a_1^* ... a_k^*$ with $a_1, ..., a_k \in A$ [5, 10],
- word-linear : $z_1^* ... z_k^*$ with $z_1, ..., z_k \in A^*$ [11].

All these results use nearly the same decision procedures as the procedure used in [7] concerning the monogeneous systems. In fact, they are based on the definition of an ordering on the reachability set (relation between the nodes of the reachability tree). This ordering must be a decidable, well ordering verifying the monotonicity property. By using this ordering, we construct a reduced reachability graph. This one is finite for the monogeneous systems. Unfortunately, we can't conclude for the word-linear systems.

In [11], the same reasoning is used. Nevertheless, a problem arises : the decoding of the contents of the channels according to the words z_i (definition 6). This leads to an ambiguous decomposition. To overcome this problem, Jéron [11] gives only a sufficient condition to decide the finiteness of the reachability graph problem by using a decidable ordering and verifying the monotonicity property, without verifying whether it constitutes a well ordering. The later is essential property to decide the halting of reachability graph construction. Consequently, the decision procedure don't allows to decide whether a given channel is bounded.

In the general case, these proofs are very laborious. They use an ordering and they must show that it is decidable, it allows to verify the monotonicity property and it constitute a well ordering.

Moreover, we have the problem to know what decision procedure (monogeneous, or linear, or word-linear, ...) we must apply. The belonging of a given system to one class is in general undecidable [7]. The set of languages of CFSM systems is the set of recursively enumerable languages.

To overcome these problems, we study the systems according to the CFSM's topology. We define the class of systems constituted of linear CFSM's.

The reader should remark that this linearity is not defined according to the input languages of the channels but according to the structure of the CFSM's.

For the systems of linear CFSM's, the input languages of the channels group together the three types already studied. Precisely, if the CFSM's are linear then the input languages of the channels are monogeneous, or linears, or word-linears. In contrast, the reverse is false.

It is easy to decide whether a system is constituted of linear CFSM's or not.

For these systems, we give an effective decision procedure concerning the boundedness detection problem of the two channels. To do this, we have introduced the morphism R that allows the reduction (anti-Dyck) of two sequences of actions and the reachability analysis.

4.2 Decision procedures

It is known that the linear CFSM is composed of some hanging circuits, we first show that the finiteness of the reachability tree and boundedness are decidable problems for systems of two CFSM's where each of which is constituted of one hanging circuit. Those results have been shown in [8] for the strictly monogeneous systems. However, the proof we used will be easily generalized to the systems of two linear CFSM's, which is not the case for the proof used for the strictly monogeneous systems.

Particular case : system of two CFSM's constituted each of which by one hanging circuit

We begin by giving a necessary and sufficient condition so that two words x^ω and y^ω, $x \in A_1^*$ and $y \in A_2^*$, communicate infinitely when the two channels c_{12} and c_{21} contain $\psi(x_1)$ and $\psi(y_1)$ respectively, where $x_1 \in E_1^*$ and $y_1 \in E_2^*$ (lemma 1).

After executing all emissions in x_1 and y_1, we have one of the following cases verified :

1) all receiving actions in x and y are executed by taking one time x and y,

2) all receiving actions in y are executed by taking one time x. To execute all receiving actions in x, we must take some time y. Taking this number of y, it is possible that some receiving actions in y are not yet executed. To execute them, we must take another time x. This implies that there exist two positive integers r and s such that all receiving actions in x^r and y^s will be executed.

3) This case is symmetrical to 2).

In the three cases, in order that $x_1 x^\omega$ and $y_1 y^\omega$ communicate infinitely, there must exist :

. y' and y'', where $y = y' y''$, such that $x_1 = {}^+y'$ and $\bar{}x$ can be written as $({}^+y''.{}^+y')$ power certain number.

. x' and x'', where $x = x' x''$, such that $y_1 = {}^+x'$ and $\bar{}y$ can be written as $({}^+x''.{}^+x')$ power certain number.

After that, we generalize the previous condition so that $x_1 x^\omega$ and $y_1 y^\omega$ communicate infinitely, where $x_1 \in A_1^*$ and $y_1 \in A_2^*$ (lemma 3). This correspond to the case where each CFSM is constituted by one hanging circuit. In fact, $\mathcal{A}_1$ contains one circuit labelled by x at which leads a path, from the initial state, labelled by x_1 and $\mathcal{A}_2$ contains one circuit labelled by y at which leads a path, from the initial state, labelled by y_1.

Lemma 1

Let x_1 and y_1 be two words on E_1^* and E_2^* respectively.
Let x and y be two words on A_1^* and A_2^* respectively.
$x_1 x^\omega$ and $y_1 y^\omega$ communicate infinitely if and only if one of the following cases is verified :

1) $R(x_1.x, y_1.y, \varepsilon, \varepsilon) = (\varepsilon, \varepsilon, w_1, w_2)$ and
$\exists\, p > 0 \quad x_1.\bar{}x = ({}^+y)^p.x_1$
$\exists\, q > 0 \quad y_1.\bar{}y = ({}^+x)^q.y_1$

2) $\exists\, r \leq s \quad R(x_1.x^r, y_1.y^s, \varepsilon, \varepsilon) = (\varepsilon, \varepsilon, w_1, \psi(y_1))$ and
$\exists\, p \geq 0 \quad x_1.(\bar{}x)^r = ({}^+y)^{s+p}.x_1$
$y_1.(\bar{}y)^s = ({}^+x)^r.y_1$

3) $\exists\, r \leq s \quad R(x_1.x^s, y_1.y^r, \varepsilon, \varepsilon) = (\varepsilon, \varepsilon, \psi(x_1), w_2)$ and
$\exists\, p \geq 0 \quad y_1.(\bar{}y)^r = ({}^+x)^{s+p}.y_1$
$x_1.(\bar{}x)^s = ({}^+y)^r.x_1$

The proof of this lemma is in [1].   ◊

By studying the contents of the channels and the words added at each time to them, by applying R, we deduce the following corollary.

Corollary 2
following the conditions of the lemma 1 :

1) c_{12} is bounded if and only if p=1.
   c_{21} is bounded if and only if q=1,

2) c_{12} is bounded if and only if p=0.
   c_{21} is always bounded,

3) c_{12} is always bounded.
   c_{21} is bounded if and only if p=0.

The proof of this corollary is an immediate consequence of lemma 1.   ◊
The following lemma generalizes the previous.

Lemma 3
Given x_1, $x \in A_1^*$, y_1, $y \in A_2^*$, $x_2 \in E_1^*$ and $y_2 \in E_2^*$,
$x_1 x^\omega$ and $y_1 y^\omega$ communicate infinitely if and only if one of the following cases is verified :

1) $R(x_1, y_1, \varepsilon, \varepsilon) = (\varepsilon, \varepsilon, \psi(x_2), \psi(y_2))$ and
$x_2.x^\omega$ and $y_2.y^\omega$ communicate infinitely,

2) $\exists p > 0 \; R(x_1.x^p, y_1, \varepsilon, \varepsilon) = (x', \varepsilon, \psi(x_2), \varepsilon)$ and
$x_2.x_3^\omega$ and y^ω communicate infinitely, where $x = x''.x'$ and $x_3 = x'.x''$,

3) $\exists q > 0 \; R(x_1, y_1.y^q, \varepsilon, \varepsilon) = (\varepsilon, y', \varepsilon, \psi(y_2))$ and
x^ω and $y_2.y_3^\omega$ communicate infinitely, where $y = y''.y'$ and $y_3 = y'.y''$.

The proof of this lemma is evident. In fact, we try to execute entirely the sequences x_1 and y_1 in order to begin the execution of x and y. After this, we are in the conditions of lemma 1.   ◊

By lemma 3, we give the following theorem.

Theorem 4
Let S be a system of two CFSM's $\mathcal{A}_1$ and $\mathcal{A}_2$ constituted each of which by one hanging circuit such that :
- $x_1 \in A_1^*$ is the word labelling the path, started from the initial state, leading to the hanging circuit labelled by $x \in A_1^*$ in $\mathcal{A}_1$, and
- $y_1 \in A_2^*$ is the word labelling the path, started from the initial state, leading to the hanging circuit labelled by $x \in A_1^*$ in $\mathcal{A}_2$.
The reachability tree of S is infinite if and only if $x_1 x^\omega$ and $y_1 y^\omega$ communicate infinitely.   ◊

Example 1

Consider the system of two CFSM's in figure 1. Each CFSM is constituted by one hanging circuit. The square denotes state and arrow denotes transition with sending and receiving messages. The initial state of each CFSM is 0. Events are basic units of communication between CFSM's and represented by the symbols associated with the state transition arcs. The event symbol takes the form (-a) or (+a), where 'a' represents the type of message exchanged between the CFSM's $\mathcal{A}_1$ and $\mathcal{A}_2$. The sign '-' denotes a transmission and the sign '+' denotes a reception. When a CFSM is in any state, the execution of event represents the traverse of the arc labelled by the corresponding event symbol and then it enters in a new state.

We have for the CFSM $\mathcal{A}_1$
$x_1=(-1)(+1)(-2)$ and $x=(+1)(-2)(+1)(-2)(-2)$
and for the CFSM $\mathcal{A}_2$
$y_1=(-1)(+1)$ and $y=(-1)(+2)$.
By lemma 3, we have :
$$R(x_1, y_1, \varepsilon, \varepsilon)=(\varepsilon, \varepsilon, 2, \varepsilon).$$
Now, we have to test if $(-2).x^\omega$ and y^ω communicate infinitely.
By lemma 1, we have :
$$R((-2).x, y^2, \varepsilon, \varepsilon)=(\varepsilon, \varepsilon, 2.2, \varepsilon).$$
In addition, we have :
$$(-2).(^-x)^r = (^+y)^{s+p}.(-2)$$
$$(^-y)^s = (^+x)^r,$$
where $r=1$, $s=2$, $p=1$, $^-x=(-2)(-2)$, $^-y=(-1)$, $^+x=(-1)(-1)$ and $^+y=(-2)$.
Then x_1x^ω and y_1y^ω communicate infinitely. Therefore the reachability tree is infinite.
Since $p>0$, the channel c_{12} is unbounded. The channel c_{21} is always bounded.

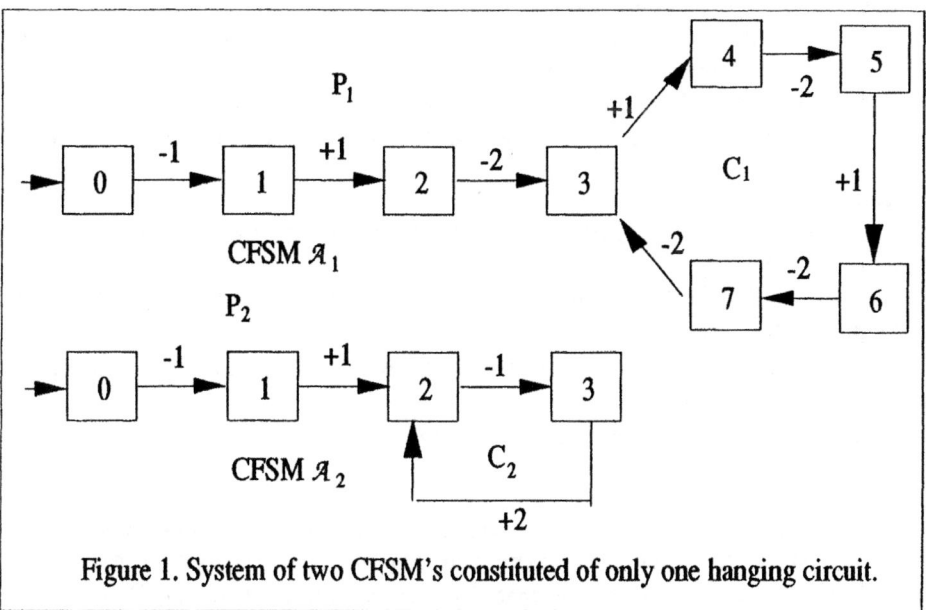

Figure 1. System of two CFSM's constituted of only one hanging circuit.

General case : system of two linear CFSM's.

Proposition 5
Let S be a system of two linear CFSM's, the reachability tree $AA(S)$ is infinite if and only if for each CFSM A_i, i=1 .. 2, there exists a sequence of actions y_i and a state q_i such that y_i is infinitely executable from q_i.

Proof
$\Rightarrow$) Suppose that $AA(S)$ is infinite. Since $AA(S)$ has a finite degree, by König's lemma, there is at least an infinite branch issued from the root labelled by e_0. Let $ch=e_0 \rightarrow^{t1} e_1 \rightarrow^{t2} e_2 \rightarrow ...e_{i-1} \rightarrow^{ti} e_i ...$ be this branch. We project ch on the states and actions executed by each CFSM, we obtain two paths :
$ch_1=q_{10} \rightarrow^{t1,1} q_{1,1} \rightarrow^{t1,2} q_{1,2} \rightarrow ...q_{1,i-1} \rightarrow^{t1,i} q_{1,i}....$ and
$ch_2=q_{20} \rightarrow^{t2,1} q_{2,1} \rightarrow^{t2,2} q_{2,2} \rightarrow ...q_{2,i-1} \rightarrow^{t2,i} q_{2,i} ...$ with $q_{j,i} \in Q_j$ and $t_{j,i} \in A_j$, for j=1..2.
First, we show that the two paths are both infinite.
Let us suppose that ch_1 is finite. In order that ch should be infinite, ch_2 must be infinite and then A_2 must contain one circuit labelled by sequence of only sending actions. Then, from the hypothesis each circuit is labelled by a sequence containing at least one receiving action. This contradicts the fact that ch_1 is finite. The reasoning is similar for ch_2. We conclude that ch_1 and ch_2 are both infinite.
Second, we show that :
$\forall$ i=1..2, $\exists$ x_i, $y_i \in A_i^*$ such that ch_i is labelled by $x_i y_i^\omega$.
Suppose that ch_i, i=1..2, is not labelled by any sequence of the form $x_i y_i^\omega$.
This implies that for all i in [1..2], ch_i is labelled by an infinite sequence of the form $x_i.y_i^\omega$. Then, this case is impossible since we consider only the linear CFSM's : if we leave some circuit we do not return to it.
Therefore, in each CFSM A_i, each sequence y_i that can be crossed infinitely is infinitely crossed, in contiguous manner.
We conclude that if the reachability tree is infinite then each path ch_i executed by A_i, is labelled by a sequence of the form $x_i y_i^\omega$ for i=1..2.
$\Leftarrow$) Inversely, it is evident that, upon executing the system, if in each A_i (i=1..2) one sequence y_i is infinitely crossed, from one state, then the reachability tree is infinite. $\lozenge$

Lemma 6
The reachability tree for a system of two linear CFSM's A_1 and A_2 is infinite if and only if there exist :
- $x_1 \in A_1^*$ word labelling the path, started from the initial state, leading to one circuit labelled by $x \in A_1^*$ in A_1, and
- $y_1 \in A_2^*$ word labelling the path, started from the initial state, leading to one circuit labelled by $x \in A_1^*$ in A_2
such that $x_1 x^\omega$ and $y_1 y^\omega$ communicate infinitely.

Proof
Suppose that the reachability tree is infinite. By previous proposition, there exist at least two paths ch_1 labelled by $x_1 x^\omega$ and ch_2 labelled by $y_1 y^\omega$ in A_1 and A_2 respectively. Therefore, if the reachability tree is infinite then $x_1 x^\omega$ and $y_1 y^\omega$ communicate infinitely.

Inversely, suppose that x_1x^ω and y_1y^ω communicate infinitely. We show easily that there exists an infinite branch in the reachability tree which has a finite degree. Then the reachability tree is infinite.

This shows the lemma.   ◊

The following theorem is an immediate consequence of the previous lemma.

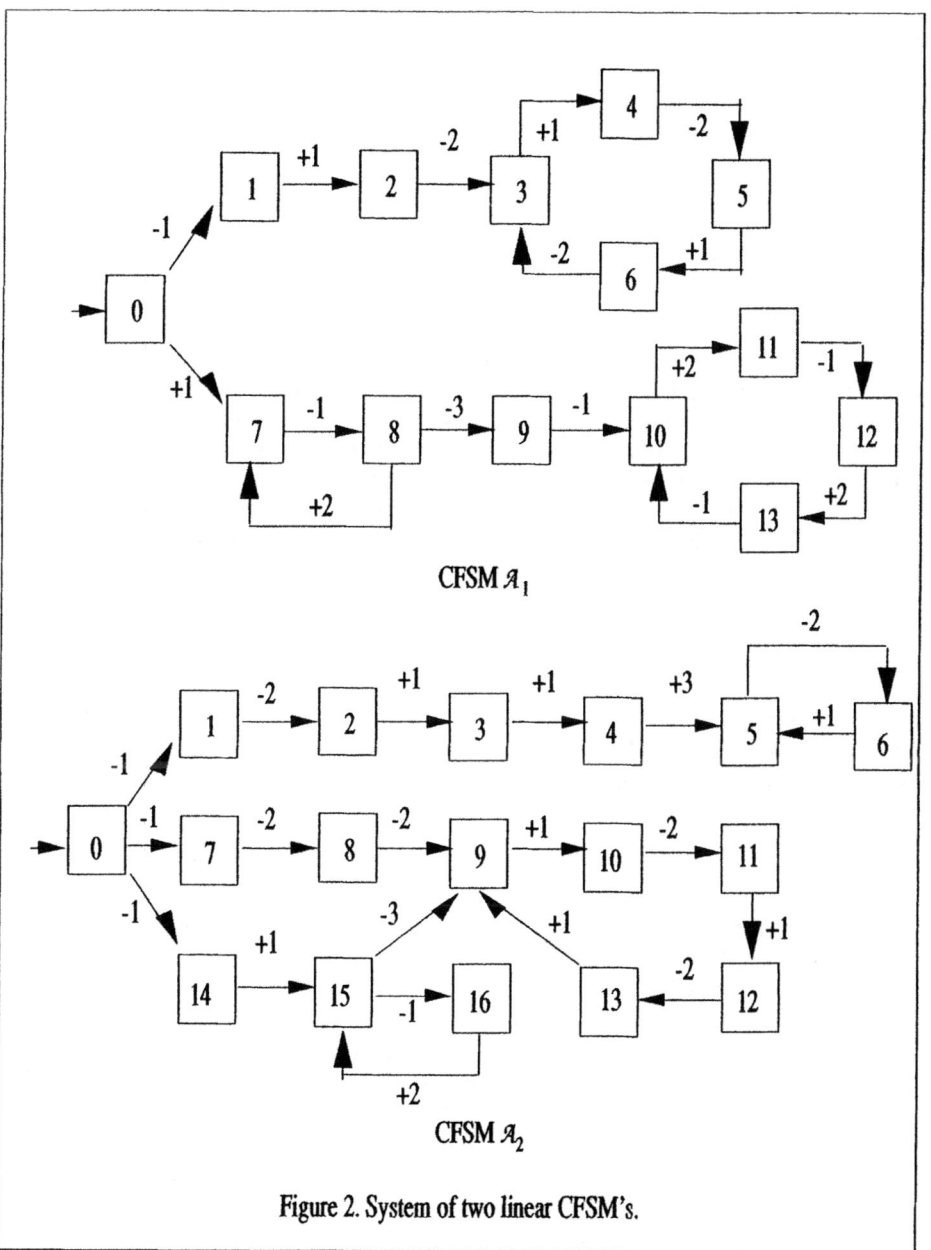

Figure 2. System of two linear CFSM's.

Theorem 7
The finiteness of the reachability tree is a decidable problem for the systems of two linear CFSM's.

Example 2
By inspecting all paths in the two CFSM's of figure 2, we find that :

1) the sequences $x_1x_2x_3x^\omega$ and y_1y^ω communicate infinitely. In fact, we take :
$x_1=(+1)$, $x_2=(-1)(+2)$, $x_3=(-1)(-3)(-1)$ and $x=(+2)(-1)(+2)(-1)$ for $\mathcal{A}_1$
and $y_1=(-1)(-2)(+1)(+1)(+3)$ and $y=(-2)(+1)$ for $\mathcal{A}_2$.

2) the sequences x_1x^ω and y_1y^ω communicate infinitely,
where $x_1=(-1)(+1)(-2)$ and $x=(+1)(-2)(+1)(-2)$ for $\mathcal{A}_1$
and $y_1=(-1)(+1)$ and $y=(-1)(+2)$ for $\mathcal{A}_2$.

From these two cases and by corollary 2, we deduce that c_{21} is bounded and c_{12} is unbounded.

5 Conclusion

Some decidability results have been shown when the input languages of the channels are considered. Therefore, to know if the input language of a given channel is monogeneous, linear, or word-linear is undecidable. For this reason, we are studied the systems by considering the topology of the CFSM's. We have introduced the class of systems of linear CFSM's. Then, we have shown that the boundedness detection problem is decidable for this class. To prove this result, we introduced a reduction relation for two communicating sequences of actions.
Some interesting extensions will be considered :
- the generalization to an arbitrary number of CFSM's,
- the implementation of the decision procedure.

Acknowledgment
We would like to thank Dr. Alain Finkel for helpful remarks and suggestions in french version of this paper. We also thank the anonymous referee for their comments.

References

1. A. Benslimane " *Contribution à la Validation des Protocoles : réduction de l'espace d'états et décidabilité du caractère borné*" Ph.D thesis of Franche-Comté University, (1993).
2. A.Benslimane " *Deciding Boundedness for systems of two communicating finite state machines*" HPDC-3 Third IEEE International symposium on High Performance of Distributed Computing, San Francisco California, IEEE Computer Society Press, (2-5 August 1994), pp. 262-269.
3. G.V. Bochmann " *Finite State Description of Communication Protocols*" Computer Networks, N°2, (1978), pp. 361-372.
4. D. Brand and P. Zafiropulo " *On Communicating Finite-State Machines*" Tech. Rep. RZ 1053. IBM Zurich Research Lab., Ruschlikon, Switzerland, (January 1981).

5.   A. Choquet and A. Finkel " Simulation of Linear Fifo Nets by Petri Nets having a Structured Set of Terminal Markings" Proc. of the 8th European Workshop on Applications and Theory of Petri Nets, Zaragoza, Spain, (June 1987).

6.   A.S. Danthine " *Protocol Representation with Finite State Models*", IEEE Trans. on Commun., Vol. COM-28, N° 4, (April 1980), pp. 632-642.

7.   A. Finkel " *Structuration des Systèmes de Transitions : Applications au Contrôle de Parallèlisme par Files Fifo*" Thèse d'état L.R.I. Orsay, (Juin 1986).

8.   A. Finkel " *A new Class of Analysable CFSMS with Unbounded Fifo Channels*" Protocol Specification, Testing, and Verification, VIII, S. Aggarwal and K. Sabnani (ed.) North-Holland Publishing Compagny IFIP, (1988), pp. 283-294.

9.   M.G. Gouda and Y.T. Yu " *Synthesis of Communicating Finite State Machines with Guaranteed Progress*" IEEE Trans. on Commun., Vol. COM-32, N° 7, (July 1984), pp. 779-787.

10. M.G. Gouda, E.M. Gurari, T.H. Lai and L.E. Rosier " *On Deadlock Detection in Systems of Communicating Finite State Machines*" Computer and Artificial Intelligence, Vol. 6, N° 3, (1987), pp. 209-228.

11. T. Jéron " *Contribution à la Validation des Protocoles : test d'infinitude et vérification à la vollé*" Ph.D thesis of Rennes University, (1991).

12. J. Pachl " *Protocol description and analysis based on a state transition model with channel expression*" Protocol Specification, Testing, and Verification, VI, B. Sarikaya and G.V. Bochmann (ed.), North-Holland Publishing Compagny IFIP, (1987), pp. 243-254.

13. W. Peng and S. Purushothaman " *A Unified Approach to the Deadlock Detection Problem in Networks of Communicating Finite State Machines*" Workshop on Computer Aided Verification DIMACS 90, (June 1990), pp. 242-252.

14. R. Razouk and G. Estrin " *modeling and Verification of Communication Protocols in SARA : The X.21 Interface*" IEEE Trans. on Comp. C-29, 12, (Dec. 1980), pp. 1052-1083.

15. L.E. Rosier and M. Gouda " *On deciding progress for a class of communicatng protocols*" Proc. of the 8th Ann. Conf. on Inf. Sci. and Syst., Princeton Univ., (1984), pp. 663-667.

16. C.A. Sunshine " *Formal Modeling of Communication Protocols*" Computer Networks and Simulation II, S. Schoemaker (ed.), North-Holland Publishing Compagny, (1982), pp. 53-75.

17. C.H. West " *Application and Limitations of Automated Protocol Validation*" Protocol Specification, Testing, and Verification, C.Sunshine (ed.) North-Holland Publishing Compagny IFIP, (1982), pp. 361-371.

18. Y.T. Yu and M.G. Gouda " *Deadlock Detection for a Class of Communicating Finite State Machines*" IEEE Trans. on Commun., Vol. COM-30, N° 12, (December 1982), pp. 2514-2518.

19. P. Zafiropulo, C.H. West, H.Rudin, D.D. Cowan and D.Brand " *Towards Analyzing and Synthesizing Protocols*" IEEE Trans. Commun., Vol. COM-28, N° 4, (April 1980), pp. 651-661.

20. J. Zhao and G.V. Bochmann " *Reduced Reachability Analysis of Communication Protocols : A new approach*" Protocol Specification, Testing, and Verification, VI, B. Sarikaya and G.V. Bochmann (ed.), North-Holland Publishing Compagny IFIP, (1987), pp. 243-254.

A Framework for Knowledge Intensive Engineering

Tetsuo Tomiyama and Yasushi Umeda

Department of Precision Machinery Engineering
The University of Tokyo
Hongo 7-3-1, Bunkyo-ku, Tokyo 113, Japan

Takashi Kiriyama

Research into Artifacts, Center for Engineering (RACE)
The University of Tokyo
Komaba 4-6-1, Meguro-ku, Tokyo 153, Japan

Abstract. This paper proposes knowledge intensive engineering that is a new way of engineering activities in various product life cycle stages conducted with more knowledge in a flexible manner to create more added value. Knowledge representation and modeling issues are discussed and a cooperative multiple intelligent agent architecture based on multiple ontology is proposed for building a computational framework for knowledge intensive engineering. Through examples, we describe the developed framework as a knowledge intensive CAD for knowledge intensive design of knowledge intensive machines. This demonstrates the power and usefulness of knowledge intensive engineering. It is also discussed that to achieve knowledge intensive engineering, systematization of knowledge is an essential process to allow intelligent agents to share accumulated knowledge.

1 Introduction

This paper proposes and exploits the concepts of knowledge intensive engineering and describes a computer-based framework for knowledge intensive engineering. Knowledge intensive engineering refers to how knowledge for various kinds of engineering activities exists and is used in an advanced computing framework. Using flexibly different kinds of knowledge is exactly the very issue concurrent engineering addresses. Concurrent engineering aims at improving manufacturing processes by taking, e.g., production issues into design consideration. Knowledge intensive engineering further elaborates the idea such that having a framework for knowledge intensive engineering allows designers and engineers to create more added value in various stages of product life cycle stages including designing, manufacturing, operations, and maintenance.

At the University of Tokyo, our group has been studying a wide variety of topics related to engineering design and trying to develop a computational framework for knowledge intensive engineering. In the early stages of the project, we aimed at building an intelligent CAD system, called IIICAD (Intelligent Integrated Interactive Computer Aided Design), that would integrate and make flexible use of various kinds of design knowledge and that would intelligently assist designers by giving advises and suggestions and by checking errors based on design process knowledge [28, 29]. The current work on knowledge intensive engineering is an extension of this intelligent CAD project, but it is a substantially expanded idea.

Note that knowledge intensive engineering is not just a new variation of knowledge engineering, but a new style of engineering based on a recognition that knowledge is the source of added value. Because societies in many developed countries have arrived at their plateaus beyond which no economic growth can be expected with increasing environmental and other problems, we need to seek for other ways for growth. Perhaps, we need to look at a new paradigm of *less production with more added value* [23]. This paradigm, *the post mass production paradigm*, requests a new way of generating added value. Since added value can be generated only from knowledge, obviously we need more knowledge in various aspects of engineering. "More knowledge is power" in this sense.

A *knowledge intensive design* environment should allow generation and flexible exchange of design knowledge and information that will be further used in various stages of the life cycle. It is not just another knowledge-based design system which provides designers with sophisticated problem solving facilities. For example, expanding knowledge and information exchanging methods can increase added value. Recently, we see various kinds of advanced mechatronics machines emerge. One of the bottlenecks in developing such *smart products* is software development. If a CAD system can provide a software engineer with design information including functional and behavioral requirements, device control information, and physical constraints regarding sensors and actuators, the software development can be greatly assisted or even automated, which we might call *knowledge intensive software development*. In case of process plants, such as power plants and chemical plants, operation monitoring and maintenance tasks can greatly change, if design knowledge and information are directly available from CAD. A *knowledge intensive CAD*, therefore, can serve as a basis for *knowledge intensive operation and maintenance*.

The present paper proposes and exploits the concepts of knowledge intensive engineering and describes a computational framework for knowledge intensive engineering. The rest of the paper is organized in the following manner. In Chapter 2, we discuss the concepts of knowledge intensive engineering from the point of view of cooperative multiple intelligent agent architecture. Chapter 3 discusses the kernel knowledge representation and modeling for such an architecture. Chapter 4 examines the concepts of knowledge intensive engineering. Chapter 5 describes the computational framework that we are currently developing. Chapter 6 illustrates examples demonstrating how knowledge intensive engineering can increase added value of various engineering activities. Chapter 7 concludes the paper.

2 The Concept of Knowledge Intensiveness

2.1 Large Scale Knowledge Bases

It is often advocated that large scale knowledge bases are useful for engineering applications including design, manufacturing, operations, and maintenance, because these activities require an extremely huge amount of and various kinds of knowledge [4]. Advanced automation is not the sole purpose of this approach. Because a knowledge base can *hard-fail* when it meets an unknown situation, building large scale knowledge bases that store more knowledge is considered crucial for advanced applications. This approach assumed that "more knowledge is power" [12], and led to a number of projects to build large scale knowledge bases [5, 13].

This approach assumed also that computational reasoning capabilities were rather simpler, more rigid than those of human. Human reasoning depends more on quality of knowledge than on quantity, and is extremely *flexible*.

Knowledge intensive engineering reflects problems and issues addressed in this discussion and it aims at both amount and flexibility of knowledge. Knowledge intensiveness refers to how knowledge exists in an intelligent environment. A single knowledge base can make inferences in a particular circumstance but it may hard-fail. Perhaps, we need more knowledge, but we also need to have different kinds of knowledge. A large scale knowledge base alone is not useful. Knowledge must be well organized and can be flexibly applied to different kinds of applications. Knowledge becomes power, in other words, when it is intensive and flexible.

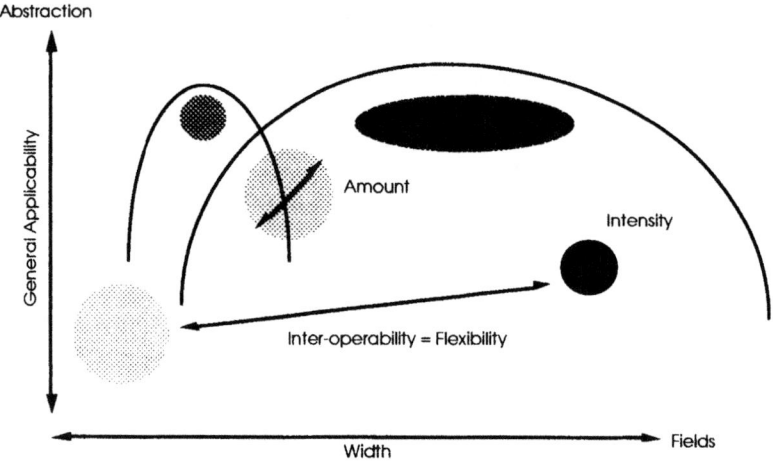

Fig. 1. Knowledge intensiveness

2.2 Knowledge Intensiveness

Figure 1 schematically depicts the knowledge intensiveness concept, in which chunks of knowledge are characterized in several ways. Certainly, knowledge intensiveness refers to the amount in a chunk of knowledge. Abstractness is an index of generality or universality of a chunk of knowledge. Width here shows domains that are covered by a chunk of knowledge. Intensity means how knowledge is well systematized within a chunk. Flexibility means ease of using a chunk in combination with others.

The difference between large scale knowledge base projects and knowledge intensive engineering resides within these explicit characteristics of knowledge. We view that knowledge intensiveness comes not only from the amount of knowledge but also from the way in which knowledge is distributed among cooperative intelligent agents. Thus, knowledge reuse and sharing become crucial issues in knowledge intensive engineering.

2.3   Knowledge Sharing and Reuse

Figure 2 compares three different types of knowledge sharing architecture. Figure 2 (1) depicts a situation with independent (and probably irrelevant) knowledge bases as a result of collecting independent *micro theories*. In this case, the strength of knowledge is just a sum of each of independent knowledge bases. Integrated knowledge bases can be represented in Figure 2 (2) in which knowledge bases can be applied to various situations and the strength of knowledge is maximum. However, this requires to have a platform with a uniform language. The Cyc project at MCC [14] is an example of this approach. In Figure 2 (3), independent knowledge bases can communicate and form an inter-operable situation, although the strength of knowledge might be weaker than that in Figure 2 (2). The entire knowledge base is a federation or a set of loosely coupled intelligent agents. For instance, the PACT [2] project employs this strategy. As a platform, they use KIF [6] and Ontolingua [7].

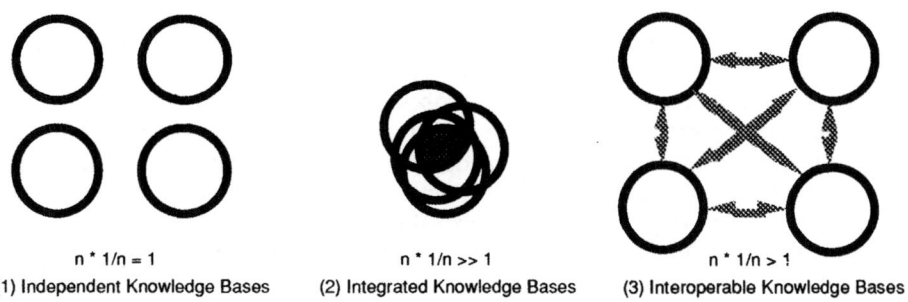

$n * 1/n = 1$	$n * 1/n >> 1$	$n * 1/n > 1$
(1) Independent Knowledge Bases	(2) Integrated Knowledge Bases	(3) Interoperable Knowledge Bases

Fig. 2. Knowledge sharing architecture

The second architecture can suffer from the lack of uniform knowledge representation. Unless carefully designed, the platform language cannot cover everything. The alternative third approach may have less problem in this regard, if the framework is enough abstract to incorporate different types of ontology about which each agent talks. However, it does not mean that we can completely avoid the problem, because communication among agents requires at least to understand what other agents are talking about.

From our previous experiences with building a large scale knowledge base for engineering design [11, 12, 22], we have found out that reasoning based on single ontology is less flexible and that describing everything in a single ontology is extremely difficult. This single ontology situation corresponds to Figure 2 (2). Instead, we need to have a system that allows reasoning based on multiple ontology in a cooperative multiple intelligent agent architecture depicted in Figure 2 (3).

2.4   Cooperative Multiple Intelligent Agents

Figure 3 depicts how we can build a cooperative architecture composed of communicating intelligent agents based on multiple ontology. Intelligent agents are computer systems or human agents of any kind including knowledge based systems (in most ordinary sense), database systems, simulation systems, and specific purpose computational tools. The simplest one could be a program that performs arithmetic.

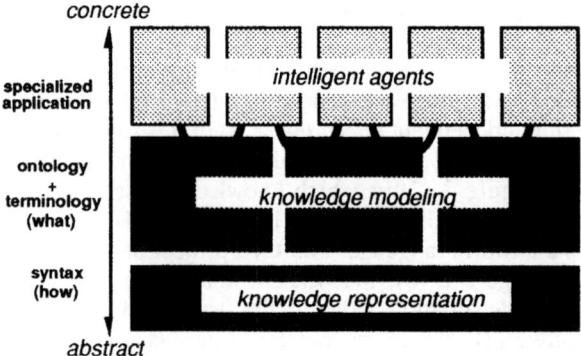

Fig. 3. Knowledge representation, modeling, and agents

At the bottom of the architecture, there is a system for knowledge representation at the syntax level. Knowledge representation at this level is a minor issue. However, the knowledge modeling level that describes *ontology* is more critical. While the knowledge representation level is more concerned with how, the knowledge modeling level represents what to be modeled and defines the basis of communication for intelligent agents. Knowledge modeling refers to such techniques as model based reasoning, qualitative physics, etc. We call this level *ontology*. The ontology at this intermediate level might include (but not necessarily be limited to) causality, spatial and temporal relationships, and actions in many cases. The ontology level describes concepts with fixed sets of concepts. These sets define *terminology*. Terminology is a network of conceptual relationships (or a dictionary), and at this moment, there are seven categories of terminology, i.e., entity, relation, attribute, physical property, physical phenomenon, physical law, and function.

Anything undefined in this level should be resolved by agents that provide problem solving capabilities. Each agent, therefore, has its own *ontology* inside it and does not have to understand other agent's knowledge and algorithms. (This usage of ontology might be different from the ontology shared at the intermediate level. Ontology at the agent level simply denotes problem solving mechanisms.) This is an advantage of this approach when compared with, e.g., the integrated knowledge representation approach in Figure 2 (2).

The intermediate layer should also contain information about what other agents know about. This includes information about the input-output relationships of an agent and information about concepts that an agent knows, although some details can and should be omitted.

3 The Kernel Knowledge Representation and Modeling

3.1 The CATS Ontology

As discussed in the previous chapter, a computational framework for knowledge intensive engineering should be based on the idea of collaborating *intelligent agents*. At the core of the framework, there is a mechanism, called *metamodel mechanism*, that provides common ontology and terminology for allowing intelligent agents to communicate each other. Intelligent agents are various kinds of modeling systems that have their own problem solving mechanisms. Since the project started as a project to build an intelligent CAD system, we primarily focused on knowledge about design objects [22]. However, the system should also cover other stages of product life cycle, such as manufacturing, operation, maintenance, recycling, and disposal.

The knowledge modeling of the metamodel mechanism is based on multiple ontology but a single terminology. Currently, there are four different types of ontology either implemented or planned. The first type is *causality* based on Qualitative Process Theory (QPT) [3]. The second type is *temporal* ontology in addition to temporal notions used by a qualitative reasoning system based on QPT. The third type of ontology is *spatial*. These two types of ontology, temporal and spatial, allow reasoning with dedicated reasoning systems. We plan to incorporate the fourth type of ontology that is about *action*. We call these four types of ontology CATS (Causality, Action, Time, and Space) ontology. The choice of the four types of ontology is reflection of the fact that the project originated from an intelligent CAD project. On top of the CATS ontology, there is terminology discussed in the next section.

Having a common set of ontology and terminology allows intelligent agents to collaborate easily. For example, if we want to describe a mechanical shaft, there is a concept *shaft* in the entity terminology that tells a shaft is a mechanical part and is a rotational object. When this shaft is associated with bearings to support it, reasoning within the causality ontology concludes that there are possible physical phenomena related to the shaft, such as supporting, rotation, and bending. Once bending is detected, we might have to worry about other physical phenomena such as stress and fatigue. This is the basic principle for intelligent agents based on multiple ontology to collaborate each other.

3.2 Terminology

Currently, terminology modeled in the framework has the following seven categories; namely, *entity, relation, attribute, physical properties, physical law, physical phenomenon,* and *function*. Each of these categories forms a static conceptual network. For instance, there is super-sub (or abstract-concrete) relationships among entities; a spur gear is a sub concept of a mechanical rotational part. Multiple inheritance is allowed; a spur gear can be a subclass of the mechanical rotational part and of the metallic object.

In the following sections, we describe the teminology of the current system in detail.

3.3 Entity

An entity represents an atomic, physical object in our representation and corresponds to an *individual* in QPT. Entities are organized in an abstract-concrete hierarchy. For instance, a spur gear is a sub concept of a mechanical rotational part. The hierarchy allows multiple inheritance, so that an entity can be categorized into more than one class; a spur gear can be a subclass of the mechanical rotational part and of the metallic object. This is done by delegation [13] that temporally adds additional properties to an entity. For instance, one can make a box to delegate an electric-conductor, so that the box can become an electric path as well. Figure 4 is a screen hardcopy of the entity browser that shows the definition of *metal*.

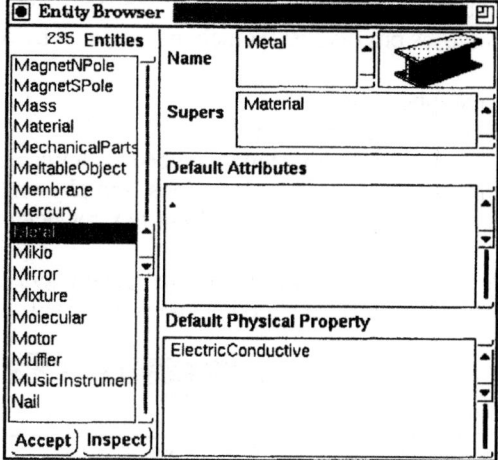

Fig. 4   The entity browser

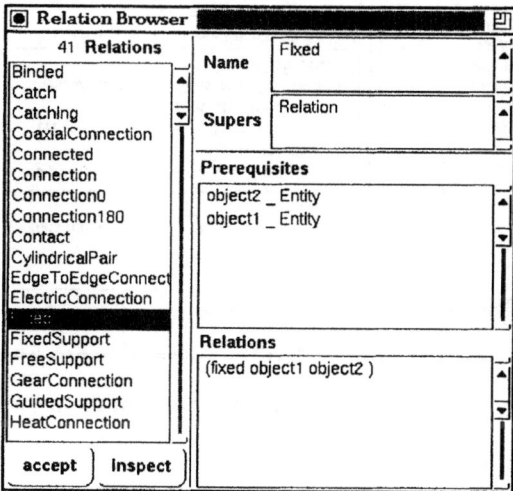

Fig. 5.   The relation browser

3.4   Relation

A relation represents a relationship among entities, such as *on*, *above*, *below*, *support*, and *connection*. In other words, it defines structure among entities. Relations work exactly in the same way as predicates in predicate logic. Relations can be hierarchically defined; for example, if electric-connection is a subclass of connection, electric-connection between A and B also implies a connection between them. Figure 5 is a screen hardcopy of the relation browser that shows the definition of *fixed relation*.

Relations are one method to introduce the spatial and temporal ontology into the system. For instance, a relation *on* between two entities, A and B, does not have any meaning, unless there is a system that can verify that A is actually on B. This is a

sort of symbol grounding problem and can be solved only by introducing an external reasoning system that can confirm if this relation holds.

3.5 Attribute and physical property

An attribute is a concept attached to entities and takes a value to indicate the state of entities, such as *position, temperature,* and *mass.* A physical property is a concept that describes generic characteristic of entities, such as *elastic* and *magnetized.* A physical property is associated with a set of attributes that indicate degree of the physical property. For example, Young's modulus indicates elasticity. The knowledge base also has knowledge about differential relations between attributes, and *has* relations between entities and physical properties. Figures 6 and 7 are screen hardcopies of the attribute browser listing *velocity* and the physical property browser listing *curved,* respectively.

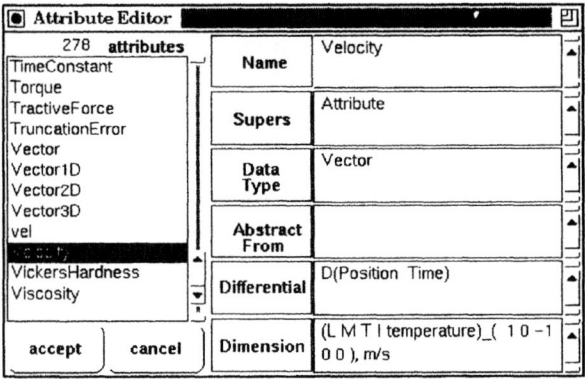

Fig. 6. The attribute browser

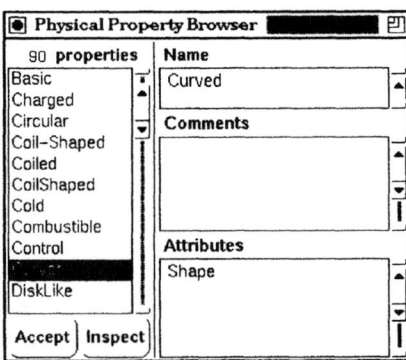

Fig. 7. The physical property browser

3.6 Physical phenomenon

A physical phenomenon designates physical laws or rules that govern behaviors of the object and corresponds to a view of QPT. Figure 8 shows a screen hardcopy of the

physical phenomenon browser defining a physical phenomenon in which a spring is reacting to external force. A physical phenomenon is defined by the following slots:

- Name of the physical phenomenon.
- Super (or abstract) physical phenomena.
- Prerequisites for the physical phenomenon to become activated in terms of entities together with their attribute definitions, relations, and physical properties. Entities are specified as pointers to link instances of entities for delegation when building physical features.
- Physical rules that define relationships among the parameters and define influences that can happen when the physical phenomenon is activated.

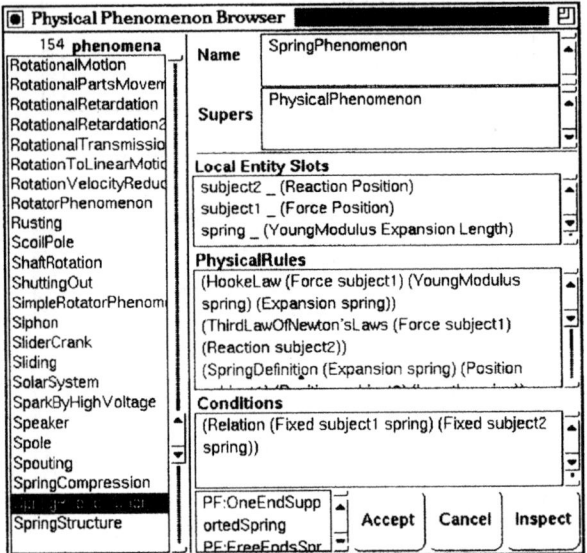

Fig. 8. The physical phenomenon browser

3.7 Physical laws

The physical law knowledge base contains two types of knowledge. One is the physical rule knowledge base that stores the unique names of physical laws and the attribute concepts used in the laws. The names of physical laws keep track of the same physical law represented in different design object models. The other is the model libraries that store typical model fragments of a design object model that relate physical concepts (such as entities, relations, physical phenomena, etc.) and design object models. For example, an equation

$$f = m\alpha$$

is a fragment of a dynamics equation model that describes Newton's second law. These fragments are used as building blocks to generate design object models. The parameters used in a fragment corresponds to attribute concepts, so that, after a design object model is generated, the metamodel mechanism can maintain relationships among the attribute concepts in the metamodel and the parameters in design object models for sharing information.

Figure 9 is a screen hardcopy of the physical rule browser that defines Newton's second law of motion.

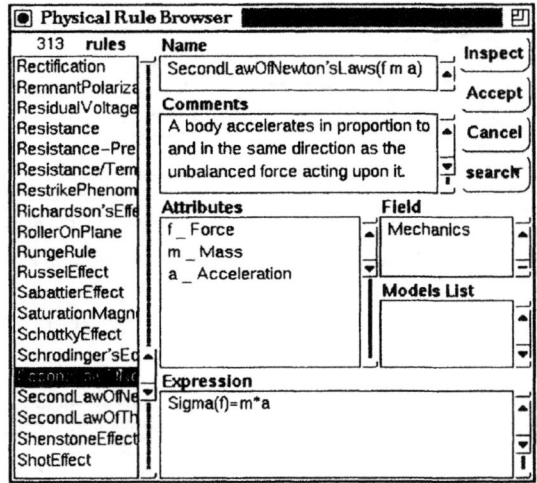

Fig. 9. Physical rule browser

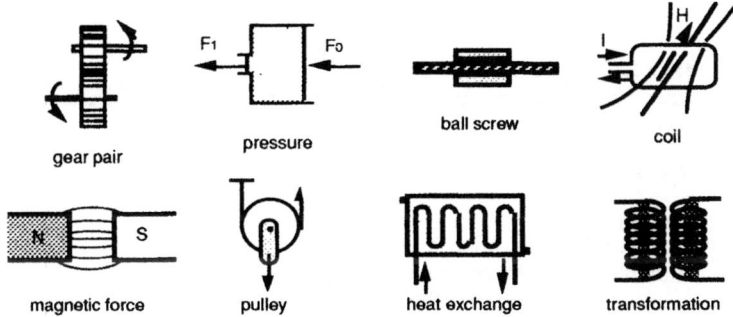

Fig. 10. Physical features

3.8 Physical features

A physical feature describes a situation in which something happens. Figure 10 depicts some examples of physical features. Figure 11 shows representation of a physical feature that includes the physical phenomenon of gear transmission. A physical feature is represented by a network of physical phenomena, entities, and relations. Links between a physical phenomenon and entities represent that the particular physical phenomenon occurs to a set of particular entities. Links among physical phenomena represent causal dependencies, which means that activation of one of the physical phenomena causes activation of others. In this sense, a physical feature is a sort of scenario of the situation; a set of causally related physical phenomena happening simultaneously or sequentially states, e.g., a situation that "fuel burns, generates heat, and warms up air." Links between a relation and entities represent that the entities have the relation.

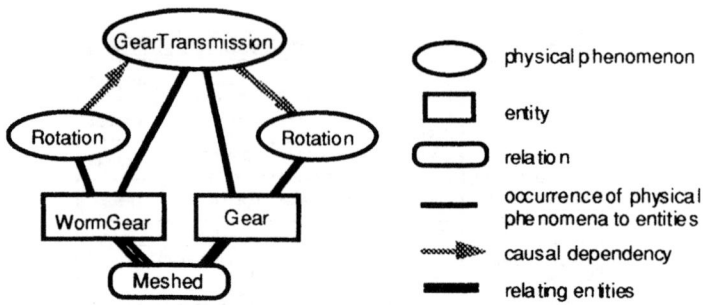

Fig. 11. A physical feature representing gear transmission

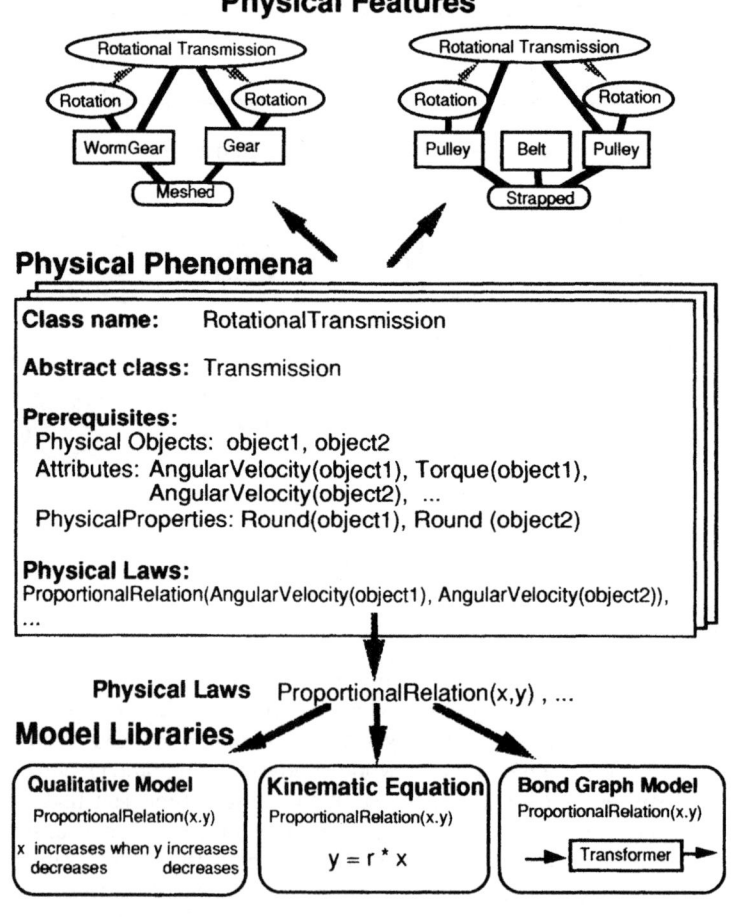

Fig. 12. Description of gear transmission

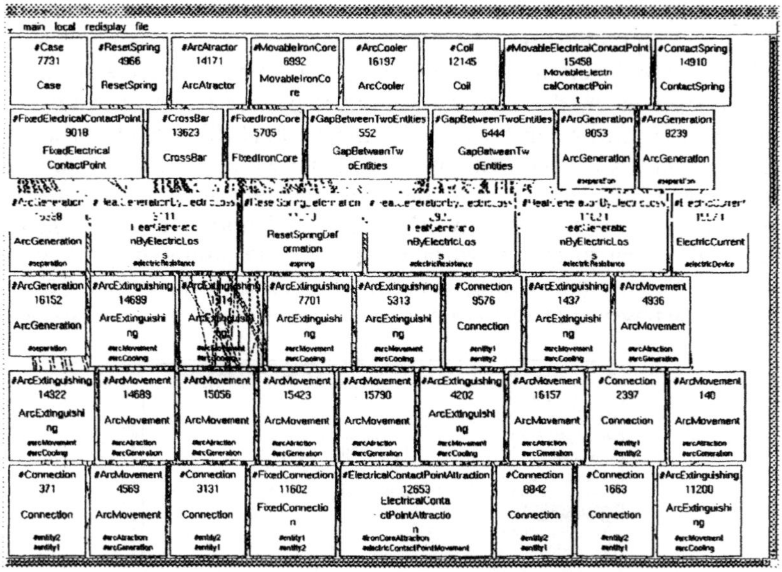

Fig. 13. A concept network

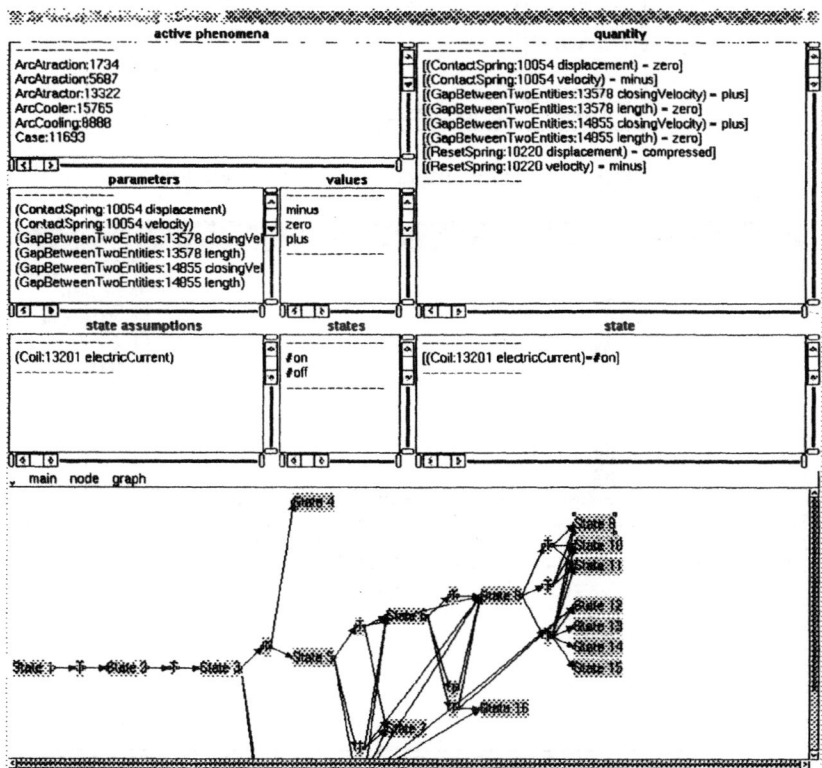

Fig. 14. State transitions

Figure 12 illustrates how the knowledge of gear transmission can be represented in our knowledge representation. We can use *rotational transmission* instead of *gear transmission*. The functional relations of gear's velocity are described in various representations in the model libraries. The qualitative reasoning system based on QPT takes physical features as an input and reasons about all feasible behaviors of the object. Its output is a concept network of relevant entities, relations, attributes, etc., (see Fig. 13) and a sequence of state transitions that represents a set of feasible behaviors (see Fig. 14). This concept network is used to identify relevant physical phenomena that might occur to the design object and that are useful information for, e.g., building models and maintaining consistency among them. Figure 15 is a screen hardcopy of the physical feature editor showing a system with a mass and a spring.

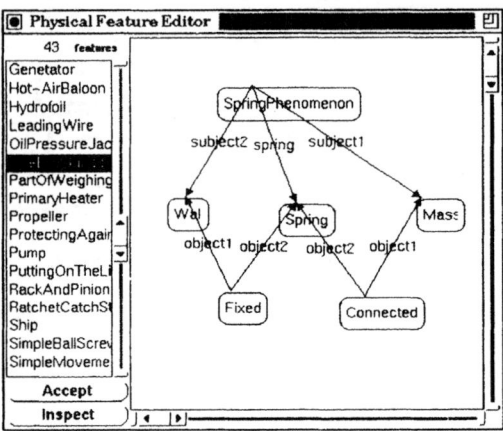

Fig. 15. The physical feature editor

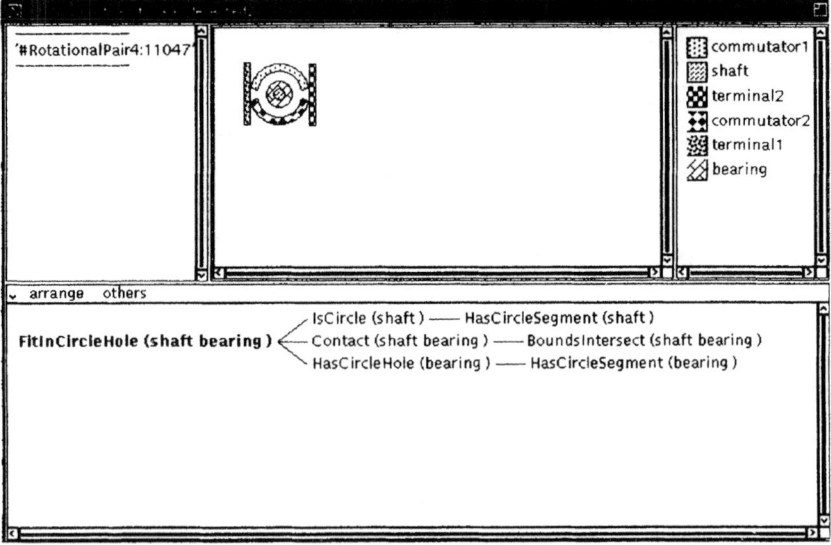

Fig. 16. Fusion of the causal ontology and the spatial ontology

The qualitative reasoning system is based on the causal ontology (i.e., QPT). However, through relations it can incorporate, for instance, spatial information that cannot be described in QPT. Figure 16 shows an example of reasoning that requires both the causal ontology and the spatial ontology. The spatial reasoning system reasons out changes of geometric contact points of a motor's commutator. This information is used by the qualitative reasoning system to reason about feasible physical behaviors of the commutator.

3.9 Function

In the context of knowledge intensive engineering activities, functional knowledge plays a fundamental role. For example, in engineering design, function is the requirement to be satisfied by the design solution. We employ the FBS (Function-Behavior-State) modeling [25] as the basis for function modeling.

We define a function as a subject description of a behavior, while a behavior is defined as a set of state transitions. An example of behavior is the state transitions shown in Fig. 14. A function is represented in the form of "to do something." Behaviors and state transitions are reasoned out by the qualitative reasoning system as described in the previous section.

A function prototype defines how this function can be hierarchically decomposed into sub functions or which behaviors (as a sequence of state transitions) can realize this function (called F-B relationship). The function hierarchy is composed of either abstract-concrete relations or whole-part relations. An example of decomposition into sub functions is that a super function "to move table" can be decomposed into three sub functions, namely, "to stop table," "to move table by motor," and "to stop table."

Figure 17 shows a function prototype for an electric switch and Fig. 18 shows its complete FBS model.

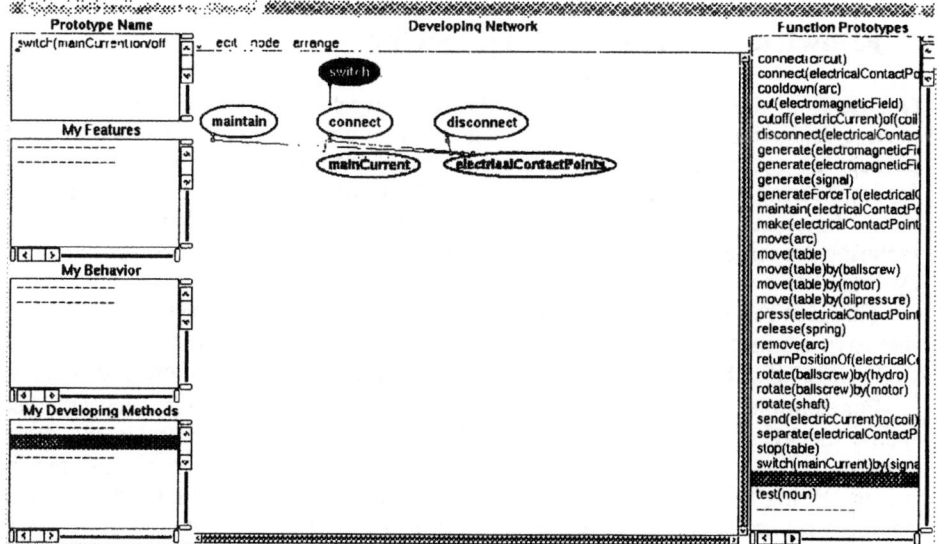

Fig. 17. The function prototype browser

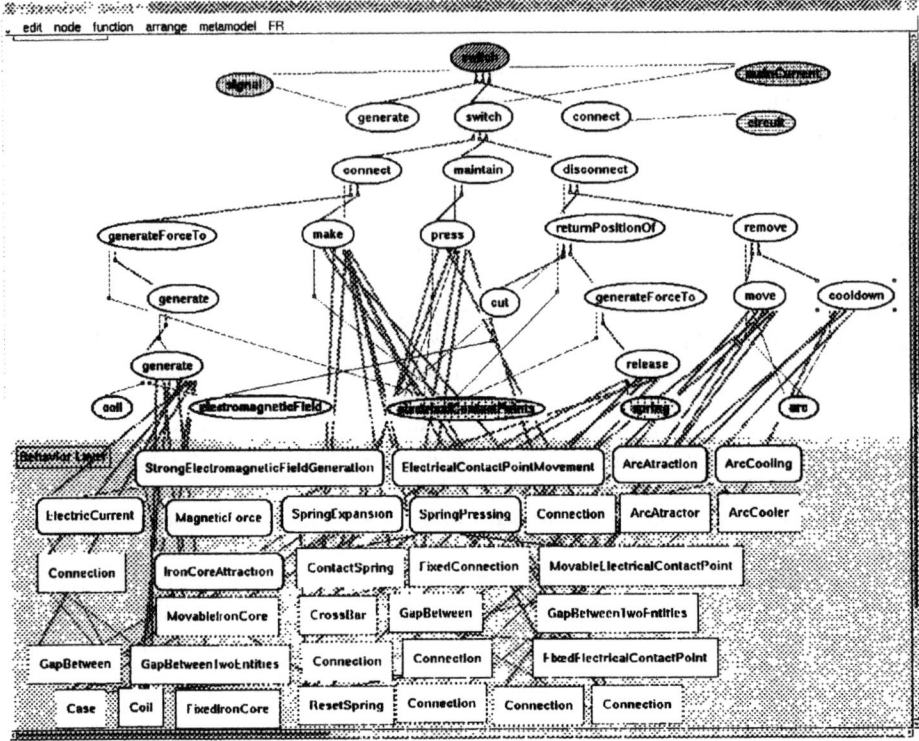

Fig. 18. The FBS modeler

4 Knowledge Intensive Engineering

Our group at the University of Tokyo has been working on various topics in engineering design research. Engineering design requires a wide variety of knowledge about product life cycle aspects. Therefore, an advanced information support infrastructure for engineering design should deal with not only design knowledge, i.e., about requirements, functions, physical phenomena, and devices, but also knowledge about manufacturing processes, operations, diagnosis and maintenance, and even disposal and recycling.

This idea resulted in the concept of knowledge intensive engineering. Knowledge intensive engineering is a new way of engineering activities in various product life cycle stages flexibly conducted with more knowledge to create more added value.

For example, a designer may need to have various kinds of *models*, such as a geometric model, a strength model, and a kinematics model, to synthesize the design solution, to analyze its properties, and to evaluate its performance against requirements. During a design process, information contained in various kinds of models are gradually refined and detailed. Models are a crucial aspect of computer supported engineering, and model-based reasoning technology is increasingly becoming important. One of the tasks of a knowledge intensive intelligent CAD system, therefore, is model management [1, 29].

Another important task in knowledge intensive design is design process management [18]. Although this feature is not further described in the paper, the framework we are developing has a design process management system that controls operations to models and navigates the design process [32].

Naturally, model management is crucial in other engineering activities as well. Operation tasks require different models from design, such as a control model, sensory data models, etc., so do maintenance tasks. For instance, knowledge intensive maintenance is a process in which various kinds of models for maintenance are used to diagnose the target machine more accurately and correctly than traditional association based diagnosis expert systems and to generate more useful information such as inspection instructions and repair plans. These should further result in feedback information to reliability design.

5 A Framework for Knowledge Intensive Engineering

Based on the ideas discussed in the previous chapters, our group at the University of Tokyo is currently developing a framework for knowledge intensive engineering. Its overall architecture is depicted in Fig. 19. The framework is developed on Smalltalk-80.

At the core of the system, we have the metamodel mechanism. Database management is done by an object oriented database system. Intelligent agents are models of an object to be modeled for various engineering activities.

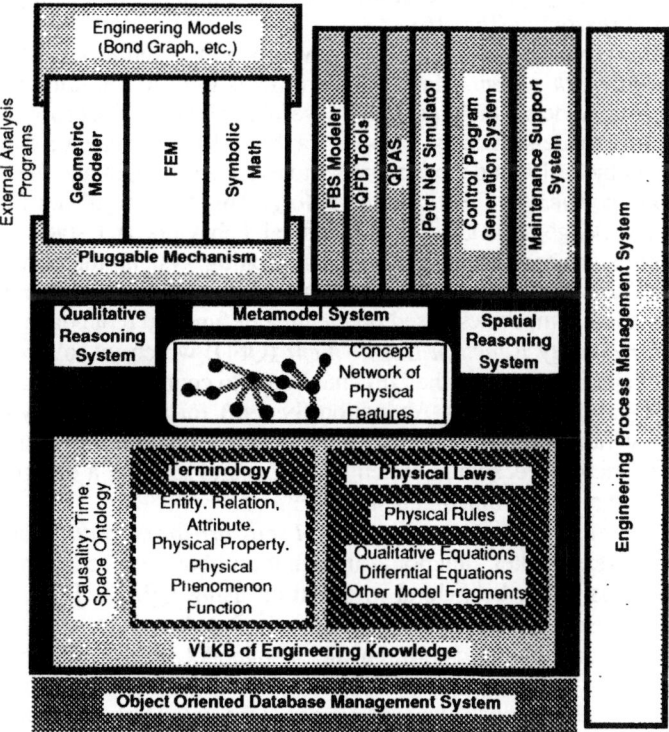

Fig. 19. The framework for knowledge intensive engineering

The metamodel mechanism [21] is based on common ontology and terminology for model management to allow intelligent agents to communicate each other. It is called metamodel, because it models various kinds of models.

The metamodel mechanism integrates various kinds of models through the concept network dynamically created by envisioning (see Fig. 13). As a result of envisioning, suppose that we know a physical phenomenon may qualitatively happen. To evaluate this, we need to generate a dedicated qualitative model concerned about that phenomenon and may further need to generate a quantitative model. The metamodel mechanism feeds information necessary to do so and automatically generates required models [11]. These generated models can share and exchange knowledge and information through the concept network. This is a substantially different approach from, e.g., product data exchange technology such as STEP [10].

So far, we have incorporated the following agents and modelers into the framework in addition to the qualitative physics based model that is dealt with by the qualitative reasoning system. However, models can have different kinds of problem solving mechanisms and different scales of abstraction. For instance, a geometric modeling system has precise geometric information based on algebraic geometry, while some other subsystem may require only topological (or connectivity) information. To absorb differences in the abstraction levels, the framework has a *pluggable mechanism* through which existing external modelers can be plugged into the framework [32].

- *A commercial geometric modeler* (DESIGNBASE from Ricoh, Co.). This system provides not only geometric design facilities but also, for example, information for spatial reasoning (Fig. 16).
- *An FEM analysis system* (MODIFY developed by Kubota *et al.* [16]).
- *A symbolic math system*, Mathematica. This is used for building various kinds of engineering models which resulted in useful applications, because many of simple engineering models, such as a Bond Graph modeler to evaluate dynamic behavior of design objects [9] and a beam model to evaluate strength and deformation [32].
- *The FBS modeler for functional modeling.* This system allows to describe functional knowledge associated with physical behaviors and states. Behaviors and states are connected to the qualitative reasoning system through the metamodel system. This system can be used for functional design including innovative design for high reliable machines based on the idea of function redundancy [24, 27].
- *A tool for Quality Function Deployment* (QFD) which in collaboration with the FBS modeler incorporates the QFD method into conceptual design [33].
- *QPAS* (Qualitative Process Abduction System) for supporting synthesis that can generate behaviors and states from functional information [8]. This system can be used for functional design in which behaviors and states of a design solution must be reasoned out from functional specifications.
- *A system for generating sequence control software for mechatronics machines* [17, 24]. This system takes functional information about a mechatronics machine and generates qualitative behavioral information using QPAS. Then, this qualitative control sequence is augmented with quantitative data, such as sensor layout and part dimensions, obtained through the metamodel mechanism from a geometric modeler, etc.
- *Systems for knowledge intensive maintenance activities.* First, there is a model based fault diagnostic system. The metamodel system can convert an object

model, described according to QPT in the metamodel, to a device oriented one for the confluence type qualitative reasoning system. This device oriented model can be used for repair planning, since it can represent faulty states and a normal state. This resulted in the development of a self-maintenance machine. Also, tools for reliability design, including Fault Tree Analysis and evaluation of reliability are incorporated.

• *A Petri net simulator for discrete event simulation.* This can be used for simulating, for instance, manufacturing processes and control logic of mechatronics systems.

6 Examples of Knowledge Intensive Engineering

6.1 Knowledge Intensive Design

The concept of knowledge intensive design is demonstrated in this section through an example of ship design [32]. Figure 20 is an FBS model of a high-speed hydrofoil boat. During the functional design, a designer can determine the working principle of such a high-speed ship from many alternatives including hydrofoil, hover craft, etc. Once its functions and realization methods are determined and described as an FBS model, the system can envision possible behaviors of the boat.

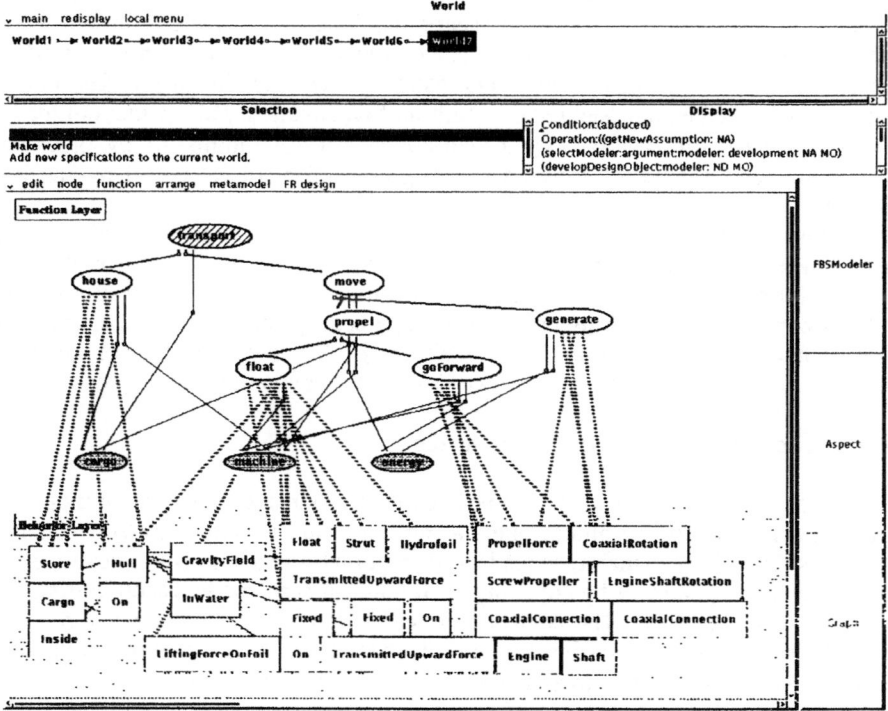

Fig. 20. An FBS model of the hydrofoil boat

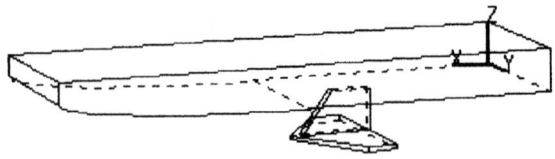

Fig. 21. A geometric model of the hydrofoil boat

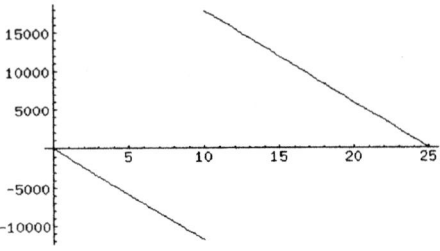

Fig. 22. Shearing force diagram of the hydrofoil boat

For instance, the qualitative reasoning system reasons out that deformation of the hull can happen and its evaluation is needed. In an early stage of design, a beam model may suffice for this evaluation. However, it might be impossible even to do so, because she does not know even rough shape and dimensions of the hull. The system then requests her to draw a rough sketch of the hull using a geometric modeling system (Fig. 21). The system then automatically feeds the dimension data into the beam model. The beam model itself is built on Mathematica. The stiffness of the boat can be assumed, for example, by a well-known rule of thumb, or by other more precise methods. Figure 22 shows a shearing force diagram calculated by the beam modeler.

In this design process of the hydrofoil boat, model management is entirely carried out by the metamodel system. Other models, such as an FEM model, are also available for further evaluation of the boat. In addition, these model operations can be entirely controlled by the design process management system. (In Fig. 20, the uppermost window shows the design process history.) This is one feature of knowledge intensive design to use design process knowledge for navigating design processes.

6.2 Knowledge Intensive Machines

Machines (or products) can also become knowledge intensive. We have developed two types of self-maintenance photocopiers as an example of such knowledge intensive machines [26, 27]. Self-maintenance is a concept that a machine continues to operate even when a fault happens by adjusting itself (control type) or by reconfiguring itself (function redundant type). The control type self-maintenance photocopier has a built-in model based reasoning system for fault diagnosing and repair planning with monitoring sensors and actuators to adjust internal states of the machine.

The function redundant type self-maintenance photocopier is a machine that can reconfigure itself so that it can maintain the most important functions while sacrificing less important functions. Function redundancy is achieved by using potential functions of existing parts in a slightly different way from the original design. This is a design strategy different from part redundancy.

In designing a function redundant machine, the FBS modeler is used to represent functions and many-to-many correspondent F-B relationships. The FBS modeler is also used to reason about alternative ways to recover the main functions with potential functions of existing normal parts. It searches for candidates of function redundancy and evaluates their behaviors using the qualitative reasoning system. It starts from a consistent FBS model of a design object and outputs an FBS model that has function redundancies (Fig. 23). In Fig. 23 that represents an FBS model of a photocopier, black rectangular nodes perform the target function "to charge the drum" in the normal state of the copier. When it is lost, a potential function of the transfer charger, which is performed by the hatched rectangular nodes, can replace the target function. The hatched function hierarchy represents function redundancy.

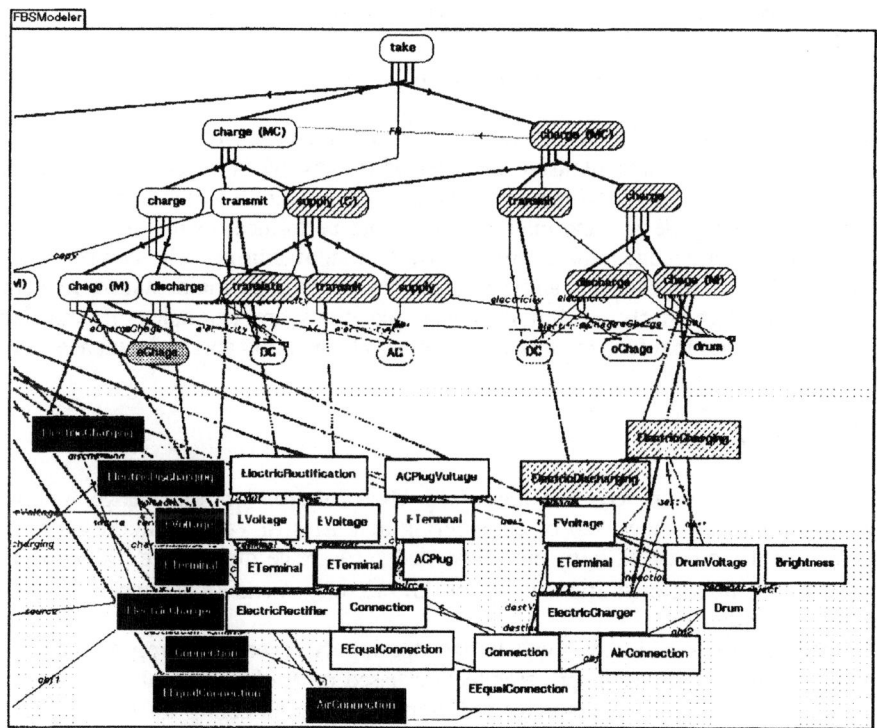

Fig. 23. Function redundant design on the FBS modeler

The FBS Modeler can support advanced (or even innovative) function design. The knowledge intensive engineering framework can even generate control software for mechatronics machines, such as photocopiers [17]. The knowledge originally stored in the framework is transferred to and embedded in a *knowledge intensive machine* that exhibits intelligent behaviors.

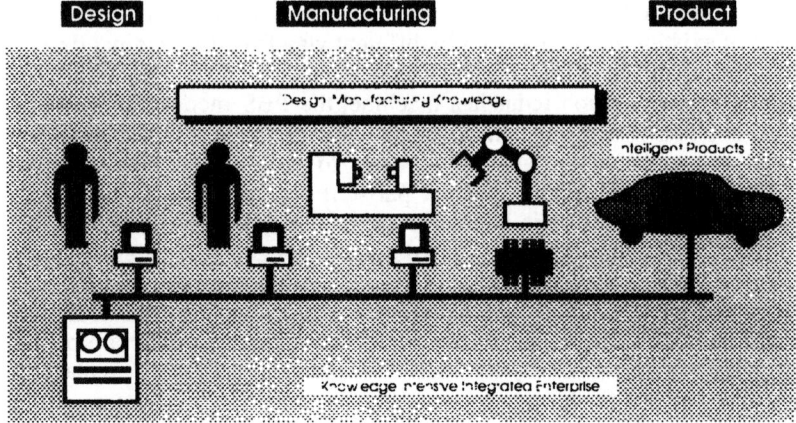

Fig. 24. Knowledge intensive integrated enterprise

6.3 Knowledge Intensive Enterprise

The concept of knowledge intensive engineering can be exercised in an integrated manner, once a knowledge intensive engineering framework is installed in a manufacturing enterprise. Figure 24 illustrates such an organization in which various kinds of activities, including design and manufacturing, are performed on the framework that allows flexible knowledge transfer from CAD to manufacturing cells and robots, and even to products. This concept can be extended to operations, maintenance, recycling, and disposal. Particularly, self-maintenance machines are examples of knowledge intensive machines that realize knowledge intensive operation and maintenance.

7 Conclusions

This paper first proposed and discussed the concept of knowledge intensive engineering that is a new way of engineering activities in various product life cycle stages flexibly conducted with more knowledge to create more added value. It then described a computational framework for knowledge intensive engineering we are currently developing.

Chapter 2 discussed issues about knowledge intensiveness and proposed the cooperative multiple intelligent agent architecture based on multiple ontology. Chapter 3 addressed the problems of knowledge representation and modeling for the framework. In Chapter 4, we examined the concept of knowledge intensive engineering and pointed out that one of the crucial issues was model management. A computational framework for knowledge intensive engineering was discussed in Chapter 5. Its architecture allows to plug in external intelligent agents. This feature is extremely useful for building, e.g., knowledge intensive intelligent CAD. Chapter 6 illustrated examples of knowledge intensive design, knowledge intensive machines, and knowledge intensive enterprise. Knowledge intensive design on the framework does not only mean advanced automation of design but also allows designers to arrive at creative, innovative design. The design of self-maintenance machines demonstrates that knowledge intensive engineering can lead to more creative, innovative design with increased added value, which is a knowledge intensive machine and knowledge intensive maintenance.

This paper reflects our continuous research effort towards knowledge intensive engineering since early 1980s. We started with a project to develop an intelligent CAD [19] and eventually arrived at knowledge intensive engineering. During the course of this research, we encountered many problems and difficulties. Among these, three issues must be particularly mentioned.

The first issue is understanding design. It is fairly easy to describe and represent design objects on computer. However, design processes are less understood and cannot be reasonably explained. Our design research [30, 20] convinced us that design study is largely design knowledge study.

We encountered the second issue during the study to investigate the feasibility of large scale knowledge bases for engineering design [11, 12, 22]. In collecting physical features, the ontology of QPT reveled many problems. These problems resulted partly from the limitation of QPT, but primarily from the consequence of relying on single ontology. In order to represent the physical world, we need to choose the most appropriate ontology for the purpose of modeling. This is one of the reasons why we arrived at the cooperative multiple intelligent agent architecture based on multiple ontology.

The third issue is also found out in the large scale knowledge base project. It can be stated in the following way: Unless knowledge is systematized, it is impossible even to collect knowledge [22, 31]. If knowledge is not systematized, the knowledge collection effort will desperately end up with unorganized, irrelevant sets of micro theories. These micro theories might be complete in their own, but are irrelevant to each other. Thus, it becomes extremely difficult to flexibly apply one knowledge chunk to another situation. This substantially decreases the value of collected knowledge. Quality of knowledge is more important than quantity of knowledge. In our view, *knowledge systematization* includes the following sub processes [22, 31].

- *Setting up a View* to define a background theory that dictates conceptual relationships among its own vocabulary and to define the scope and focus of attention.
- *Articulation* to identify instances of concepts that belong to the view and to give them representations of the background theory.
- *Codification* to find out structural relationships among instances articulated in the previous stage based on the background theory by codifying them. This process will result in generating pieces of factual knowledge. The articulation and codification sub processes correspond to actual knowledge collection process.
- *Crystallization* to generate general, abstract knowledge from purely factual knowledge, which can be called a theory. This theory will be tested and verified against the background theory and can be improved, abandoned, or taken as it is.
- *Reusing and sharing of knowledge.* These are the goals of knowledge systematization and can be achieved by having not only a common knowledge description format but also terminological, taxonomical, and ontological level standardization. The net result includes not only sharable knowledge but also a knowledge standard.

In building a computational framework for knowledge intensive engineering, knowledge systematization is essential. The sub processes listed above are indispensable to find out mechanisms to explain how engineering knowledge is dynamically formulated, how it is stored and shared, and how it forms more abstract knowledge that can be called a theory. However, this is exactly the same as the first issue of understanding design. After all, research on design theory and methodology turned out to be a critical path for knowledge intensive engineering.

Acknowledgment

This paper represents and summarizes a long research effort of our group at the University of Tokyo. We are grateful to Professor Hiroyuki Yoshikawa (currently President of the University of Tokyo) who outlined and initiated the research two decades ago. We would like to thank members of our group, especially, Masaki Ishii and Masaharu Yoshioka, who greatly contributed to the implementation of the system and helped us to prepare this paper. Many other colleagues and collaborators of our group contributed to the development of the ideas and systems. Professor Akihiro Kubota (RACE, the University of Tokyo) kindly allowed us to use the MODIFY system and provided us with the ship design example.

The research was sponsored and supported by a number of funding agencies and companies. In particular, we would like to acknowledge the supports from Nippon Steel Corporation and Mita Industrial, Co.

References

1. V. Akman, P.J.W. ten Hagen, and T. Tomiyama: A fundamental and theoretical framework for an Intelligent CAD System. *Computer-Aided Design* **22**(6), 352-367 (1990).
2. M. Cutkosky, R. Engelmore, R. Fikes, M. Genesereth, T. Gruber, W. Mark, J. Tenenbaum, and J. Weber: PACT: An experiment in integrating concurrent engineering systems. *IEEE Computer* **26**(1), 28-37 (1993).
3. K. Forbus: Qualitative process theory. *Artificial Intelligence*, **24**, 85-168 (1984).
4. K. Forbus: Intelligent computer-aided engineering. *AI Magazine* **9**(3), 23-36 (1988).
5. K. Fuchi and T. Yokoi (eds.): *Knowledge Building and Knowledge Sharing.* Tokyo: Ohmusha and Amsterdam: IOS Press, 1994.
6. M.R. Genesereth and R. Fikes: *Knowledge Interchange Format, Version 2.2 Reference Manual.* Technical Report Logic-90-4. Stanford, CA, USA: Computer Science Department, Stanford University, 1990.
7. T.R. Gruber: *Towards Principles for the Design of Ontologies Used for Knowledge Sharing.* Technical Report KSL 93-04. Stanford, CA, USA: Knowledge Systems Laboratory, Stanford University, 1993.
8. M. Ishii, T. Tomiyama, and H. Yoshikawa: Synthetic reasoning method for conceptual design. In: M.J. Wozny and G. Olling (eds.): *Towards World Class Manufacturing.* IFIP Transactions B-17. Amsterdam: North-Holland, 1994, pp. 3-16.
9. M. Ishii, T. Sekiya, and T. Tomiyama: A very large-scale knowledge base for the knowledge intensive engineering framework. In: N.J.I. Mars (ed.): *Towards Very Large Knowledge Bases.* Amsterdam: IOS Press, 1995, pp. 123-131.

10. ISO/TC184/SC4: *STEP (Standard for The Exchange of Product model data)*. Draft Proposal. ISO, 1990.
11. T. Kiriyama, T. Tomiyama, and H. Yoshikawa: The use of qualitative physics for integrated design object modeling. In: L. Stauffer (ed.): *Design Theory and Methodology —DTM '91—*. New York: ASME, 1991, pp. 53-60.
12. T. Kiriyama, T. Tomiyama, and H. Yoshikawa: Building a physical feature database for integrated modeling in design. *Working Papers of the Sixth International Workshop on Qualitative Reasoning about Physical Systems*. Edinburgh, UK, 124-138 (1992).
13. D.B. Lenat and E.A. Feigenbaum: On the thresholds of knowledge. *Proceedings of IJCAI-87*, 1173-1182 (1987).
14. D.B. Lenat and R. Guha: *Building Large Knowledge-Based Systems*. Reading, MA: Addison-Wesley, 1989.
15. H. Lieberman: Using prototypical objects to implement shared behavior in object oriented systems. *Object Oriented Computing 1986*. New York: ACM, 1986, pp. 189-198.
16. H. Ohtsubo, Y. Kawamura, and A. Kubota: Development of the object-oriented finite element modeling system — MODIFY. *Engineering with Computers* **9**(4), 187-197, (1993).
17. Y. Shimomura, T. Sakao, K. Ohmichi, T. Widmer, Y. Umeda, T. Tomiyama, and H. Yoshikawa: Model-based automatic generation of control sequence from design information. *Proceedings of the Eighth Annual Meeting of the American Society of for Precision Engineering*. Seattle, WA, USA, November 7-12, 1993, Raleigh, NC, USA: ASPE, 495-498 (1993).
18. H. Takeda, P. Veerkamp, T. Tomiyama, and H. Yoshikawa: Modeling design processes. *AI Magazine* **11**(4), 37-48 (1990).
19. T. Tomiyama and H. Yoshikawa: Requirements and principles for intelligent CAD systems. In: J.S. Gero (ed.): *Knowledge Engineering in Computer-Aided Design*. Amsterdam: North-Holland, 1985, pp. 1-23.
20. T. Tomiyama and H. Yoshikawa: Extended general design theory. In: H. Yoshikawa and E.A. Warman (eds.): *Design Theory for CAD*. Amsterdam: North-Holland, 1987, pp. 95-130.
21. T. Tomiyama, T. Kiriyama, H. Takeda, D. Xue, and H. Yoshikawa: Metamodel: A key to intelligent CAD systems. *Research in Engineering Design* **1**(1), 19-34, (1989).
22. T. Tomiyama, D. Xue, Y. Umeda, H. Takeda, T. Kiriyama, and H. Yoshikawa: Systematizing design knowledge for intelligent CAD systems. In: G. Olling and F. Kimura (eds.): *Human Aspects in Computer Integrated Manufacturing*. IFIP Transactions B-3. Amsterdam: North-Holland, 1992, pp. 237-248.
23. T. Tomiyama and Y. Baba: Artifactual engineering and the post-mass production paradigm. *Proceedings of the First International Symposium on Research into Artifacts*. Tokyo: RACE, the University of Tokyo, 16-23 (1993).
24. T. Tomiyama and Y. Umeda: A CAD for functional design. *Annals of CIRP* **43**/1, 143-146 (1993).
25. Y. Umeda, H. Takeda, T. Tomiyama, and H. Yoshikawa: Function, behaviour, and structure. In: J.S. Gero (ed.): *Applications of Artificial Intelligence in Engineering V*, Vol. 1. Berlin: Springer-Verlag, 1990, pp. 177-193.
26. Y. Umeda, T. Tomiyama, and H. Yoshikawa: A design methodology for a self-maintenance machine based on functional redundancy. In: D. Taylor and L.

Stauffer (eds.): *Design Theory and Methodology —DTM '92—*. New York: ASME, 1992, pp. 317-324.

27. Y. Umeda, T. Tomiyama, H. Yoshikawa, and Y. Shimomura: Using functional maintenance to improve fault tolerance. *IEEE Expert, Intelligent Systems & Their Applications* **2**(3), 25-31 (1994).

28. B. Veth: An integrated data description language for coding design knowledge. In: P.J.W. ten Hagen and T. Tomiyama (eds.): *Intelligent CAD Systems I — Theoretical and Methodological Aspects*. Berlin: Springer-Verlag, 1987, pp. 295-313.

29. D. Xue, H. Takeda, T. Kiriyama, T. Tomiyama, and H. Yoshikawa: An intelligent integrated interactive CAD— A preliminary report. In: D.C. Brown, M. Waldron, and H. Yoshikawa (eds.): *Intelligent Computer Aided Design*. IFIP Transactions B-4. Amsterdam: North-Holland, 1992, pp. 163-192.

30. H. Yoshikawa: General design theory and a CAD system: In: T. Sata and E.A. Warman (eds.): *Man-Machine Communication in CAD/CAM*. Amsterdam: North-Holland, 1981, pp. 35-58.

31. H. Yoshikawa: Systematization of design knowledge. *Annals of CIRP* **43**/1, 131-134 (1993).

32. M. Yoshioka, M. Nakamura, T. Tomiyama, and H. Yoshikawa: A process model with multiple design object models. In: T.K. Hight and L.A. Stauffer (eds.): *Design Theory and Methodology —DTM '93—*. New York: ASME, 1993, pp. 7-14.

33. M. Yoshioka, M. Oosaki, and T. Tomiyama: An application of quality function deployment to functional modeling. Accepted for publication in: *Proceedings of the IFIP WG 5.2 Workshop on Knowledge Intensive CAD*. Helsinki, October 25-29, 1995.

2 CAST Methods

Multiparadigm (Knowledge-Based and Numerical) Continuous Simulation Environments: Architectural Issues

Tuncer I. Ören
Ottawa Center of the McLeod Institute
for Simulation Sciences
Computer Science Department, University of Ottawa
Ottawa, Ont., Canada K1N 6N5
oren@csi.uottawa.ca

Nasser Ghasem-Aghaee[1]
Computer Engineering Department
University of Isfahan
Isfahan, Iran

Abstract

A multiparadigm approach for the solution of ordinary differential equations is proposed. The approach consists of augmenting a numerical continuous simulation environment with a knowledge-based system that can provide the solutions and integration constants of ordinary differential equations. The new approach is called M-Dif (multiparadigm –knowledge-based and numerical– continuous simulation environment). As a typical example of M-Dif, the architecture of M-Gest (multiparadigm Gest) is described.

1 Introduction

In continuous simulation, i.e., in simulation of systems described by a set of ordinary differential equations (ODEs), the fundamental issue is solving the differential equations which represent a dynamic model. Conventional continuous simulation relies on numerical techniques to solve the differential equations. In multiparadigm environments, more than one solution generation paradigm such as knowledge-based and numerical paradigms can be used. In this article, the abbreviation "M-Dif" is used to denote multiparadigm –knowledge-based and numerical– continuous simulation environments.

A multiparadigm continuous simulation environment, called M-Gest (Multiparadigm Gest) is being implemented. M-Gest is based on two tools implemented recently; they are GestCell [1] which is a contemporary implementation of Gest [2] and KbDif (Knowledge-based differential equation solver) [3]. In the sequel, after brief reviews of the characteristics of GestCell and KbDif, an architecture for multiparadigm Gest simulation environment (M-Gest) is presented. However, the concepts are general and applicable to other M-Difs (multiparadigm environments). Other con-

---

[1]This research has been done while Dr. Ghasem-Aghaee was spending his sabbatical leave at the Ottawa Center of the McLeod Institute of Simulation Sciences.

tinuous simulation environments may be augmented with KbDif, or other similar knowledge-based systems, to put in practical use the knowledge about several types of thousands of ODEs [4] within the realm of continuous simulation.

2 GestCell

GestCell version 1.0 is implemented by Keon [1] under the supervision of the first author. The architecture of GestCell is shown in Figure 1. The system consists of (1) a knowledge-based interface, (2) files to represent specifications of model, parameters, and experiments as well as model behavior, (3) a program generator, and (4) programs. The knowledge-based interface is implemented in Microsoft's Visual Basic Application. With the assistance and guidance of GestCell, a user can edit a continuous model which is stored in a Microsoft Excel spreadsheet. A program generator accepts a Gest model specification and generates a table-driven simulation program written in C. The front-end interface is used to specify model parameters as well as experimental and behavior generation conditions which are stored in tables (in Excel).

The simulation program which is a conventional numerical simulation program is compiled only once; it reads the appropriate tables and executes a simulation run or study. Through the back-end interface of GestCell, a user can specify behavior processing and display conditions. A user does not need to maintain the simulation program. All he/she needs to do is to edit an existing model or to create a new one with the assistance and guidance of the front-end interface of the GestCell system. If only the model parameters or experimental conditions or behavior generation conditions change, the new set of values are stored in tables which are read by the table-driven simulation program.

3 KbDif - Knowledge-Based Differential Equation Solver

The basic philosophy in the development of KbDif is that "numerical approach can be complemented with a knowledge-based approach to take advantage of the situations where the analytical solutions of the ordinary differential equations are known" [3]. KbDif is an expert system to provide known analytical solutions and integration constants of ordinary differential equations. Therefore, it can be used as a separate tool.

In this study, KbDif is considered a module of software implementing M-Dif in general and M-Gest in particular. As a module, the characteristics of KbDif can be summarized by its inputs and outputs. When used as a module, the inputs to KbDif are: (1) an ordinary differential equation and (2) the initial values of the independent and dependent variables; i.e., the initial condition. If the ODE is recognized within the knowledge base of KbDif, the solution as well as the integration constant are provided by KbDif. If the ODE is not recognized by KbDif, a message indicates the fact to the user.

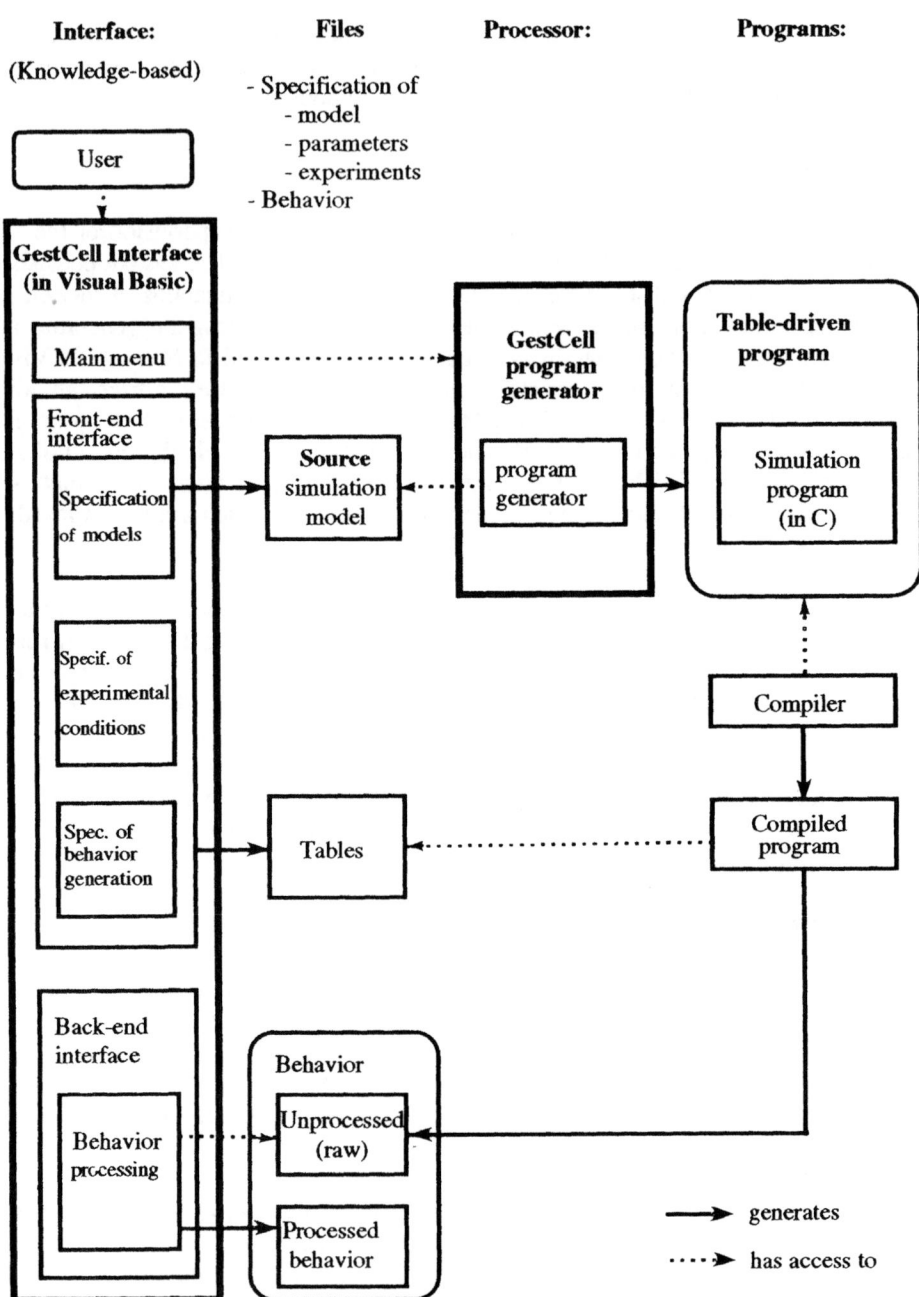

Fig. 1. Architecture of GestCell

3.1 Knowledge Base of KbDif

Currently, the knowledge base of KbDif contains about 60 first and second-order linear and non-linear ODEs. KbDif can provide explicit and implicit solutions. Only the explicit solutions are considered for M-Gest and M-Dif. There are two possibilities for solutions where one or several expressions can be provided one of which to be selected based on the values of some parameters.

Example 1. A non-linear ODE with *one expression* to represent the solution

Problem:
$$x' = \frac{a(t^n . x^{m+1} + b . t^n)}{t^{n+1} . x^m + c . x^m} \qquad x(t_0) = x_0$$

Input to KbDif:
$$x' = a*(t^\wedge n * x^\wedge (m+1) + b * t^\wedge n)/$$
$$(t^\wedge (n+1) * x^\wedge m + c * x^\wedge m)$$
$$x(t_0) = x_0$$

Solution:
$$x = [(k . \frac{1}{t^{n+1} + c^{-a(m+1)}})^{\frac{1}{n+1}} - b]^{\frac{1}{m+1}}$$

Integration constant:
$$k = (x_0^{m+1} + b)^{n+1} . (t_0^{n+1} + c)^{-a(m+1)}$$

Example 2. A first-order linear ODE with *several expressions* to represent the solution

KbDif checks the values of relevant parameters to provide an appropriate solution. In this example, the solution and the integration constant depend on the values of "a" and "b." If numerical values of "a" and "b" are given and constants, then only the applicable solution is provided by KbDif. Otherwise all applicable solutions are given.

Problem:
$$x' = \frac{a . x}{t} + c . t^b \qquad x(t_0) = x_0$$

Input to KbDif:
$$x' = a * x / t + c * t^\wedge b \qquad x(t_0) = x_0$$

If $a = b$
Solution:
$$x = c . t^{a+1} + k . t^a$$

Integration constant:
$$k = x_0 . t_0^{-a} - c . t_0$$

If $a \neq b$, $b - a = 1$

Solution:
$$x = t^a . (\ln|t^c| + k)$$

Integration constant:
$$k = t_0^{-a} . x_0 - \ln|t_0^c|$$

If $a \neq b$, $b - a \neq 1$

Solution:
$$x = \frac{c}{b - a + 1} . t^{b+1} + k . t^a$$

Integration constant:
$$k = \frac{1}{t_0^a} . (x_0 - \frac{c}{b - a + 1} . t_0^{b+1})$$

4 M-Gest and M-Dif

M-Gest (Multiparadigm Gest) is a typical M-Dif (Multiparadigm –Knowledge-Based and Numerical– Continuous Simulation Environment).

4.1 Architecture of the M-Gest environment

As seen in Figure 2, the architecture of the M-Gest system, is an extension of the architecture of the GestCell system and consists of (1) a knowledge-based interface, (2) files to represent specifications of model, parameters, and experiments as well as model behavior, (3) processors, and (4) programs. In Fig. 2, the modules which need to be added to GestCell are represented by dashed rectangles.

The knowledge-based interface of M-Gest is almost the same as the interface of GestCell. A *source simulation model* can be specified by a user under the assistance and guidance of the modelling environment. A *statement analyzer* of the M-Gest system scans every statement of the source simulation model. If the scanned statement is not a differential equation, then it is copied to the target simulation model. However, if the statement represents a differential equation, then a query is formulated to the KbDif module with the ODE and its corresponding initial condition. If the ODE exists in the KbDif system, the solution and the integration constant are put in relevant places of the target simulation model. Otherwise the ODE is copied to the target simulation model.

There are three possibilities so far as the target simulation model is concerned; i.e., all, some, or none of ODEs of the source simulation model are replaced with their analytical solutions.

In the first case, (pure knowledge-based approach) all the ODEs of the source simulation model are replaced with their analytical solutions. In this case, the target simulation model does not contain any ODE. Therefore, a *program generator* can transform the target simulation model into an non-simulation program necessary to

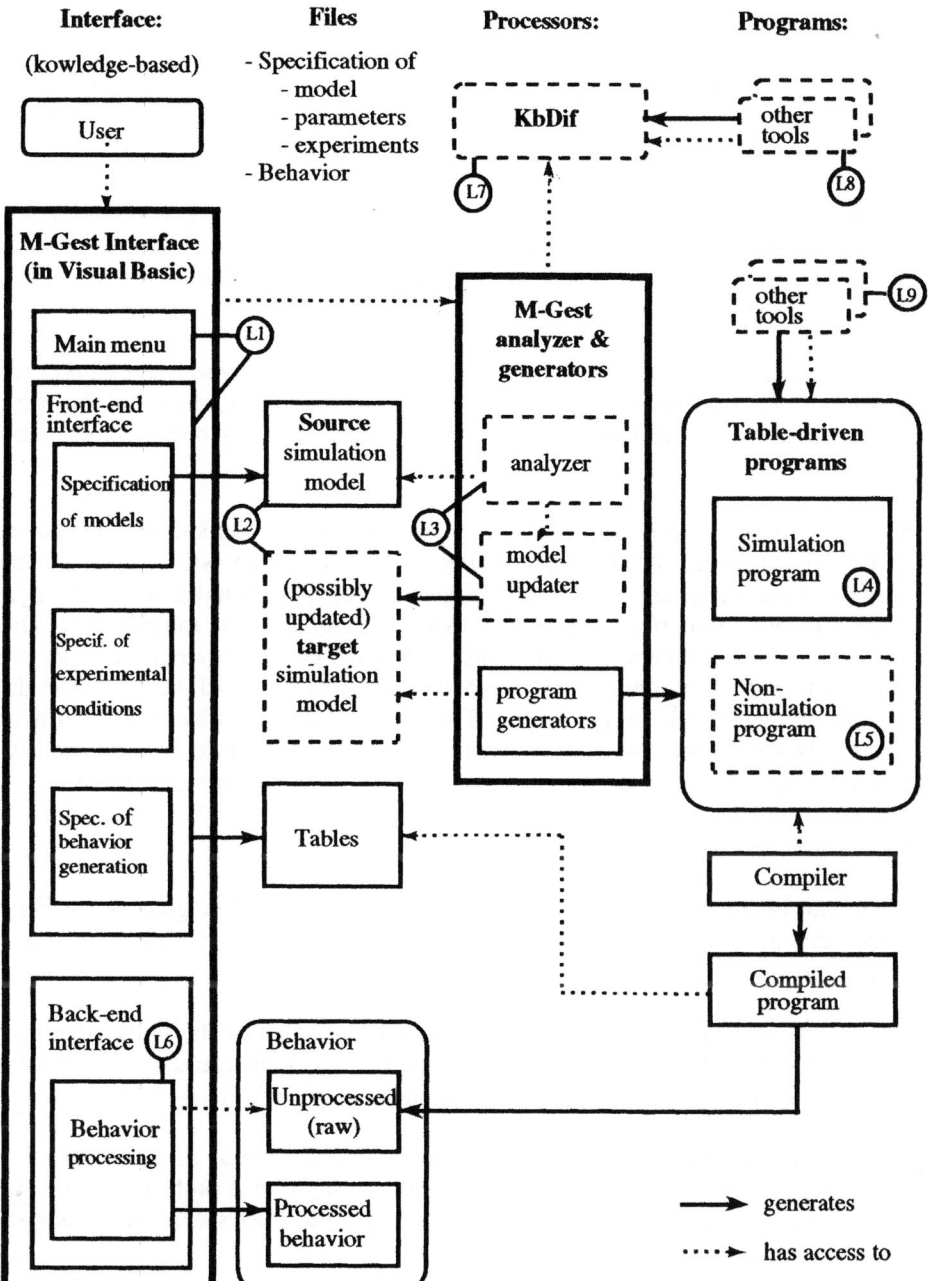

Fig. 2. Architecture of M-Gest (Multiparadigm Gest), a typical M-Dif (multiparadigm –knowledge-based and numerical– continuous simulation environment) (⌐¬ are extensions to GestCell)

generate the model behavior. A an non-simulation program does not require a numerical techniques to solve the differential equation(s) which were present in the definition of the original problem. A *table generator* can generate tables to be read by the non-simulation program to specify the conditions under which the model behavior has to be generated.

In the second case, some of the ODEs of the source simulation model are replaced with their analytical solutions. In this case , the target simulation model contains less ODEs than the source simulation model. In the third case, the target and source simulation models are identical.

Both in the second and third cases, a *simulation program generator* and a *table generator* can generate, respectively, a table-driven simulation program and the necessary tables to hold the model parameters, experimental and behavior generation conditions.

The interface of the M-Gest system has two major parts for the specifications of *behavior generation* and *behavior processing.* Based on the knowledge about the target simulation model and the simulation methodology, the behavior generation part of the user/system interface may assist a user to specify and then to update the tables about the behavior generation conditions. Then the compiled version of either the simulation or the non-simulation program can be run to first read the tables and then to generate the behavior of the model. The behavior generated by the system is unprocessed (or raw) behavior. The behavior processing interface of M-Gest can assist the user to specify behavior processing, i.e., behavior compression and display.

4.2 Languages in M-Dif and M-Gest

M-Dif can be implemented in different ways by using several languages. In Figure 2, the languages which are involved in an M-Dif environment are represented as Li (I=1, ..., 9).

(L1) The main menu as well as the front-end interface of the knowledge-based modelling and simulation environment can be implemented in several languages. In M-Gest, Visual basic Application is used.

(L2) The language of the source simulation model as well as the language of the possibly updated model. L1 is normally, a high level problem specification language. In M-Gest, an updated version of the Gest specification language is used [2].

(L3) The implementation language(s) of the analyzer as well as program and table generators of M-Dif.

(L4) The language of the generated simulation program. L4 can be a high level programming language such as an object-oriented, procedural, or functional language; or it can be another simulation language. In M-Gest, C language is used. A switch set from the main menu allows generation of C programs compatible with either Microsoft Visual C++ or with Borland C++ (depending on the available C++ system).

(L5)   The language of the of the generated non-simulation program. (This case occurs when all the differential equations have known solutions in the knowledge base of KbDif). In M-Gest, C language is used.

(L6)   The language used to process model behavior generated by the compiled program. L6 can be a structured language. The raw model behavior can also be processed by a package having necessary statistical and data display features.

(L7)   The implementation language of KbDif (Arity Prolog is used in this particular version).

(L8)   Implementation language(s) of tools that may be implemented for KbDif. These tools may support KbDif; however, they do not need to be integrated within M-Dif or M-Gest systems.

         Currently, two tools are being implemented for KbDif: a data acquisition tool and a knowledge-base documentation tool. The data acquisition tool for KbDif is to synthesize Prolog rules from specifications of differential equations, their solutions, and the integration constants. The knowledge-base documentation tool is to prepare structured and two-dimensional documentation of the rules of KbDif for the auditors of KbDif.

(L9)   Implementation language(s) of tools that may be implemented for the simulation and non-simulation programs. These tools may be integrated within M-Dif or M-Gest systems. As an example, a new version of VISAG (Visual Intelligent Software Agent Gauges) [5] is being implemented (in Microsoft Visual C++) to permit a user to attach virtual instruments on any of the observable variables of the programs.

In promoting the multilingual aspect of M-Gest and especially similar other multi-paradigm environments, we want to point out that, each one of the languages L1-L9 can potentially be a different language. Of course, a given language may (and should) be used in implementing several software modules to avoid unnecessary proliferation of the languages. However, to take advantage of the abilities of the languages, we should aim to use the language most appropriate for the implementation of different software modules. Therefore, we should be ready for multilingual environments.

5 Advantages of Multiparadigm Approaches and Conclusions

Exact solutions of over 5000 ordinary differential equations of several types (linear / nonlinear; first, second, third, forth, and higher order) are known [4]. If an ODE and its solution are in the knowledge base of KbDif, then numerical techniques can be avoided in a simulation study, for this particular ODE. M-Gest and similar M-Dif environments, once fully implemented, may thus reduce the number of differential equations which need numerical techniques for their solutions. Multiparadigm (knowledge-based and numerical) simulation environments may therefore take advantage of the existing knowledge about solutions for a very large number of types of ODEs.

Several advantages exist for such a multiparadigm approach: (1) elimination of errors in numerical integration, (2) speeding the process of the calculation of the trajectories of state variables, especially for faster than real-time computations in process control, (3) reducing the degree of parallelism for parallel ODE solvers, and (4) possibility of avoidance of the stiffness issue. By possibly reducing the number of ODEs in a simulation study, the degree of parallelism and hence the complexity of the associated problems would also be reduced. For some step wise adjustment procedures for parallel ODE solvers see [6]. An example taken from Gear [7] of the possibility of the elimination of stiffness was given by Ghasem-Aghaee and Ören [3]: The eigenvalues of the following initial value problem:

$y' = -y$, $y(0) = 1$ and $z' = -1000z$, $z(0) = 1$ are -1 and -1000; which gives the ratio of stiffness $s = 1000$. KbDif can provide the solutions $y = e^{-t}$ and $z = e^{-1000 \cdot t}$. With these solutions for y and z the stiffness problem is eliminated.

With these advantages, it is hoped that some of the advanced environments for continuous simulation will be multiparadigm environments. It is hard to justify the use of a single computer language to implement different modules of a comprehensive simulation environment such as a multiparadigm environment. Hence, multiparadigm environments may be, by necessity, multilanguage environments.

A multiparadigm approach for the solution of ordinary differential equations is feasible. It can make mathematical knowledge directly and conveniently usable through a knowledge-based system such as KbDif.

Acknowledgment

This work has been done while the second author was spending his sabbatical leave (1993-94) at the Ottawa Center of the McLeod Institute of Simulation Sciences, at the Computer Science Department of the University of Ottawa, Ottawa, Ontario, Canada. Both authors appreciate the support provided by the University of Isfahan to the second author, which made possible this co-operation that resulted with several finished and on-going research projects.

References:

1. N. Keon: Simulation of proton exchange membrane fuel cell: dynamic modelling and an environment. Master's Thesis, Systems Science Programme, University of Ottawa, Ottawa, Ontario, Canada, 1995

2. T.I. Ören: Gest - A modelling and simulation language based on system theoretic concepts. NATO ASI Series, vol. F10, Simulation and model-based methodologies: An integrative view. Berlin: Springer 1984, pp. 281-334

3. N. Ghasem-Aghaee, T.I. Ören: KbDif: A knowledge-based approach for continuous simulation. Transactions of the SCS, 12:4, 287-301 (1994)

4. A.D. Polyanin, V.F. Zaitsev: Handbook of exact solution of ordinary differential equations. Boca Raton, Florida: CRC Press, 1995, 720 p.

5. B. Abdullah, T.I. Ören: Integration of visual intelligent software agent gauges (VISAG) into a simulation environment. In: Proc. of the 1995 Summer Computer Simulation Conf., Ottawa, Ont, July 24-26. San Diego, CA: SCS, pp. 269-274

6. L.G. Birta, M. Yang: Some step size adjustment procedures for parallel ODE solvers. Transactions of the SCS, 12:4, 303-324 (1995)

7. C.W. Gear: Numerical initial value problems in ordinary differential equations. Englewood Cliffs, NJ Prentice-Hall 1971

A Development Methodology for Systems Engineering of Computer-Based Systems and its Environmental Support

Markus Voss   Gerhard Schweizer

Universität Karlsruhe
Institut für Mikrorechner und Automation (IMA)
Karlsruhe, Germany
mvoss@ira.uka.de

Abstract. Systemobject-Oriented Structured Analysis and Service-Oriented Design (SOOSA/SOD) is a coherent development methodology for computer-based systems developed by IMA at the University of Karlsruhe, Germany. It is based on strong system theoretic principles and is formal in a way that allows for execution of the specifications by simulation and partially automated implementation. It defines clear interfaces to the encompassing engineering process by covering all the engineering steps of an underlying tailorable engineering process model. Both, the development methodology and the process are supported by prototype tools forming an integrated computer-based systems engineering environment.

Keywords: Engineering of Computer-Based Systems, Development Methodology, System Theory, Modeling Theory, Design Theory, Engineering Process, Engineering Environment

1 Introduction

We will fist summarize the state and technical contents of the discipline of engineering of computer-based systems and illustrate a few problems that we address in our methodological framework for both governing the engineering process and performing the operational task of development of computer-based systems (CBS). The major point we make is that the application of system theoretic principles or in other words the 'concept of the system' [1] helps a lot in organizing the work.

1.1 ECBS

It is widely agreed, that the development of large CBS calls for a new discipline at the systems engineering level that goes beyond state-of-the-art engineering practice. This discipline is called Engineering of Computer-Based Systems (ECBS, [2]). While most engineering disciplines like mechanical or electrical have developed engineering standards to be followed in planning, realizing, and operating systems, which reflect the basic system theory of the domain and are understood by the entire community, the same does not hold for the computer-based systems (CBS) domain. That domain is dominated by many different standards developed by numerous different organizations to serve as guidelines for the engineering processes, there are even more (hundreds of) methods and supporting tools to aid in very specific areas of the engineering work, and still a great many heuristics.

The CBS domain lacks a *basic system theory* understood and agreed upon by the entire community. This is very much a result of *fragmentation* and *isolation* in research and development of computer science. Fragmentation manifests itself in focusing on programs of narrow declared goals and wrestling with the internal technical and management problems. Isolation means regarding the transfer of results into practice and market as secondary to the achievements of the internal goals.

A discipline of ECBS is an attempt to overcome these tendencies. Its promotion requires the definition of a generally accepted *terminology*, the development of fundamental but practical *procedure, information, and architecture standards*, *product standards* for baseline and application architectures, and the definition of *tool standards*. Further, it requires curricula and materials for training and education and a public relation campaign explaining the need, the use, and the benefits of ECBS to industry operating organizations of CBS and government agencies responsible for R&D programs.

The impacts and advantages of ECBS for the future markets to be expected are increased efficiency in planning and developing CBS, increased efficiency in operating and maintaining these, increased mobility of systems engineers, a harmonized market and increased transparency for both tools with accepted tool standards and products in soft- and hardware for basic and application architectures with accepted product standards, increased transparency for procurers and suppliers, a harmonized market and increased transparency for complete CBS, and increased transparency for the technical and legal processes of procurement including project management, system development, configuration management, and quality assurance.

It is obvious that regulations concerning procedures, products, and tools have to be defined for all activity classes within ECBS processes. Besides the activities directly concerned with the systems life cycle those from supporting management tasks have to be explicitly emphasized also in order to arrive at a complete framework. In [3] it is shown how these regulations can be organized within *engineering standards* by investigating generally valid activity classes and emphasizing multi-dimensional activity class space decomposition resulting from inherent organizational principles as an analysis tool for standards evaluation, evolution, and operationalization.

1.2 Issues

It is problematic only to regard ECBS as a discipline of applying given systems engineering concepts (e.g. [4]) to the development of CBS or to overstate the need to maintain continuity with the systems engineering culture [5]. Although ECBS is a system level discipline it is also about learning from modern software and hardware engineering practice to avoid incompatibilities at the interfaces. Talking about modern software engineering note, e.g., the fact that the modeling of any system when strong system theoretic concepts are applied is of *object-oriented nature*, in the sense that object is the *modeling term* for system or subsystem. Therefore it is highly useful to integrate the benefits of object-oriented development (encapsulation, abstract data typing, and generalization to support reuse within every engineering step) with current systems engineering practice.

Mathematically correct system representations (system models) are not only a precondition for satisfying all sorts of certification issues but almost automatically lead to 'better' tools. A problem here is the gap between current knowledge about 'formal' techniques for system specification and its presence in most of the popular development methods actually used in practice. In a way this corresponds to the observation that development methods often not explicitly refer to an underlying theory and that, on the other hand, existing theories are not made applicable in terms of standard processes to follow and standard notations. Emphasizing that, we address the topic of distinguishing between methods and formally defined modeling paradigms as proposed in OSA e.g. [6]. We propose starting with a mathematical theory (being a sort of generally valid meta method) and then building concrete system development methods in terms of procedures and notations upon it.

It is necessary to support 'ergonomic' ways to formulate knowledge acquired during the development process. For applicability, as we have stated in [7], it must be possible to get graphic-oriented ways based upon the theory of CBS model building applicable for practitioners by 'adequate packing' of the theoretical knowledge without loss of accuracy. Sure this is done in some areas already but it has not yet been made a paradigm for all of the systems engineering process.

All of this is of only limited value if there is no common agreement about the technical contents of the core engineering steps to form a tailorable process model (model of the project organization as a system), the underlying information models, and their representation within procedure, information, and architecture standards. Considerable work towards defining these has been done already (as in [8],[9] for the ECBS community) but what is often neglected is the fact that there is a standard architecture of a project organization resulting from the system properties of the product under developmental consideration and sorts of problems to be solved within the ECBS domain. Having generally valid information acquisition steps in hierarchical modular modeling, e.g., yields a way of structured project management, namely building and maintaining the project organization according to the product's structure.

SOOSA/SOD is concerned with all these issues. This is why we call it a methodology rather than a method. Within the next chapters we emphasize general system, modeling, design, and engineering process theory for ECBS, sketch ways of mapping notations and procedures, and investigate possibilities for tool support.

2 System and Modeling Theory

2.1 Systems and System Models

Modern system theory distinguishes between a system's *structure* and *behaviour*. Structural information contains all the knowledge about *what* components a system is built from called the *component hierarchy*, and about what *further associations* exist between its parts or the parts properties describing *how* the system is assembled together from its components including static cooperation structures sometimes called *systems coupling* [10]. Unlike in traditional system theory, this is contrasted with classification properties of the systems parts incorporating the concept of generalization resulting in the so-called *class hierarchy*.

Behavioural information, on the other hand, represents the knowledge about the system's property in terms of mapping its *input* (influence) to its *output* (reaction).

That system function's interpretation depends on the system's *state*. It is clear that, given the system's parts, its behaviour can be deduced by the knowledge about the parts behaviour and the way the components interact via their coupling structure. This can mathematically be captured by modeling the system as a set of *communicating extended finite state machines* (EFSM, e.g. [11]) or *discrete event systems* (DEVS, e.g. [12]) both modeling system behaviour as state sequences.

The most important statements to make up a theory of system modeling are listed below. By these the classes to constitute a *model of system models* and their relations are defined (system meta-model). Every instance is called an *object*. Note however that these are not meant to be an axiomatic specification of what a system is.

- A system has a unique identity and is of a sort.
- A system may be regarded as aggregated from components which are also systems. All aggregation relations constitute a hierarchy (component hierarchy).
- A system may be further described by attributes.
- Apart from aggregation relations there may exist further structural relationships between a system's components which determine the system's attributes from the components' attributes.
- A system has a state.
- A composite system's state is an abstraction of its components' states
- A system may have inputs and outputs.
- A system has a behaviour.
- The behaviour of a system as a whole can formally be captured by functions

  f_{state} : *input* $\times$ *state* $\rightarrow$ *state*

  and f_{output} : *state* $\rightarrow$ *output*.
- Behaviour of a composite system $s = \langle s_1,..,s_n \rangle$ results from the componets' behaviour and input/output relations which can formally be captured by functions

  f_{input,s_i} : *input*$_s$ $\times$ *output*$_{s_1}$ $\times$ .. $\times$ *output*$_{s_n}$ $\rightarrow$ *input*$_{s_i}$ $i = 1,..,n$

  and $f_{output,s}$: *output*$_{s_1}$ $\times$ .. $\times$ *output*$_{s_n}$ $\times$ *output*$_s$ $\rightarrow$ *output*$_s$
- Systems may be classified according to common attributes, behaviour, component classes, inter-component structural relations, inter-component input/output relations or purpose.
- All classification relations constitute a hierarchy (class hierarchy).
- The minimal system classes constitute system sorts.
- A system's identity results from its components and inter-component relations.

2.2 Abstraction and Knowledge Representation

The modeling process is very much driven by some hierarchical structure which evolves during model development to keep track of the model's complexity and to serve as a media for abstraction. Different methods organize system modeling according to different hierarchies (objects, classes, functions). To incorporate the concept of *abstraction*, modeling can be decomposed into system class definition and system architecture definition both organized in formal *knowledge representation schemes* (KRS).

A *system class definition* for some system S is captured within a KRS, which is a 3-tuple

$$Class_S = \{A, K, R\}$$

- *A* is a set of types for objects (called classes), attributes, states, inputs, and outputs
- *K* is a set of elements being of a type included in *A*.
- *R* is a set of relations over *A*, which define relations both structural and behavioural.

Class$_S$ is also known as a system's *signature*. All information from above syntactically map into that scheme (both identity and class map into the KRS's syntax). It is used as an organizing framework (some sort of template) for the knowledge acquired during the modeling process. For a certain abstraction, *A* contains all types of a systems components, attributes, state, inputs, and outputs.

Abstractions between one higher-level and *n* lower-level abstractions have to be kept consistent which means some sort of *abstraction mapping* must be provided for each level of decomposition according to the underlying hierarchy. We call this process a *model extension*, because it extends 1 + *n* models into one multi-dimensional model, called the *extended model*. The model to result contains its own plus the sub models' types, elements, and relations as well as additional relations modeling the abstraction mapping. The process of model extension is performed by defining the higher-level system's architecture.

A *system architecture definition* is captured within a KRS, which is a 4-tuple

$$Architecture_S = \{A, K, R, Fo\}$$

- *A* is a set of types as in *Class$_S$*. It is recursively defined as the union of *Class$_S$* and all sets *A* of *Class$_C$* with *C* being component.
- *K* is a set of elements like in *Class$_S$* with the same set building principle.
- *R* is a set of relations over *A* as in *Class$_S$* with the same set building principle plus a set of relations for inter-class properties like aggregation, coupling, cooperation, etc.
- *Fo* is a set of axioms (formulas) defined upon the interpretation (values) of *A* and variables from *K*. They define the actual values (semantics) of the elements and relations.

A system model is given by the interpretations that fulfil the axioms. This abstraction principle yields a structure called the *model organization structure (MOS)* to be applied to the system information defined above. That structure is to be exploited in project management as will be shown later. The MOS also corresponds to so called mega-object structures used as extensions of 'flat' object-oriented modeling for 'engineering-in-the-colossal' [13].

2.3 Use of Models within the Process

Systematic systems development is characterized by building different sorts of models of the system. It usually starts with building a model of the system, which abstracts from implementation details used as a media for requirements analysis. We call these sorts of models *conceptual models*. In system design, these are transformed into models of the actual physical system by taking these details into account, called *design models*. We will investigate system design within the next chapter. With respect to critiques towards special life-cycle forms imposed on the development process like the waterfall model [14] or spiral model [15] it should be noted at this point, that incremental processes of development are supported by the 'breakpoints' given with the MOS. Conceptual and design work can be done nearly in parallel as shown below.

The ways of acquiring the knowledge to be put into the models is yet another issue to be discussed. For conceptual work in terms of 'picking the objects' a lot of different heuristics like problem domain analysis (e.g. [16]), use-case (scenario-oriented) analysis (e.g. [17]), or functional approaches like event partitioning (e.g. [18]), or problem decomposition have been proposed (for a good abstract see e.g. [13]). We do not go into detail on these approaches but refer to the next passage about the process for a framework for all these techniques to be applied resulting from general problem classes to be solved. It shows that following this framework conceptual models are created from more than one system.

At the first tier of decomposition of the engineering process there is always requirements definition, conceptual modeling, design modeling by extending the conceptual model in terms of 'imperfect technology' and allocating conceptual objects or functions to physical system components, trade-off analysis to select suitable designs, build and test plan generation, detailed hard- and software components construction, integration and test as well as validation tasks and supporting technical management, quality assurance, and configuration management tasks. The next tier of the development process definition though is guided by the system theoretical concepts described above and the following idea about general problem classes.

Although the purposes and shapes of CBS seem unlimited, there is only two reasons *why* CBS are built. One is that the future behaviour of some system is to be controlled and the other is that the future behaviour of some system is to be predicted or planned. That reality leaves us with two 'kinds' of systems or problems that we have to deal with in modeling (obviously these are not disjoint classes).

In case of *automation problems* the structure and the required behaviour of some plant to be automated is regarded as a system. The model built represents the conceptual model of the fully automated system. From that the conceptual model of the automation system can be deduced which specifies the relations between the automation demands, the measurable states and the controls required to put the plant

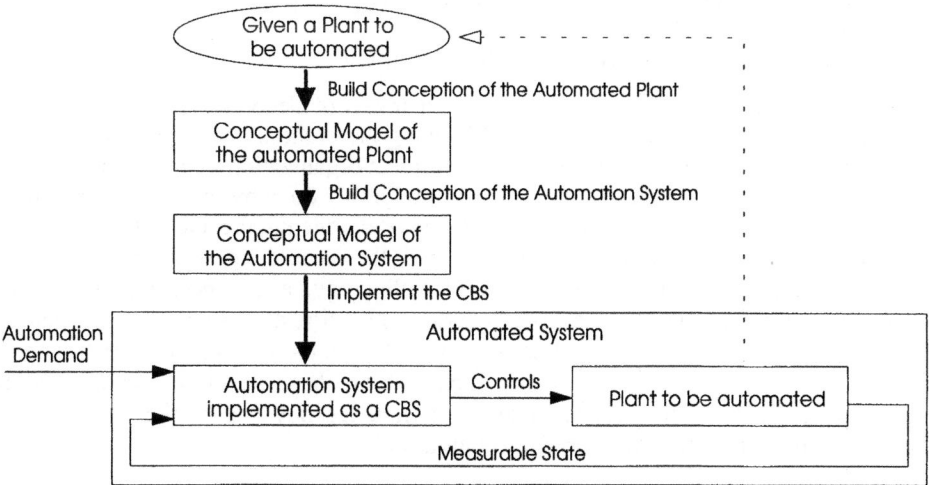

Figure 1: Engineering Automation Systems

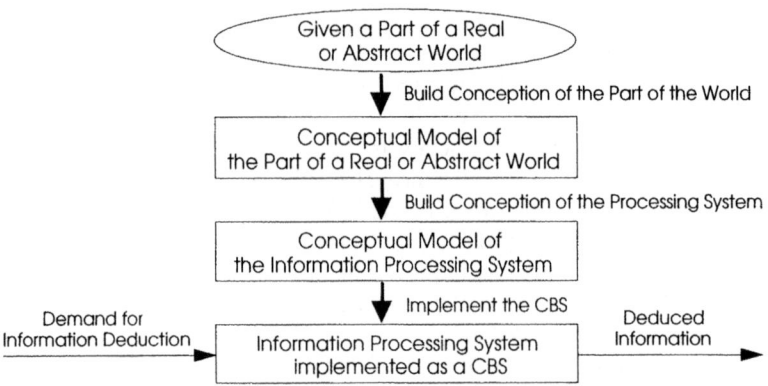

Figure 2: Engineering Information Processing Systems

into the specified state. The later is then to be transformed into a design model for the automation system to be implemented. Figure 1 illustrates this fact.

In case of *information processing problems* a part of the world to which the information belongs and from which required information must be deduced is regarded as a system. The model built represents the conceptual model of this system. From that the conceptual model of the processing system is deduced. The later is then to be transformed into a design model for the processing system to be implemented. Figure 2 illustrates that fact.

As a consequence, the hierarchy resulting in the abstractions always corresponds to a component hierarchy of some system existing in reality, which imposes a clear structure on analysis (hence the structured analysis in SOOSA).

Now building all these models follows a MOS as defined above.

It results in process steps or activity classes in modeling. It gives us *units of work* separately manageable as a basis for concurrent modeling efforts. Three major units of work classes to be deduced from the above ideas are:

- System class definition for a system or subsystem
- System decomposition
- Model extension by system architecture definition

This separation principle is valid through all levels of decomposition and therefore gives the project manager many ways of 'braking the process into pieces' according to both the project's technical contents and the project organization's characteristics.

System class definition is a unit of work preferably done by experts in the domain the model should define the abstractions for. Referencing the ideas from the next chapter, concerning models for baseline services these could e.g. be specialists in GUI or OS design or physical process control technology, and concerning models for application services these are specialists in the application domain or its sub domains.

System decomposition should be performed by people having the overall view. It can actually be regarded as a management activity because of its impacts on the resulting project organization. A major consequence however is the need to train managers in systematic development techniques.

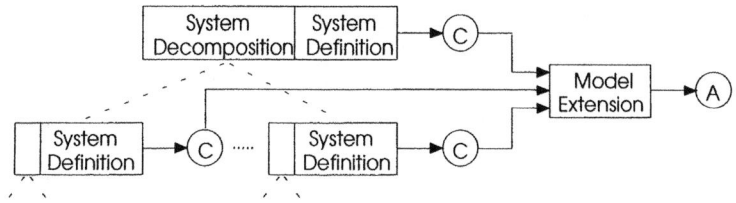

Figure 3: Process Model for Modeling

Model extension consequently can be regarded as a management activity, too. It has quality assurance character because it is concerned with consistency checking within the modeling process.

The sequence of the process steps execution is arbitrary. Only the system and subsystem modeling steps from one level of decomposition must be run before a process step towards model extension. All process steps combined must assure consistent extended models of the system under consideration.

The separation principle is illustrated in figure 3 through a kind of generic process model for modeling.

2.4 Notational Representations

Notations are now applied within the frame set by the MOS. The canonical notation is given by the KRS's elements and it is appealing explicitly to use this notation in some contexts possibly combined with other notations (mixed representation).

A structure model's graphic-oriented notational representation is always closely related to those given in object-oriented modeling (object models, e.g., from OMT [19]) or hierarchical modular modeling (system entity structures from multifacetted modeling [12]) as different as these approaches may seem. Note that the differences between these approaches to modeling do not result from what can be expressed within the notations. Object models of oo-modeling are very well capable of expressing hierarchy, only that oo-methods make no use of it. Behavioural modeling there is flat at the leave level of the component hierarchy. The structure model yields object-identity as well as the 'carriers' of the behavioural functions. SOOSA makes use of a notation similar to OMT.

A behaviour model's graphic-oriented notation must be capable of expressing the state machines associated with the objects. SOOSA defines a behavioural modeling concept with *hierarchical behaviour diagrams (HBD)* as its representation. These have a well defined interface to the structure models and allow for explicit specification of functional as well as dynamic properties. We will only go into a little detail on these diagrams and their associated semantics. For further information we refer to [20] and [21].

Having objects as carriers for behavioural definition allows for different ways of utilizing these. For a certain level of abstraction one can always do a *behavioural classification* of the objects into *active* and *passive* ones. Active objects are the ones with a 'relevant' behaviour and are represented by *function symbols* (rectangles) within the HBDs representing their *hierarchical process*. Passive objects on the other hand are represented by *information symbols* (circles) representing their states and attributes.

Active objects either cooperate via typed *channels* also represented by information symbols, which are given by the objects input/output coupling structure and in contrast to traditional object-oriented methodology may hide different sorts of *communication protocols* or by accessing passive objects. Every access of that kind corresponds to a sort of event which in consequence can be assigned to an *action* defined *on* the passive object. This makes a new function symbol to be used within the active objects internal process description representation. Both kinds of cooperation relations are represented by *information flow symbols* (arrows connecting function and information symbols). These can be attributed with access protocol specification information represented by *protocol symbols*.

Upon information symbols representing state one may define functional relations with other information symbols representing *modes* (of operation) Structural relations known from the structure models may also be represented by information symbols. Both kinds of information may be used within the process descriptions. The process descriptions themselves are represented by actions attributed with duration values ordered by control flow information represented by *control flow symbols* (arrows connecting function symbols representing sequences) and special *control constructs* for iteration, concurrency, and selection .

With the HBDs symbols other established ways of behavioural descriptions can be expressed like SDL component and process diagrams [22], RDD behaviour diagrams [5], Real-Time Networks [23], and functional and dynamic models from object-oriented analysis e.g. [19]. The proposed notation therefore is a sort of intermediate form for behavioural specification.

3 Design Theory

3.1 Services

System design is the activity of mapping a conceptual model's information into a physical structure of cooperating components in order to define an architecture that satisfies the users needs in terms of both functional and non functional requirements. The system is assembled together from physical components of defined functionality. This work covers both the system's architectural definition as well as the detailed, internal design of the components constituting this architecture. Figure 4 illustrates this general meaning of design.

SOD organizes system design around the term service. A service is a physical component of a system, that can be ordered from outside, this is from other system

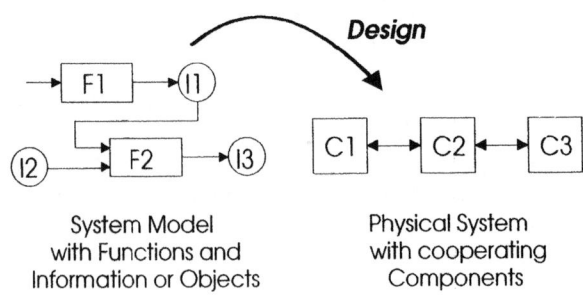

System Model
with Functions and
Information or Objects

Physical System
with cooperating
Components

Figure 4: Nature of Design

components, to carry out some sort of functionality to assure the performance of some part of the system's overall functionality. A component of that kind consists of *hardware and/or software* parts. Software parts of a service are linked data structures and algorithms (*modules*) which assure wanted functionality from the software engineering point of view. Hardware parts of a service are both components directly assuring the needed functionality or components the software parts can be installed and executed upon.

Especially in automation where temporal and performance requirements tend to be very important the system designer should be given the highest degree of freedom to decide upon which technological solution is the most suitable for his problem. Services may be dedicated hardware components, intelligent subsystems of mixed hardware, software and control parts or software for standard components and operating systems.

Once specified in terms of its functionality a service may manifest itself in a *variety of products* of different designs, each with its own specific non functional quality measures. Design itself then has the character of choice between different alternatives which is a prerequisite to satisfy design's inherent nature of being a matter of trade-off decisions. The service is the entity of managerial consideration within the design process both in terms of having non functional quality measures and re-use properties assigned. For more details on the nature of SOD we refer to [24].

3.2 Allocation

The design mapping from above can be partitioned into several steps or distinct tasks. The first way of partitioning design work is in terms of a design's *location dependant and location independent* characteristics. This is done by separating issues concerned with 'logical' unit building like process structure definition or decisions in terms of parallelism from issues concerned with physical structure building like choosing sorts of physical components and interconnection media. The 'logical' units will from now on be referred to as *Design Units* (DU). They constitute a *location independent system design*.

Building a location independent system design is very often called *system-level design*. All sorts of languages are proposed for performing that task (SDL, StateCharts, Lotus, Estelle, OCCAM, CSP-like languages etc.). These have in common that one models the system as a (possibly hierarchically ordered) set of parallel and interacting extended finite state machines (EFSM). The means of communication may be abstracted as happening via channels and the models communication protocol assumptions is where the approaches differ the most. That observation calls for a sort of intermediate form for system-level design description general enough to model all forms of interaction from the established notations. We already talked about this when investigating HBDs.

One may regard executing both tasks mentioned above as building two distinct sorts of system structures represented within two distinct system models. The first model has DUs at its objects, the second physical components and interconnection media. Resulting from separating these issues is a third step of integrating these two structures again for the system structure of cooperating services by mapping design units onto physical components. Physical components therefore can be regarded as *places* whereon to *allocate* defined functionality in terms of design units. We call this process system design according to the *allocation principle*. It is illustrated in figure 5.

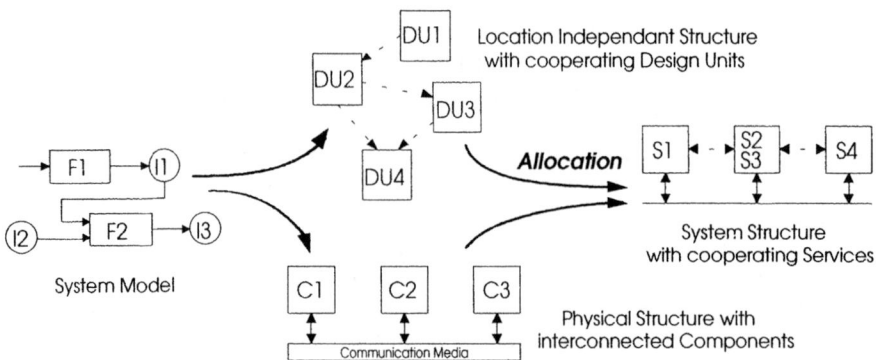

Figure 5: System Design by Allocation

In software engineering this separation corresponds to a separation between hardware and software issues. Design units correspond to software modules and physical components correspond to the hardware the software can be installed and executed upon. In mixed hardware/software systems engineering the physical components may be both components directly assuring the needed functionality or components which are hosts for software. On a sufficient level of detail these are *either one or the other*. If this level of detail is matched, the physical structure of interconnected components not only covers what may be called the architecture interconnection structure, as e.g. described in architecture interconnect diagrams (AID) from RT/SD [25], which is a system-level information, but also hardware/software partitioning decisions.

These decisions consequently also apply for the location independent structure of design units. On a sufficient level of detail (in terms of the hierarchy of design units stated above) this structure not only covers what may be called the architecture flow structure (described in architecture flow diagrams, AFD, in RT/SD), which also is a system-level information, but the same hardware/software partitioning decisions addressed above.

Note, that going from system-level issues down to hardware/software design issues following these ideas is only a matter of abstraction or *level of detail*. The conclusion is, that using the introduced abstractions, one can seamlessly proceed from the system-level view to the hardware/software view.

3.3 Service Access

When working on the software side one can easily abstract from the underlying hardware and communication implementation details. Service modules may be regarded as only having access to other services via a transparent *service access* using functionality provided by a service access *interface*. That service access is the media for all communication among services and realizes conceptual communication protocols to be used by the service modules. This abstraction principle is illustrated in figure 6.

The service access can be regarded as a kind of 'logical system bus' into which services can be 'plugged in' for system integration or configuration purposes. That abstraction effectively separates the design issues as described above and allows for a most natural understanding of the integration or configuration aspects. An

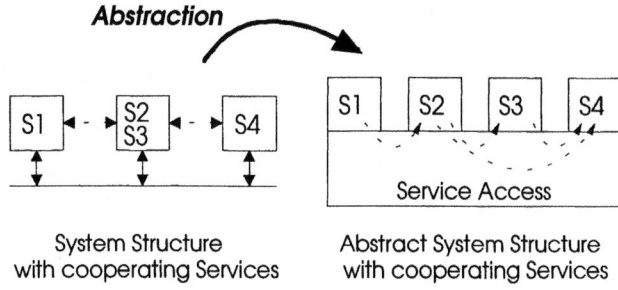

Figure 6: Abstraction from Hardware and Communication

implementation of the service access as a basis for software system development can be realized by an extended operating system to be present on all physical components and being accessible (in whatever syntactic form) from within the programming language [24].

In mixed hardware/software system design we naturally have to care about the communication details. Due to service-orientation however, one can utilize the same abstraction from above concerned with software system design to form a sound theoretical basis for mixed hardware/software system design. Services cooperate by accessing the service access which serves as some sort of co-processor responsible for inter-service communication execution. It realizes the channels the design units use for communication as well as the actual physical links.

3.4 Hardware/Software Codesign

Hardware/Software Codesign is about systematically proceeding from system-level issues down to hardware/software specific issues. The abstractions defined above yield a sound backbone for hardware/software codesign. The most important ideas to proceed towards this direction are (1) regarding the physical structure of interconnected components as instantiated from a generic architecture framework, (2) partitioning the service access into components according to sets of the single channels of communication between design units, (3) making these correspond to special communication units within the structure of cooperating design units, (4) partitioning this structure to match the physical structure, and (5) allocating these by performing communication synthesis (channel binding and channel mapping). For more information we refer to [26].

3.5 Service Characteristics

Most important to note is the possibility to classify cooperation structures between system components. Very often there are client-server relations where the client component uses the server component's functionality within the frame of performing its own job. This structure is found on all levels of system design detail. There is a hierarchy in terms of utilization.

Usually this principle is extended in a way, that special components are made responsible for specific classes of jobs. These can then be used by many components in need to perform a service of the specific kind within their own job. The term service as defined above reflects this form of cooperation. The reasons for building a service usually is the need to allocate dedicated functionality and abstractions at a very specific location within the system.

The term of a service therefore is not only a media for design organization but also reflects a style of thinking or conceptual framework. Systems designed according to that framework tend to have five major characteristics which are regarded essential for governing the system life cycle (in terms of complexity, efficiency, quality). These are hierarchy, abstraction, modularity, encapsulation, and concurrency. For a further discussion about why these characteristics are met, we refer to [24].

3.6 Service Reuse

When a service is integrated into a system, seen from the view of the other components the available functionality of the system is increased. One may regard service integration as an *extension of the system interface*. Therefore it is possible to make the definition of a new service with the highest possible use of given services functionality a *design paradigm*. This paradigm has to be supported by preparing the designer with as many and as mighty services as possible.

The definition of a *new* service therefore always has to be judged in terms of its general solution potential for a class of problems. If a service is general enough it can be added to a *infrastructure of reusable components*. Working according to that principal yields policies to be applied in configuration management.

With respect to a given infrastructure all services comprising a system can be classified into baseline services, the ones within the infrastructure, and application services.

With proposals like the ECMA reference model, architecture templates from RT/SD, and our experience within the field of process automation in mind, we propose defining the following three services as baseline:

- Data Handling Service (DHS): Realizes the abstractions from data base technology like databases, relations, etc. plus a set of pre-defined operations like searching e.g.
- Man Machine Interaction Service (MMS): Realizes the abstractions from GUI technology like interfaces, windows, menus, presentation elements, etc. plus a set of pre-defined operations like display or monitoring.
- Physical Process Control Service (PPS): Realizes the abstractions from measurement and control technology like channels etc. plus a set of pre-defined operations like timed measurement e.g.

The main idea in defining these services as baseline is having pre-defined and fully tested components at hand that cover most of the issues arising when going from analysis to design, which otherwise tend to consume most of the time. The philosophy is being able to solve as many problems as possible concerned with data handling, man machine interaction or physical process control by simply reusing the baseline services.

This philosophy can only work if it is complemented with constantly trying to incorporate additional knowledge from those cases where reuse is impossible into the baseline services by extending the realized abstractions. Note that the infrastructure can be regarded as a collection of evolving prototypes. As long as these prototypes do not reach a stable state resulting in a product standard, it is necessary to be able to base the baseline service's abstractions and functionality on the use of lower level standards and realize it by mapping to lower level 'off the shelf' components. Concepts and prototype implementations of all three services (both pure software and mixed hardware/software solutions) are described in internal papers of the IMA.

The service access itself is baseline, too. Although it is the media for service definition, it also has service character because it defines the abstractions necessary for service-oriented design and fulfils the criteria of encapsulation (by hiding the communication implementation details) and modularity. The service access and its prototypical realization is also described in internal papers of the IMA.

4 Engineering Environments

We conclude this paper with a short passage about infrastructures for that kind of engineering methodology. Infrastructures contain libraries of reusable specifications (model bases) and components (service infrastructures) for efficient development as well as tools to aid in the engineering work. We talked about model bases and service infrastructures before. Now we briefly investigate the tools.

The prototype engineering environment consists of two toolsets, one for project management and one for development related tasks. It is organized in services to aid in performing the engineering steps. It contains services for front-end modeling supporting the sketched notations, simulation of the specification for dynamic verification and timing analysis, partially automated implementation (in the prototype for C++ software components) implementation simulation utilizing a special simulation kernel to replace the run-time kernel of the service access implementation, partially automated system configuration as well as model and component infrastructure management and documentation and report generation. These services can be used to configure a complete ECBS environment which is a design of a project organization as described above based upon the tool services.

References

[1]   Schweizer, G.: The Concept of the System as a Key to Systems Engineering in the Information Technology Domain. Proc. 3rd CBSE Workshop, University of Maryland, College Park, MD, Mar. 1992
[2]   Lavi, J.Z., et.al.: Formal Establishment of Computer Based Systems Engineering Urged. IEEE Computer, 24 (3), pp. 105-107, 1991
[3]   Voss, M., Hummel, H., Wolff, T.: The Role of Process Standards within ECBS Work. Proceedings of 6th Intern. Workshop on Engineering of Computer-Based Systems (ECBS), Stockholm, Sweden, Computer Society Press, 1994
[4]   Blanchard, B.S., Fabrycky, W.J.: Systems Engineering and Analysis. Prentice-Hall, 1990
[5]   Alford, M.: Stregthening the Systems Engineering Process. National Council on Systems Engineering (NCOSE) Workshop, Chattanooga, Tenn., Okt. 1991
[6]   Embley, D.W., Kurtz, B.D., Woodfield, S.N.: Object-Oriented Systems Analysis: A Model Driven Approach. Prentice Hall, 1992
[7]   Schweizer, G., Voss, M., Blank, M.: Towards Standardization of CBS Modeling. Position Paper, IEEE Task Force on CBSE 5th Workshop, Lawrence, Kansas, Aug. 1993
[8]   Pyle, I., et.al.: Real-Time Systems. Investigating Industrial Practice. Wiley, 1993
[9]   Oliver, D.W.: A Tailorable Process Model for CBSE. GE Research & Development Center, 1993

[10]   Ören, T.I.: GEST - A Modelling and Simulation Language Based on System Theoretic Concepts. In: Ören, T.I., Zeigler, B.P., Elzas, M.S. (Eds.): Simulation and Model Based Methodologies: An Integrative View. North-Holland Pub. Co., Amsterdam, 1984

[11]   Shaw, A.C.: Communicating Real-Time State Machines. IEEE Transactions on Software Engineering, 18 (9), S. 805-816, Sep. 1992

[12]   Zeigler, B.P.: Multifacetted Modelling and Discrete Event Simulation. Academic Press, 1984

[13]   Booch, G.: Object-Oriented Analysis and Design with Applications, 2nd Edition, Benjamin/Cummings, 1994

[14]   Royce, W.W.: Managing the Development of Large Software Systems. Proc. WESTCON, Ca., USA, 1970

[15]   Boehm, B.: A Spiral Model for Software Development and Enhancement. Software Engineering Notes, 11(4), 1986

[16]   Shlaer, S., Mellor, S.: Object-Oriented Systems Analysis: Modeling the World in Data. Englewood Cliffs, 1988

[17]   Jacobson, I., et.al.: Object-Oriented Software Engineering, Addison-Wesley, 1992

[18]   Yourdon, E.: Modern Structured Analysis. Yourdon Press, 1989

[19]   Rumbaugh, J., et. al.: Object-Oriented Modeling and Design. Prentice Hall, 1991

[20]   Voss, M.: Algebraic Interpretation of Hierarchical Systems. University of Karlsruhe, IMA, 1993

[21]   Voss, M.: Systemobject-Oriented Structured Analysis (SOOSA). University of Karlsruhe, IMA, 1993

[22]   Recommendation Z.100. CCITT Specification and Description Language (SDL). CCITT, 1992

[23]   Simpson, H.: A Data Interaction Architecture (DIA) for Real-Time Embedded Multiprocessor Systems. RAE Conference on Computing Techniques in Guided Flight, Boscombe Down, 1990

[24]   Voss, M.: Service-Oriented Design and its Support by a Service Access System. University of Karlsruhe, IMA, 1993

[25]   Hatley, D.J., Pirbhai, I.A.: Strategies for Real-Time System Specification. Dorset House, 1988

[26]   Voss, M., Ben Ismail, T. Jerraya, A.A., Kapp. K-H.: Towards a Theory for Hardware/Software Codesign. Proceedings of: 3rd Intern. Workshop on Hardware/Software Codesign - Codes/CASHE '94, Grenoble, France, IEEE Computer Society Press, 1994

An Approach to the Design of Complex, Heterogeneous Hardware/Software Systems

Christoph Schaffer

Systems Theory and Information Engineering
Institute of Systems Science
Johannes Kepler University
Linz/Austria

Abstract. In this paper an approach will be presented that allows the development of complex systems in a way that a high degree of portability to any heterogeneous systems topology is achievable. By introducing a layered design approach systems behavior will be split into atomic-, channel- and port behavior. This approach allows significant abstraction and results in highly reusable components.

1 Motivation

Nowadays systems engineers have to cope with the problem of an increasing effort for the development of complex heterogeneous systems, whereas the time the development has to be finished within, decreases dramatically. To reduce the development time it is worth striving for reusing already developed components within new designs.

The components have to be described at a very abstract technology independent level, because these abstract components have a higher added value than directly coded ones [4]. This is because the designer has the ability to reuse these component descriptions and realize them within different realization technologies (e.g. hardware (HW) and/or software (SW)).

This will work quite well if all the components of the system will be realized by the same realization technology. But it will cause a lot of problems if some of the components have to be realized in software and some of them in hardware because here we have the problem that a lot of HW/SW-interfaces will be brought into the system that have not been defined within the abstract description. Up to now it is very usual that at this state the behavioral description of the components will be changed as it is required by the new HW/SW-interface (requirements due to the realization technology). But having changed the behavioral description the reusability of this component is restricted dramatically.

Another big problem is coming up if the components have to be realized within different heterogeneous systems. One can imagine that the whole system can be realized within one single computer or ASIC (application specific integrated circuit) if all the components have a negligible geographic distance to each other (fig. 1 a)). Another scenario is given in figure 1 b) where there

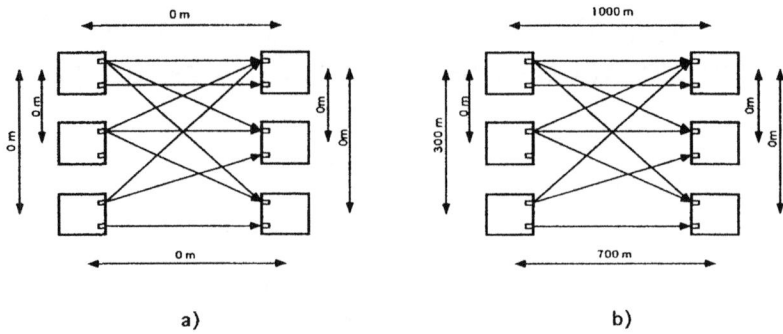

Fig. 1. a) Overall system behavior realized within one single computer b) System behavior split among several distributed computers

is a geographic distance between the components of e.g. several hundred meters. Here we cannot realize the whole system within one computer. We have to use a special communication medium like twisted pair, fiber optic, radio etc. to exchange data between these coupled components. Consider a small example where two components, each realized in SW, have to exchange data. If they can be realized within one computer the exchange of data can be achieved by using the computer memory. But if we have to realize the same components within two different computers we will have a lot of additional interfaces (SW-HW-...-HW-SW) that have to be designed.

It should be stressed that the overall system behavior should not change if the geographic distance or the realization technology of a special component changes.

When executing the systems design the topological structure (distances between the different components, communication media etc.) is mostly regarded as being fixed. This causes the designer to model all the behavior (protocols etc.) needed to communicate with another component over a given communication medium within the components themselves. This again will restrict the reusability of these components.

In the following a formalism will be presented that supports the development of components in a way that they provide a high degree of reusability and the ability to be adaptable to a wide range of different system topologies.

2   Splitting Components Behavior

As mentioned above the reusability of components will be reduced dramatically if the designer adds some interface and communication aspects to the behavioral description of a component. To avoid this situation the behavioral description of a component has to be split into (figure 2):

- *atomic behavioral* aspects
- *port behavior* and
- *channel behavior.*

This split leads to a well structured design that guarantees a high level of reusability.

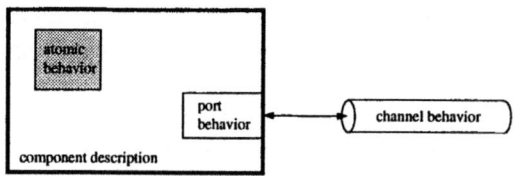

Fig. 2. Split of behavior

2.1 Atomic Behavior

Atomic behavior defines the behavior of a component supposing optimal communication channels (infinite bandwidth, no losses, able to transport all kinds of data, etc.), where the output of one component is routed directly to the input of another component without using any real communication media. In this case we can also speak of a *virtual channel (VCH)*. The description of the atomic behavior will not change if the communication media changes.

2.2 Port Behavior

Ports can be defined as being the interfaces of the component to its environment. Communication with the environment only takes place through ports. If a component is coupled with another component by a VCH, the types of both ports have to be the same. So we can imagine that the designer decides to realize one component in SW and the other component in HW. Here it is obvious that the "HW-signals" have to be transformed into "SW-signals". Another scenario is given when we want to reuse two already defined components that have different input-port, output-port types. Here again we have to cast the types. The problem occurs again when we want to realize a component in HW. Here some output driver, etc., have to be added to the output ports of the component, which is also some kind of type conversion.

2.3 Channel Behavior

Channels are needed to couple two or more components of the system and to transport data between them. Coupling itself is well developed and is discussed in a number of publications [5], [11], [10], [6], [7], whereas the concept of channels and ports lacks fundamental theory. Only some aspects are mentioned in the literature [11], [1].

As mentioned above communication channels can be realized with a lot of different realization technologies (twisted pair, ethernet, dual ported memory, optical fiber, etc.). The following requirements are common to all these channels:

- Reliable communication: the output to the channel should be the same as the input. If this requirement is not fulfilled automatically safety mechanisms have to be built in (e.g. protocols). If a protocol is used special synchronization mechanisms have to be built in too.
- In general, communication channels can transport only a special type of data. So a data-type conversion might be necessary.

2.4 Design Requirements

Because we would like to have a well structured view of the overall system, and because communication channels can be very complex, the principle of hierarchy has to be introduced to define communication channels and ports. So it should be possible to specify a complex communication bus as a virtual channel as well as to give a detailed description of each single communication line. Additionally, bottom-up as well as top-down design should be possible because this is an essential requirement in systems design [8].

3 Port Definitions

Before starting with the formal definition of ports we have to investigate the principal functionalities that have to be fulfilled by a port. First of all it is a means to connect (couple) one component to one or more other component(s). Here the following cases can occur (figure3): Whereas case 1) is very simple, cases 2) and 3) need more investigation. Case 2 shows that a single internal signal is split into multiple signals needed to supply all the input ports of the other components this one signal is connected to. This kind of port is called a *split-port*. Case 3 is the most complicated, because here more than one signal will write to a single input port. Here we have to set up the question which of the signals applied to this port will determine the value passed to the component. A port like this will be named as *joint-port*.

Because ports are the interfaces of a component to the environment they have to have some type of conversion facilities. They have to convert the internal data type to data types required by other components or the transmission channel. A combination of these two considerations will be shown in figure 4.

Now after having seen the principal needs we can proceed to give a formal port definition.

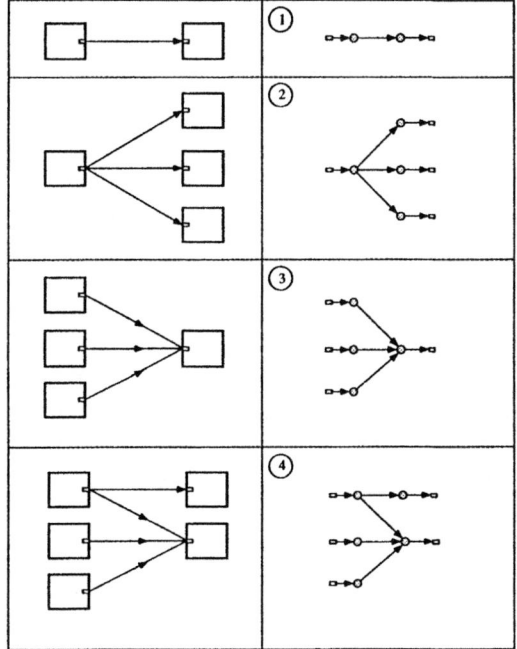

Fig. 3. Types of port coupling 1) one to one (1:1) 2) one to many (1:n) 3) many to one (m:1) and 4) many to many (m:n)

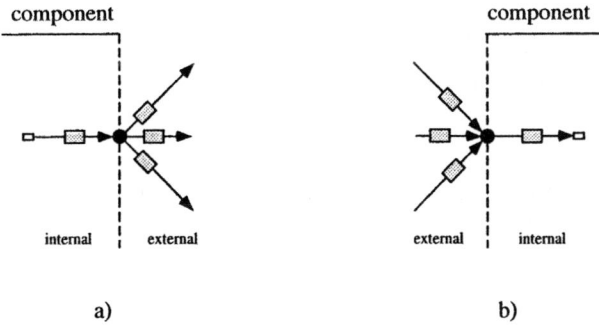

Fig. 4. a) Split-port and b) joint-port in combination with type conversion facilities (shaded boxes)

3.1 Atomic Ports

An *atomic port* $P_{i,k}$ of a component or a communication channel X is defined by a five-tuple

$$P_{i,k}(X) = (P_{int}, P_{ext}, F_{conv}, f_{res}, dir), where \tag{1}$$

$$\{i : i \in N \wedge i \leq |P(X)|\}, and$$

$$\{k : k \in N \wedge k \leq |P_{ext}|\}$$

$P_{i,k}(X)$ is interpreted as being the i-th port of component X, where $|P(X)|$ specifies the number of atomic ports assigned to this component. The second index, namely k, identifies the external part of the port needed to connect this special port to a port of another or even the same component. P_{int} is a tuple used to specify the part of the port that is internal to the component, whereas P_{ext} is a set of tuples needed to specify the external part of the port (figure 5).

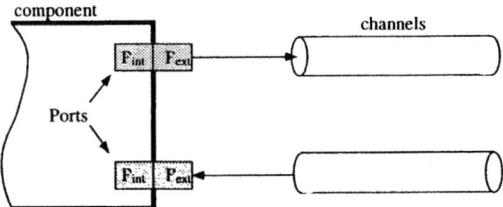

Fig. 5. Definition of P_{int} and P_{ext}

P_{int} consists of an entry that defines the name of the port (N_{port}) and the realization of this port internal to the component.

$$P_{int} = (N_{port}, R_{int}) \tag{2}$$

P_{ext} is a set of tuples consisting again of entries that specify the name and the realization of the external port.

$$P_{ext} = \{(N_{ext}(j), R_{ext}(j)) : j \in N \wedge j \leq M\} \tag{3}$$

Here M is the number of external ports needed. It will be given by the number of ports this special port is connected to. So we will need four external ports if the port has to be connected to four other ports within the overall system. $N_{ext}(j)$ is needed to differentiate between the different external ports. The name used here should express some special characteristics of the port given due to the definition of $R_{ext}(j)$. It should support the readability of the resulting description.

R_{int} and $R_{ext}(j)$ are elements of $\hat{R}$ ($R_{int} \in \hat{R} \wedge R_{ext}(j) \in \hat{R}$) which is the cartesian product of the type (TYPE) of the port, the range (RANGE) of the data flowing through this port and the realization technology (R_TECH) of this port.

$$\hat{R} = (TYPE \times RANGE \times R\_TECH) \tag{4}$$

Considering the two realization technologies hardware and software, R_TECH can be defined in the following way:

$$R\_TECH = \{HW, SW\}. \tag{5}$$

This definition can be extended if e.g. a mechanical realization of a component should be possible too.

Having the possibility to specify and to use different realization for the internal and the external ports we can define a set of functions that can be used to convert data from one representation to the other.

$$F_{conv} = \{f_{conv}(j) : \hat{R} \rightarrow \hat{R} \mid j \in N \wedge j \le |P_{ext}|\} \tag{6}$$

In figure 4 we have seen that we have to distinguish between split and joint ports. Split ports are used to broadcast data from one port to all the other ports connected to it. Whereas this kind of port will cause no major problems the joint port needs a more detailed investigation. This is because here more than one port wants to write to one single port. But which of the writing ports will determine the resulting value? To solve this problem the designer has the ability to specify the so called *resolution function* f_{res}. Here he has to define in which way the values of the different ports writing to the single input port will influence the value at the input port itself. So the designer can assign a priority to each port, use an AND-gate, define some kind of time-multiplex etc. Figure 6 shows the realization of a resolution function.

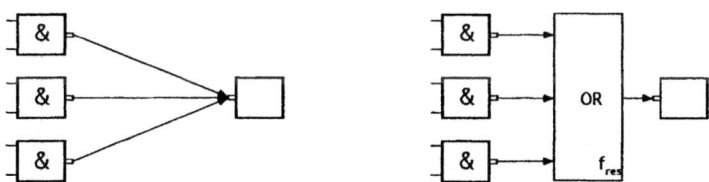

Fig. 6. Realization of a resolution function

$f_{conv}(j)$ as well as f_{res} should be described at an abstract level too, because in this way it is possible to generate e.g. C- or VHDL-code out of the description. So both software and hardware components can be used to do data type conversions.

The last element of the five-tuple is *dir* which specifies the direction of data passing through this port. *dir* is an element of DIR ($dir \in DIR$) which is a set of all possible directions.

$$DIR = \{In, Out, InOut\}. \tag{7}$$

Example To demonstrate the application of the definitions given above a simple example is given in (figure 7):

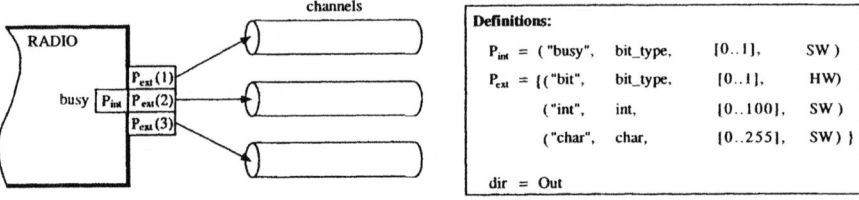

Fig. 7. $P_{2,3}(RADIO)$ is the second port of component RADIO with the name "busy". $f_{conv}(3)$ of this port has to convert the internal data type "bit_type" to the external data type "char". Here both ports are realized in software, so no conversion from SW to HW, as it is necessary at port $P_{2,1}(RADIO)$, is needed.

3.2 Compound Ports

Sometimes it is not necessary to depict all the ports of a component in a flat way. So it should be possible that e.g. a bus-connector, which consists of a lot of ports, is represented by a single port only in a higher layer of abstraction. In this case, we are speaking of a *compound port*. A compound port is a means to achieve a better structure of a component. It is defined by

$$CP_{i,j}(X) = (N, dir, COLL), where \qquad (8)$$

$$i, j \in N \wedge i, j \leq |P(X)|$$

Here i is the ith compound port of the component X at abstraction layer j (figure 8). N specifies the name of the compound port and dir again is an element of DIR ($dir \in DIR$) which has been defined in equation (7). $COLL$ is the set of all atomic and compound ports $CP_{m,k}(X)$ (where $k < j$) assigned to the compound port $CP_{i,j}(X)$. This definition will lead to a hierarchical presentation of all the ports of the component X. To avoid confusion the following definition has to be fulfilled:

$$\forall_{(j \in N)} \forall_{(i,k \in N \wedge i \neq k)} COLL_{CP_{i,j}(X)} \cap COLL_{CP_{k,j}(X)} = \emptyset. \qquad (9)$$

At the same hierarchical level there is no port that is assigned to two or more compound ports.

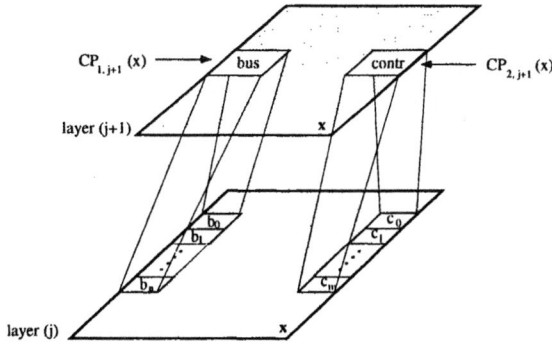

Fig. 8. Abstraction layers of compound ports

4 Channel Definition

A channel is a mean to transport data from one port to another port. At the highest level of abstraction a channel is a simple single line. Here we are only interested in the fact, that two components have to exchange some data. How this will be realized will be out of the scope at this level. But when we want to realize this system available in an abstract description we need the ability to bring the constraints given due to a special realization technology into our channel description.

Again we will start with the definition of an atomic channel which is the most detailed description of a channel.

4.1 Atomic Channel

An *atomic channel* is defined by a 6 tuple:

$$CH_i = (N, P_{1,j}(CH_i), P_{2,k}(CH_i), CON, dir\_ch, pr) \qquad (10)$$

N specifies the name of the channel. $P_{1,j}(CH_i), P_{2,k}(CH_i)$ are the two interfaces (ports) of the channel (figure 9). The index i specifies the number of the channel within the overall system. CON is a set of tuples that define which port is connected to which port over the channel CH_i.

$$CON = \{(x, y) : x, y \text{ are atomic port } \wedge x, y \rightleftharpoons CH_i\} \qquad (11)$$

The first element in the tuple represents the port of the component connected to the port $P_{1,j}(CH_i)$ of the channel, whereas the second element represents the port connected to the the port $P_{2,k}(CH_i)$ of the channel CH_i. "$\rightleftharpoons$" is the short form for "being connected with". So the definition $CON_{CH_i} = \{(P_{1,3}(RADIO), P_{5,2}(STATION))\}$ defines that the port 1 (external port 3) of the component RADIO is connected to port 5 (external port 2) of the component STATION.

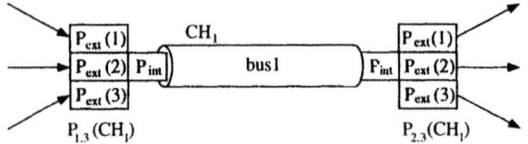

Fig. 9. Definition of an atomic channel

The definition of the direction of the data flow through the channel dir_ch which is an element of DIR_CH ($dir\_ch \in DIR\_CH$) differs from the definition given in (7). DIR_CH will be defined in the following way:

$$DIR\_CH = \{Simplex, Halfduplex, Duplex\} \tag{12}$$

Here $Simplex$, $Halfduplex$ and $Duplex$ are defined as usual in telecommunication. Simplex means that data flow will be allowed in one direction only. Duplex operation mode will allow a data flow in both directions at the same time, whereas Halfduplex will allow a data flow in both directions too, but not at the same time. The last entry within the 6-tuple is pr which defines the protocol used for doing data transmission over the channel CH_i. Here pr is an element of PR

$$pr \in PR \tag{13}$$

which is a pool of available protocols. The kind of protocol used will depend on many different factors like realization technology, speed, geographic distance, error probability, costs, etc. The selected protocol should ensure that the data to be transmitted from one component to the other can be detected correctly at the receiver. It should be stressed that this kind of protocol has nothing to do with the protocol used between the components themselves, e.g. to synchronize some activities. This component related protocol has to be modeled within the atomic behavioral description of the component. It will not change if the system topology changes. The communication protocol pr on the other hand will change if e.g. the communication media or the data rate changes.

4.2 Compound Channel

Here again we use the concept of the compound channel as a means to achieve a better structure of the system, and to allow top down as well as bottom up design. When merging two or more different channels to one compound channel we have to distinguish between two different cases. The channels can be connected in series or in parallel. So we have to define two different types of compound channels:

Compound Channel Serial (CCHS) This type of channel will be used when two or more channels have to be connected in series and the designer wants to group them to one single compound channel (figure 10). This is a typical case of a serial composition. The order in which these channels are connected to each other is a essential information. This is because the channels can be realized by different technologies, or the TYPE or RANGE definition can change from one channel to the other. So we may have, e.g., the following situation. Two HW-gates have to be connected by an optical fiber. Here we need a copper wire, that transports the data to and from the HW-components to/from the optical fiber channel. This will lead to a serial composition ($\ominus$) of a copper-wire an optical-fiber channel and again a copper-wire. It should be stressed that the order the channels are concatenated has no influence to the transmission function of the whole compound channel but it is, as we have seen, a essential information to do the corresponding type conversions. A *Compound Channel Serial* (CCHS) is defined in the following way:

$$CCHS_{i,j} = (N, SCC, dir\_ch) \qquad (14)$$

Here the index i specifies the number of the channel within the overall system at abstraction layer j. N again specifies the name of the compound channel. SCC is an ordered set of all channels that are connected in series. At abstraction layer $j=2$ a set of L serial connected channels can be defined in the following way:

$$SCC = \overset{L}{\underset{i=1}{\ominus}} CH_i \qquad (15)$$

When connecting more channels in series the following condition has to be fulfilled:

$$\forall_{(j \in N \wedge 1 \leq j < L)}(P_{2,x}(CH_j) \ominus P_{1,z}(CH_{j+1}) \Rightarrow x = z) \qquad (16)$$

Assuming a higher level of abstraction $(j > 2)$ also CCHS and atomic channels or also CCHP can be connected in series.

dir_ch is used again to specify the direction of the data flow through the channel $(dir\_ch \in DIR\_CH)$.

Compound Channel Parallel (CCHP) The parallel compound channel will be used if the designer wants to group two or more parallel channels to one single compound channel. A typical example for an application of this type of channel is a bus system. Here the designer wants to group many single channels to one common bus. This will lead to a better structure of the overall system. A *Compound Channel Parallel* will be defined in the following way:

$$CCHP_{i,j} = (N, PCC, dir\_ch) \qquad (17)$$

All the indices and entries of the triple will be handled in the same way as defined in (14). The only difference will be the definition of *PCC* which is the set of all

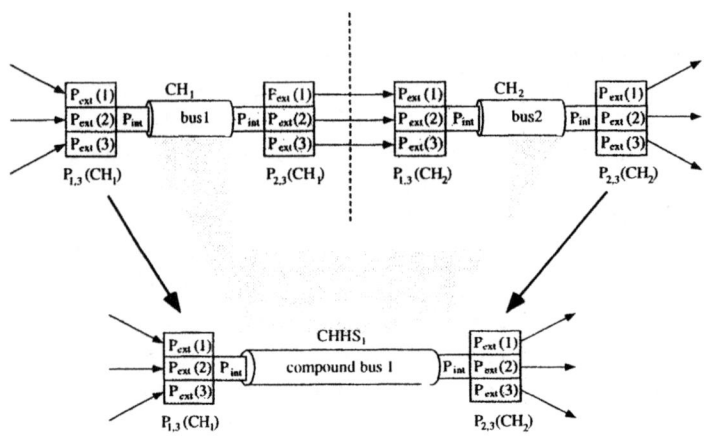

Fig. 10. Compound channel serial

parallel compound channels grouped within $CCHP_{i,j}$. At abstraction layer j=2 a set of L parallel connected channels can be defined in the following way:

$$PCC = \overset{L}{\underset{i=1}{\big\Vert}} \ CH_i \qquad (18)$$

5 Communication Meta Language (CML)

As it was mentioned above the atomic behavior of a component will describe all the behavioral aspects that will not change if e.g. the communication medium or the realization technology changes. This atomic behavior has to be described at a very abstract level. But also at this abstract level we will have the need to exchange data between the different components. This will be achieved by sending data to and receiving data from a special port of a component. These data (messages) will be passed to the channel and transported to all connected components. The format (type) of the messages being sent to or received from a port are given by the definition of P_{int}.

To express the need for communication with the environment the designer needs only two commands.

1. **send(port_id, data):** this command will be used if a data item has to be passed to a specific port. The port will be specified by the port_id. The command send($P_{3,j}(X)$, test) will send the data "test" to the third port of component X. It is not necessary to specify the index of the external port j, because the external ports hold the same information as the internal port. There is only a difference in the representation.

2. **receive(port_id, data):** this command is used to receive data from a specific port. The parameters are used in the same way as they are used in the send() command.

The communication protocol *pr* of a channel will convert this send and receive messages to the corresponding protocol commands. This should be done automatically.

6 Modeling HW/SW systems

When doing systems design using the formalism introduced above the design steps that have to be executed will have the following appearance (figure 11). While executing the step *"modeling of the atomic behavior"* the designer will

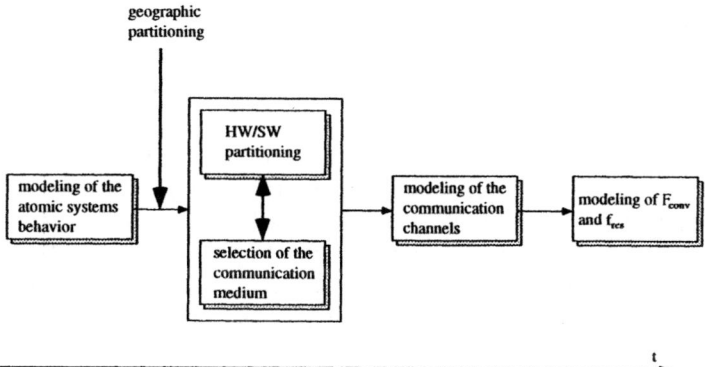

Fig. 11. Design steps that have to be executed

identify components and define the necessary couplings between them. Additionally he will define (model) the atomic behavior of these components. This description has to be done at a very abstract level by using, e.g., SDL [9], statecharts [2], [3] or some other formal description technique. Simulation and formal verification should guarantee that the resulting systems description fulfills all the requirements.

Having finished this step the designer has an abstract system description, and simulation data of the overall systems performance (e.g. data load on the virtual channel etc.).

Now the designer can proceed with the *"geographic partitioning"* of the overall system. This step is depicted in figure 12 where all these components are grouped (A, B, C) that have a negligible distance between them. This step is

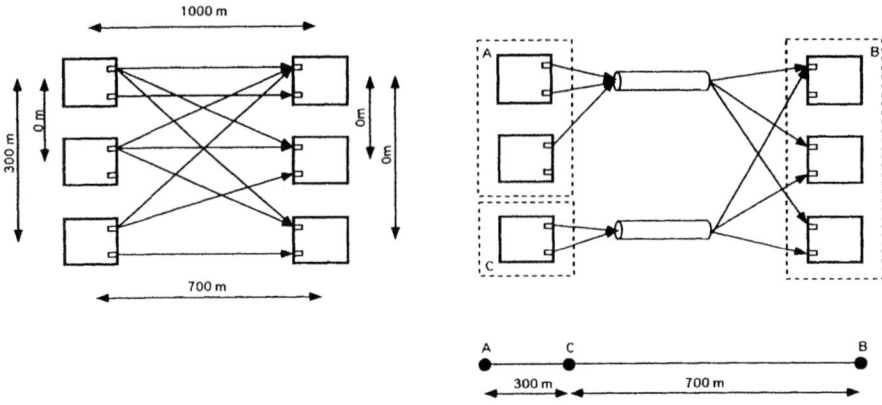

Fig. 12. geographic partitioning of the system to be designed

necessary because we have to identify the communication medium one specific channel has to be realized in. The distance between two components will be an essential parameter in the selection process. So it is a fact that two components realized in SW cannot communicate over longer geographical distances without using HW channels (e.g. twisted pair, optical fiber etc.).

Having finished this partitioning process we can start with the selection of an appropriate communication channel. Here the selection process will be influenced by a lot of parameters like geographic distance, available data rate, error probability, costs, etc., to mention only a few of them. Because the realization technology of the components connected to this channel will also influence the communication medium the HW/SW-partitioning process should be executed in parallel. Here it is possible that a lot of iterations are needed to achieve an optimal selection. Up to now we have only selected the medium to transport the data. Now we have to select a protocol that can be used in conjunction with the selected communication medium. One aspect of this protocol will be to guarantee an error free transmission of the data.

Now we need a realization of the communication channel using the selected communication medium. Here two different scenarios are imaginable:

1. A realization is available on the market.
2. The channel has to be designed by the designer himself. In this case he can regard the channel as a system again and he can use all the concepts introduced above.

Having finished modeling of the channels we can proceed with the modeling of F_{conv} and f_{res}. These functions should be seen as sub systems and should be described at an abstract level too. If these functions are seen as systems too, we again can apply the formalism introduced above.

7 Conclusion and Outlook

An approach was presented that allows the design of highly reusable components. The partitioning process of the overall system into HW- and/or SW-components as well as the reuse of these components in different heterogeneous environments will be supported quite well.

But it is clear that this work is not finished with this paper. So it will be of interest to use this approach in conjunction with an expert system. This combination seems to be very powerful especially when, e.g., selecting the communication channel. Here we have a lot of different parameters that will determine the communication medium and the protocol to be used. The selection process itself is straight forward.

Currently the development of a real system is investigated. Here the formalism is used as a methodology. The system itself will be modeled with SDL.

References

1. William L. Chapman, A. Terry Bahill, and A. Wayne Wymore. *Engineering Modeling and Design*. Systems Engineering Series. CRC Press, 1992.
2. David Harel. Statecharts: A visual formalism for complex systems. *Science of Computer Programming*, 8:231–274, 1987.
3. David Harel. On visual formalisms. *Communications of the ACM*, 31(5):514–529, 1988.
4. Jonah Z. Lavi, Michael Winokur, Reuven Gallant, and J. Kudish. The reuse of generic subsystem specifications in multisystem cbs development. In *The Proceedings of The Sixth Israel Conference on Computer Systems and Software Engineering*, pages 49–55, June 1992.
5. M.D. Mesarovic and Y. Takahara. *General Systems Theory: Mathematical Foundations*. Mathematics in Science and Engineering Series. Academic Press, Inc, 1975.
6. Franz Pichler. *Mathematische Systemtheorie: Dynamische Konstruktionen*. Walter de Gruyter & Co., 1975.
7. Franz Pichler and Heinz Schwaertzel. *CAST Computerunterstuetzte Systemtheorie*. Springer Verlag, 1990.
8. Christoph Schaffer and Herbert Praehofer. On requirements for a cast-tool for complex, reactive system analysis, design and evaluation. In F.Pichler and R.Moreno Diaz, editors, *Computer Aided Systems Theory-EUROCAST 93*, number 763 in Lecture Notes in Computer Science. Springer Verlag, 1994.
9. Kenneth J. Turner, editor. *Using Formal Description Techniques: An Introduction to ESTELLE, LOTOS and SDL*. Wiley Series in Communication and Distributed Systems. John Wiley & Sons, 1993.
10. A. Wayne Wymore. *A Mathematical Theory of Systems Engineering: The Elements*. John Wiley & Sons, 1967.
11. A. Wayne Wymore. *Model-Based Systems Engineering*. Systems Engineering Series. CRC Press, 1993.

A Strategy for Realizing Traceability in an Object-Oriented Design Environment

Jean-Pierre Corriveau and Craig Hayashi

School of Computer Science, Carleton University
Colonel By Drive, Ottawa, CANADA K1S 5B6

Abstract. Object Oriented Software Engineering has focused mainly on the notations, models and tools to be used in order to carry out the analysis, design and implementation of systems. *Iteration* has been emphasized as a solution to the problems of scalability and maintenance, but little attention has been paid to the management of iterations through the use of *traceability*. We first contend that traceability constitutes a fundamental requirement for OOD. We then argue that traceability reduces to the need to *hyperlink* the different models of OOD between themselves, back to requirements, and onto the code. Several hyperlinked CASE systems for the structured paradigm have been developed. However, no comparable system exists for the object oriented paradigm. And yet we suggest that the properties of object oriented software development cause an even greater need for hyperlinks than with structured development. We conclude with an overview of a prototype for such a tool.

1 Introduction

The design and development of large complex software systems constitutes the fundamental task addressed by software engineers [30]. However, it has become apparent in the last ten years that traditional approaches to software development do not scale up gracefully to the complexity of today's problems [5,27]. With hardware technology rapidly improving, computers are increasingly called upon to implement larger and more complex applications. Examples of this are systems to control nuclear power plants, the space shuttle, or bank account transactions for millions of accounts. The resulting strain on the software development process has been addressed by different generations of languages and paradigms. Recently, the emergence of object-oriented programming (OOP) [4] and object-oriented software engineering (OOSE) [10,19] has led to the development of a multitude of 1) object-oriented programming languages, 2) object-oriented design methodologies and 3) corresponding object-oriented (OO) CASE tools.

However, OOSE has focused almost exclusively on the design notations and tools to be used by a single developer *in one iteration* in order to carry out the analysis, design and implementation of systems. More precisely, it has been suggested [5,19,27,30] that object-oriented design (OOD) reduces to the production of a combination of 1) a *structural* model (which captures objects and relationships between them), 2) a *scenario* model (which captures the details of interactions between objects), 3) a *behavioral* model (which captures the internal workings of an object, and 4) functional model (which specifies the flow of data between objects).

Considerable effort has been invested in explaining how these models are to be constructed *iteratively* and serve, at least in principle, as the basis for coding and testing. Indeed, iteration has been emphasized as a solution to the problems of scalability and maintenance [17]. Roughly put, it is assumed large complex systems will merely *evolve* from smaller (and therefore more verifiable) systems and maintenance will reduce to the introduction or modification of details on top of existing models. With respect to verification, this design philosophy rests mainly on the notion of executable specifications, that is, on the ability of the developer to *simulate* the system by executing these models. (Simulation, in this context, merely implies the capacity to send messages between objects and run the finite state machines that typically define the behavior of objects.) But simulation does not tie models back to requirements and is thus insufficient by itself from a verification viewpoint.

In this paper, we want to argue that an approach to OOD that only focuses on models not only downplays the important role iteration is supposed to play in OOD, but also oversimplifies if not ignores the maintenance issue. Following Jacobson [19], we instead suggest that *traceability* be viewed as a most fundamental requirement for an OOD process. Semantically, traceability means establishing a causal relation between the different models used during the evolution of a system from requirement specification down to implementation and testing. Operationally, we argue that traceability reduces to the need to *hyperlink* the different models of OOD between themselves, back to the requirements, and onto the code produced by the developer, regardless of the specific notations and tools used. Our claim is that only in such a semantically-rich framework can maintenance be realized as an evolution of an existing set of models. Conversely, the availability of the proposed hyperlinks allows for the development of principled OO development process that addresses the whole lifecycle. A detailed description of this process lies beyond the scope of this paper (see [9]). We merely want to emphasize here that a key characteristic of this process is that, because we recognize and in fact rely on the importance of iteration in OOD, hyperlinks are viewed as versionable entities in their own right. Such a strategy allows the process to be centered around the notion of a *design history*, which corresponds to the rich interpretation of the idea of evolution in the context of an iterative development process.

2 The Object Oriented Life Cycle

The software lifecycle refers to the activities that take place during the creation, use, and evolution of the software system. These activities can be divided into the categories of analysis, design, implementation, testing, and maintenance. Analysis should capture the requirements the system must fulfill, knowledge learned about the application domain, and how the system will interact with the problem domain [18]. The results of object oriented analysis include a description of the objects in the problem; identification of classes and their attributes, behavior, and relationships; and documentation of the manner in which the objects interact with each other [22]. From the system described during analysis, a solution to realize the system is produced in design. The resulting artifacts are language-independent and similar to those produced in analysis: classes and interactions between instances are again described. The design will include objects that are not explicitly in the real world problem domain but are required in order to realize the software system. Notwithstanding, many of the objects identified in analysis form part of the design solution. Implementation concerns the language specific code to implement the system on a particular hardware platform. The level of detail of the design phase will determine how closely design and implementation match. At any rate, implementation should augment the design with language or hardware specific constructs that optimize the system's performance and implementation.

Testing activities take place in all three of the previously discussed phases. Roughly put, testing should be done to ensure that each phase correctly accomplishes its goals and proceeds from previous phases and iterations [14]. More specifically, the results of each phase must be closely tied together and testing should be centered around the notion of traceability (see [9]). But currently, such requirements-based testing has been downplayed in favor of ensuring, through simulation of the models, that the solution that was developed will work: the pervasive role of traceability during OO testing remains to be emphasized in literature.

Finally, any activities occurring after the delivery of the system are termed maintenance. This can include bug fixes, evolving the system due to changes in the problem domain, or extending the system to give more functionality [16]. For OO, maintenance can be viewed as *evolution*, that is, as a mini lifecyle which encompasses the tasks of analysis, design, implementation, and testing. The added difference is that the existing system must be understood in order for it to be modified [6].

3 Object Oriented Methodologies and CASE Tools

Various methodologies have been developed to support the OO lifecycle [5, 19, 23, 27, 28,32]. A CASE tool is often available to give automated support for the methodology. In general, a methodology will include a *process* and *representation* [22]. The process gives a technique on how to develop the system while the representation provides a notation to describe the system. As an illustration, consider Booch's methodology [5]. His macro development process includes the following activities.
- *Conceptualization*: establishing the requirements
- *Analysis*: modeling the system's behavior
- *Design*: creating an implementation architecture
- *Evolution*: successively refining the implementation
- *Maintenance*: managing the post-delivery evolution

This essentially corresponds to the phases of the lifecycle. The micro development process consists of the following steps:
- identification of the classes and objects
- identification of the behavior and attributes of the classes and objects
- identification of the relationships and
- implementation of the classes and objects

These activities are performed by the system developers during the various phases of the macro development process. Booch also includes a graphical notation to depict classes, class relationships, state machines, and interaction diagrams.

The Booch method was chosen as an example because it represents one of the more complete methodologies. However, close examination reveals weaknesses in his and most other methodologies. The two major weaknesses are a naive view of iteration and the lack of support for the complete software lifecycle. Let us elaborate.

It is well established that the waterfall view of software development is not realistic [30]. Instead, OO development encourages the successive iteration and refinement of objects. All methodologies advocate iteration but they never provide ways to capture or manage it. In the examples of using a particular methodology, the various classes, attributes, behavior, and interactions are quickly identified and recorded. The examples do not account for the added classes, re-arrangement of hierarchies, modification of old interaction diagrams, or abandoned efforts that characterize any practical iterative development.

With respect to the second weakness, most methodologies only handle the analysis and design phases with any great detail. The implementation phase is assumed to be covered by literature on a specific language such as C++ or Smalltalk [13]. And code generation is typically limited to the production of code 'skeletons' [9]. Thus, there is generally a discontinuity between design and implementation. Furthermore, testing, we repeat, is not adequately dealt with [9, 11]. Consequently, no methodology includes testing across the phases as an integral part of the process. Furthermore, maintenance is only briefly mentioned as an activity that will occur after the system is developed. It is assumed that maintenance will be similar to the initial development effort and will involve analyzing and designing new features for the system.

This lack of attention to the maintenance phase is disturbing. Recall that the object oriented paradigm was motivated from two promises. The first one was the ability to handle the development of large systems and the second one was the improvement in the maintenance of the systems. We should not be blinded by the increase in productivity *possibly* resulting from the use of object oriented methods: If maintenance is ignored, object oriented systems will become unmaintainable legacy systems like those that exist from the structured revolution. In our opinion, an object oriented process must effectively support iteration and maintenance if it is to be successful in the long term. Let us briefly elaborate.

From studies of systems developed using the structured paradigm, maintenance is known to consume between 60 and 70 percent of the life cycle cost [16, 21,29]. Of this effort, 40 percent of maintenance is performed to add features to the system [21]. This means the system will grow in terms of code and analysis and design documentation. Finally, 47 to 60 percent of the maintenance effort is spent on understanding the software [16]. This is because the original developers are almost never involved in the maintenance for the system they developed.

As a result of iteration, documents can become out of date during the initial development of a system. This often makes code the only reliable document for the system. However, even if all analysis and design documents were up to date, maintainers would gain additional benefit from information *not recorded* during the development effort. This information includes the various iterations (i.e., the design history) that lead to the final solution. An analogy can be drawn between a scientist's lab diary tracing the discovery of a new phenomenon to a text book describing the phenomenon. A good understanding of the phenomenon can be attained from the text book but a better understanding can be found in the diary which details the thought process in finding the phenomenon.

Thus it should be clear that capturing the iterations during development should help with maintenance activities. Furthermore, because of its great costs, maintenance must be an integral part of the system's development and documentation.

4 Proposed System

4.1 Proposal

In order to address the issues of iteration and maintenance, we propose combining hyperlinks to OOD. In the following discussion, we assume the existence of:

A methodology: The methodology should be relatively thorough and cover as much of the life cycle as possible. It should give a process that will result in the description of the system to develop and how it will be developed. The artifacts that are created can be text, diagrams, or code. They can be used to represent requirements, analysis, design, implementation, test data, metrics, cost estimates, project schedules, etc.

A software repository: The artifacts produced during the creation and life of the system are stored on-line in a database. This database is called a software repository [15]. The database serves as an archive that stores prior and current versions of a particular artifact.

CASE Tool(s): One or more CASE tools should provide sufficient support to develop an application. At the minimum this means support for analysis, design, and implementation. It can take the form of a CASE tool to support a methodology and programming environment for the implementation language. Additional tools to help in management or project planning can also be used.

Hyperlink capability: Within the repository, artifacts can be linked together. A link between two artifacts is bi-directional. Additionally, there exists mechanisms to view the links, navigate (via links) through the system, and manage the links. This essentially forms a *hypertext* system [8].

Taken together, these entities provide a way to develop software *and* account for maintenance. The combination of a methodology, repository, and CASE tools is nothing new. The methodology gives an organized plan to develop software systems. To develop any industrial sized system, a methodology is required to manage the resources and phases of development. Furthermore, CASE tools and repositories are increasingly used to help manage the size and complexity of new systems. Hypertext technology has also been around for a long time. It has traditionally been used for on-line books, education, and help systems.

The originality of the proposed system comes from the combination of hypertext and CASE for object oriented software development. Several hyperlinked CASE systems for the structured paradigm have been developed. However, no comparable system exists for the object oriented paradigm. And as it will be argued later, the properties of object oriented software development cause an even greater need for hyperlinks than with structured development.

The proposed system will operate as follows. Developers will follow the methodology to proceed from analysis to design to implementation. The information that is discovered (e.g., classes and class interactions) will be captured through the CASE tool and stored in the repository. The important point is that related artifacts will be linked using hyperlinks. In addition, the links themselves will contain useful information. This will allow the evolution of the application to be captured and organized. The various iterations leading to the final system can be traced by following the links. This makes it easier to understand the system by viewing the iterative growth from analysis up to implementation. In turn, the maintainability of the system is improved.

As it was mentioned, the hyperlinks will hold meaningful information. This can include the following:

- the time and date of creation
- a type telling the relationship between the linked artifacts (e.g., depends on, related to, implements, cooperates with)
- the iteration number (e.g., this represents the third refinement of this class)
- design annotations (e.g., rationale for implementing things a certain way or any assumptions made)

This information increases the power of the hyperlinks in that it gives a person greater flexibility when exploring the system through the links. Using this information, some of the ways the links can be traversed are as follows.

- **Time**: By sorting the links by creation time, it is possible to 'replay' the development of the system.
- **Type**: Depending on the user's needs, different types of links from an object can be traversed. For example, starting with a class in design, the user can follow the 'cooperates with' link to see how the class is used in the system. Once this is understood, the user can travel the 'implements' link to see the class' implementation.
- **Iteration**: The various refinements of a particular object can be traced with the iteration number of the link. For example, the first attempt at designing a class specified in analysis will be represented by iteration number one. A second refinement on the design class will be connected by the iteration number two.

Deciding exactly which artifacts to link will boil down to the judgment of the developer. Just as Smalltalk programmers write good method comments to help others understand their code, the users of this system will learn to make meaningful links to help maintainers understand the design of the software. Some general guidelines as to what to link are given below:

- link analysis classes back to the requirements
- link design classes to the analysis classes
- link implementation classes to the design classes
- link messages in interaction diagrams to their implementation
- link classes in class hierarchies to the class definition
- link classes in class relationship diagrams to the class definition
- link pseudo code to the method implementation

In other words, anything relating an isolated artifact to the developed application will help in understanding the application.

In summary, the proposed system uses hyperlinks to tie together relevant artifacts created during the lifecycle. In the next section, a prototype of the system is discussed in order to better understand how the system works and what advantages it gives.

4.2 Prototype

In order to determine how such a system would benefit the development and maintenance of an object oriented software project, a prototype system was implemented. The system is not a CASE tool that allows the linking of artifacts created during the project's lifecycle. The prototype represents an 'end result' of a project developed using the envisioned system. The various analysis, design, and implementation artifacts that are related to the project were manually linked together. By using the prototype, the developed system can be explored by traveling the links. The prototype will be used to answer several questions below (e.g., how did the links affect the development of the system, how much more difficult would it be to develop software in an environment where the artifacts had to be linked together, did the links help during the development process, how did the links affect the maintenance of the system, how did they help with understanding the system, how hard would it be to link in new artifacts created, how easy would it be to understand the maintenance additions).

The prototype was implemented in Supercard[1] and dealt with the development and maintenance of a weather monitoring application. This is a simpler version of the one found in Booch's book [5]. A generic methodology guided the development. It only covered the phases of analysis, design, and implementation. The methodology contained the minimum descriptions in order to describe a system. In particular, English text, hierarchy diagrams, interaction diagrams, and pseudo code express the analysis and design objects. These notations are used to represent the class attributes, class behavior, class relationships, and interactions between classes.

Two activities were performed to simulate the manner in which a real project is developed. First, the requirements for the system were given. These outlined the features the system needed (see [9] for full details and screen captures of this prototype). From these requirements, the weather monitoring application was analyzed, designed, and implemented. The artifacts created by this process were captured and linked in the prototype. Secondly, a maintenance enhancement (namely "Add ability to periodically save the displayed measurements to a file") was performed after the system had been developed.

When each card was created, it was manually linked to any other cards deemed important. For example, analysis classes were linked to the requirements they addressed. Clicking on a requirement listed the relevant analysis artifacts. Similarly, analysis cards were linked to design cards and design cards were linked to implementation cards. An example of viewing the links from a design card is shown in figure 1.

Each other type of card has a particular link association. Cards containing pseudo code are linked to the implementation. Cards containing class hierarchies are linked to the class descriptions. Finally, cards with interaction diagrams are linked to the class descriptions and method definitions contained in the card.

---

[1] Supercard is a rapid prototyping tool based on the card metaphor. A Supercard application is composed of many cards. A card contains information in the form of pictures, fields with text, or buttons. Users can interact with the card through the fields or buttons. They can also navigate between different cards by clicking particular buttons or using built in navigational tools. The cards as well as Supercard's scripting language provide a basic hypertext system. This was used to create the hyperlinked prototype.

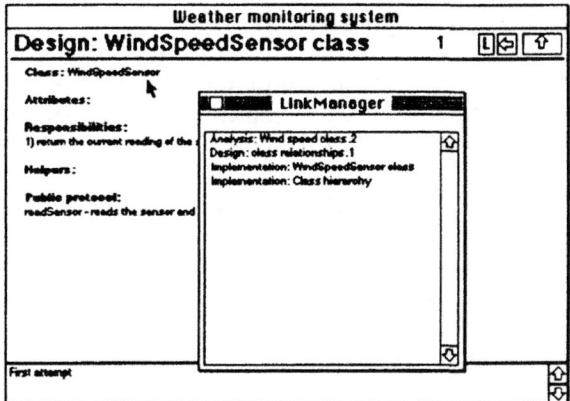

Figure 1: Links from the class name

The link sequence window (shown in figure 2) provides a global navigation facility. It contains a list of all analysis, design, and implementation artifacts. Each list is sorted by the order of creation. Clicking on a name displays the associated card.

In order to perform the enhancement, the system was examined to determine what would have to be modified. Then, a new iteration was created for each existing card that had to be changed. Any new cards that were required were also added. To finish the maintenance, a list of the cards that were added and changed during the maintenance was placed in the link sequence window. This provided a record of the maintenance performed. The window is shown in figure 3.

The initial development and maintenance tasks concluded the implementation of the prototype. From the experience gained, a number of open issues were uncovered and are presented in the next section.

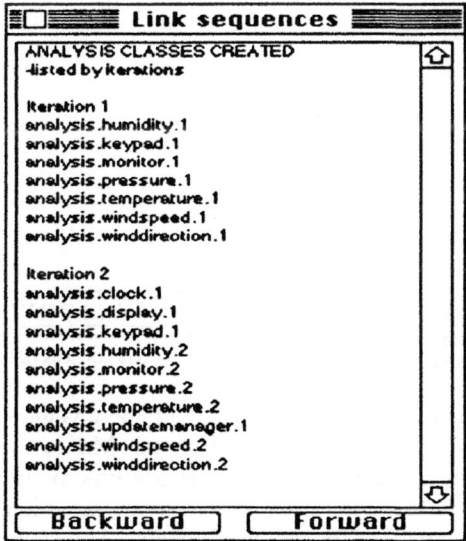

Figure 2: The link sequence window

4.3 Implementation of a hyperlink facility

We now give a sample of the features whose need was uncovered through experience with the prototype. Obviously, an actual implementation of the system would have to address a great deal of additional technical issues.

Navigation. Providing navigational abilities will help prevent users from becoming 'lost' as they explore the links. This is a common problem in hypertext systems and is the focus of a great deal of research [8]. Some features that help are:
- list the last x visited cards
- list all direct links going to and coming from an object
- list all links that are available x steps from an object
- list all the links available from a card and how they are accessed
- keyword search on the link type, name, or comment

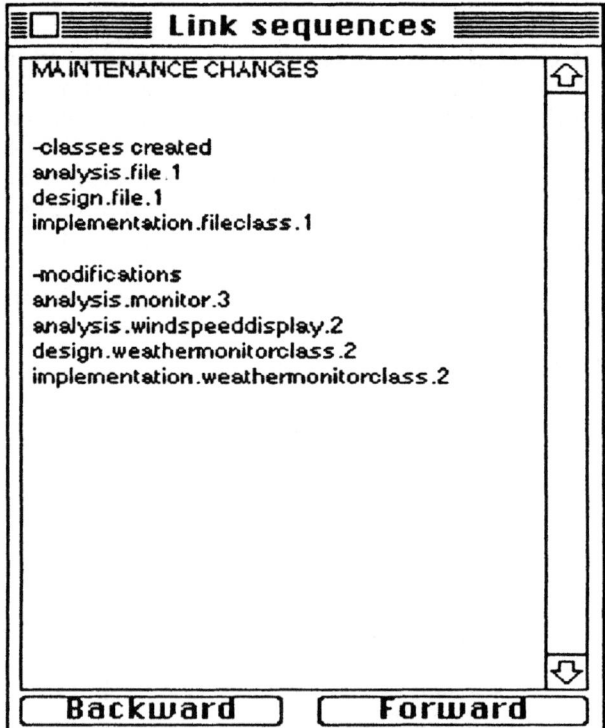

Figure 3: Cards affected by the maintenance

Versioning. It should be possible to highlight the changes from one iteration to a next. This makes it easier to see what has changed. Moreover, because iterations may only differ by a slight amount, the ability to hot-link or inherit cards is desirable. Using this feature, a different iteration will have the prior iteration as a parent. The new iteration will inherit all of the parent's objects. Any changes in the new iteration will over-write the inherited objects. This way, the amount of copying is reduced.

Cognitive overhead of creating links. When a new iteration is created, it should be automatically linked to all the places the prior iteration was linked to. It should be possible to perform keyword searches on the names of the objects in the system so the appropriate objects to link can be found. A facility to add audio comments to links or objects would make it quicker to record rationale or assumptions.

Controlling the number of link types. A link repository to list all of the available types of links in the system should be provided. This will limit the number of analogous links (e.g. a link with the type 'implementation' and a link with the type 'code'). Users will select links from this (user-growable) list instead of entering their own link type specifiers.

Beyond these identified requirements for the hyperlink OOD tool, the prototype also raised several issues:

The difficulty of creating the links during development. This question is harder to address in the prototype since it did not provide automated support for the creation of links. Furthermore, because of the weather monitoring application's small size it was possible for one person to develop it. This meant the entire system could be understood at once and it was easier to determine which objects to link together. In a larger project with multiple developers, the tool will have to provide features to make it easy for the developer create links and determine which artifacts to link.

The links benefit during development. The links were found to be very helpful during the development of the application. Even with the application's small size, it was still possible to forget the exact details of certain objects. Fortunately, the links could be used to quickly find the relevant information. For example, it was easy to use the links to locate the pseudo code or interaction diagram that specified how a method was to be used or implemented. The advantage of links (specially for reuse [26]) should become even more apparent in a large system where different people perform the analysis, design, and implementation phases.

The links usefulness to understand the system for maintenance. When the initial development of the weather monitoring application was completed, the system was not used for several days. After this time, the maintenance enhancement was performed. The first step involved understanding the system. Using the links, it was easy to understand specific parts of the system and trace the development from requirements to analysis to design to implementation. However, it was harder to get a sense of the system as a whole. This is probably because there was not an organized manner in which to read the information. The link sequence window helped but it was not powerful enough. Taking the effort to make an organized path of exploration to guide people on a sequential exploration of the system would have been beneficial. This would be analogous to creating a system document that would read like an ordinary book.

The difficulty of linking in maintenance modifications. This relates to the same issues covered in the question on creating links during the initial development. There must be automated support for this activity. It is even more critical in maintenance because the modified cards have to be linked to the same places that the original cards are. Doing this in the prototype was very hard because there was no way to automatically search for all references to a particular card. The linking had to be done manually. During the time since the initial development it was difficult to remember the link relationships that were present.

The benefit of using links to understand the maintenance enhancements. By creating a list of added or modified cards as a result of the maintenance, it was very easy to understand what was changed. Additionally, by using the links in the new cards it was relatively simple to see how the changes related to the existing system. The tracking of maintenance activities was the greatest benefit the links provided.

In summary, the prototype gave an indication of some of the advantages, disadvantages, and issues that a hyperlinked CASE tool would encounter. A better prototype will be developed and more controlled usage experiment will be performed in order to further understand the needs of the proposed system.

4.4. On existing Hyperlinked CASE Tools

Several such hyperlinked CASE tools have been developed. The majority of the tools are for use with the structured paradigm and all of them are research systems. They provide valuable information regarding pitfalls of such systems:
- creation of links is time consuming but critical for efficient navigation [12]
- the navigation architecture (i.e., links) must be browsable if not visual [3,15]
- the repository must store a wide variety of artifacts [2]
- versioning even of partial solutions is important but difficult [30]
- beyond models and code, design assumptions and decisions must be captured [1]
- traceability of code back to requirements is feasible [7]

4.5 General Advantages of Hyperlinks

As mentioned above, the prototype was used to uncover some of the advantages that hyperlinking design artifacts provided. These general advantages and other envisioned advantages are given below.

The links help the CASE tool mimic the developer's mental process of software development. One of the advantages of objects is that they close the gap between the human's and computer's representation of the problem [25]. Both representations are based on interacting objects. Due to the functioning of the human mind, related mental objects are linked together. Furthermore, associated objects are instantaneously retrievable and it is possible to mentally navigate between the objects. By abstracting the design into different levels and evolving the problem through successive iterations, the final solution is attained. In other words, the use of mental links is the key to this problem solving process. By providing object oriented CASE tools with a hyperlinking feature, they will further match the problem representation and solution process of people and computers.

Iteration is an important part of the object oriented development process and links provide both support and management for it. Links give an audit trail of development iterations. If a current stream of iteration is found not to work and is abandoned, the links can be used to trace back to the point of departure. From this point, another iterative attempt can be started. By keeping a record of the failed attempt, important information as to decisions or rationale can be obtained. A hyperlinked CASE tool recognizes that change is a natural part of object oriented software development and actively tries to capture and record it.

Treating the links between entities as first class objects rightfully makes the links as important as the entities themselves. The links can be used to capture rationale, show tradeoffs, and indicate dependencies. Listing the links in different sequences can be used to give preset paths of exploration [24]. For example, paths can be created to show the iterative development of a class from analysis through to implementation. The links thereby form an integral part of the documentation of the system.

Links give some leeway with respect to keeping information up to date. One disadvantage of iteration is that new iterations can quickly outdate existing information. However, links can be used to tie related pieces of information together. This makes it easier to find the information to update. Moreover, if the change is minor, it may not be necessary to take the time to update the information. A comment can be associated with the link to the new iteration recording the fact that the old information was not updated. Since the links capture the iterative evolution, minor discrepancies between the documents can be tolerated.

By sequentially tracing the development of the system, links make it easier for maintainers to understand the system. They can trace the actual development process by following the links in the order of creation. This makes it easier to understand the system by watching it build up incrementally from the start. The capturing of iterations and rationale also helps with comprehension. Finally, the links make it easy to selectively understand a small part of the system and the effect that any modifications will have.

During maintenance of the system, the changes and additions are linked with the existing information. This provides a record of a particular maintenance activity. It makes it easier to separate the original system from the maintenance changes and also separate different maintenance tasks. This helps to organize the maintenance and hopefully prevent the system from becoming incomprehensible and unmaintainable as it gets older.

5 OO Development and Links

In this section, the properties of object oriented software development that make links particularly necessary are discussed. To begin, the relevant features of objects and object oriented development are given.

Features of objects. One of the consequences of encapsulation is that the control flow of an object oriented program is distributed across several objects [31]. In a similar fashion, inheritance can cause a class' state and functionality to be spread across a hierarchy of objects [20].

Features of OO development. The boundaries between analysis, design, and implementation are said to be blurred in the object oriented paradigm [10]. A consequence of this fact is that many objects found in analysis persist through the design and implementation phases. Another feature of object oriented development is the many different but related views that are produced [11]. Each of the views (e.g., interaction diagrams, class descriptions, or collaboration diagrams) uniquely contributes to the understanding of the system. Finally, as it was already mentioned, iteration is encouraged in object oriented development.

From these points, it can be concluded that the object oriented paradigm produces discrete artifacts which are highly related to each other. These include both the documentation and objects in the system. Therefore, links between the artifacts are essential to capture the various relationships. Only then can a true understanding of the system be attained by using the links to move between different objects and versions of objects. *Links help organize the objects and thereby facilitate understanding.*

Understanding is the key to maintenance because the failure to understand the objects and interactions between objects in a system leads to maintenance error [25]. Using links during the development and maintenance of a system will help facilitate understanding and future maintenance efforts. *This will increase the life of the system and make maintenance easier.*

6 Conclusions

The current methods and tools used to develop software in the object oriented paradigm have been examined. From this, it has been suggested that iteration and maintenance issues have been somewhat neglected. The use of hyperlinks to connect related development artifacts was described. It was shown how the links complemented the existing methodologies and tools to better account for iteration and maintenance. The very properties of objects and object oriented development were used to argue for the links necessity. As a result, links were deemed to be an essential component to further advance the object oriented paradigm.

References

1. Arango, G., Bruneau, L., Cloarec, J., and Feroldi, A. (1991) A Tool Shell For Tracking Design Decisions, *I.E.E.E. Software*, March 1991, p. 75-83
2. Bigelow, J. (1988) Hypertext and CASE, *I.E.E.E. Software*, March 1988, p. 23-27
3. Blum, B. I. (1988) Documentation for Maintenance: A Hypertext Design, *1988 I.E.E.E. Conference on Software Maintenance*, p. 23-31
4. Budd, T. (1991) *Object-Oriented Programming*, Addison-Wesley.
5. Booch, G. (1994) *Object-Oriented Analysis and Design with Applications*, Benjamin/Cummings
6. Chapin, N. (1988) Software Maintenance Life Cycle, *1988 I.E.E.E. Conference on Software Maintenance*, p. 6-13
7. Cimitile, A., Lanubile, F. and Visaggio, G. (1992) Traceability Based on Design Decisions, *1992 I.E.E.E. Conference on Software Maintenance*, p. 309-317
8. Conklin, J. (1987) Hypertext: An Introduction and Survey, *Computer*, September 1987, p. 17-41
9. Corriveau, J-P. and Hayashi, C. (1995) Traceability and Object-Oriented Software Development, submitted to *Object-Oriented Systems*.

10. de Champeaux, D., Lea, D., and Faure, P. (1993) *Object-Oriented System Development*, Addison Wesley
11. Firesmith, D. G. (1993) Testing Object-Oriented Software, *TOOLS '93*
12. Fletton, R. (1990) A Hypertext Approach to Browsing and Documenting Software, *Hypertext: State of the Art*, R. McAleese and C. Green (eds.), Ablex, p. 193-204
13. Goldberg, A. and Robson, D. (1983) *Smalltalk-80: The Language and its Implementation*, Addison-Wesley
14. Graham, D. R. (1992) Testing and Quality Assurance – the Future, *Information and Software Technology*, Vol. 34, No. 10, p. 694-697
15. Gulla, B. (1992) Improved Maintenance Support by Multi-Version Visualizations, *1992 I.E.E.E. Conference on Software Maintenance*, p. 376-383
16. Hall, P. A. V. (1992) Overview of Reverse Engineering and Reuse Research, *Information and Software Technology*, Vol. 34, No. 4, p. 239-249
17. Henry, S. M. and Humphrey, M. (1990) A Controlled Experiment to Evaluate Maintainability of Object-Oriented Software, *1990 I.E.E.E. Conference on Software Maintenance*, p. 258-265
18. Høydalsvik G. M. and Sindre. G. (1993) On the Purpose of Object-Oriented Analysis, *OOPSLA '93*, p. 240-255
19. Jacobson, I, *et al.* (1992) *Object-Oriented Software Engineering - A Use Case Driven Approach*, ACM Press/Addison Wesley
20. Lejter, M., Meyers, S., and Reiss, S. P. (1991) Support for Maintaining Object-Oriented Programs, *1990 I.E.E.E. Conference on Software Maintenance*, p. 171-178
21. Lientz, B. P. and Swanson, E. B. (1981) Problems in Application Software Maintenance, *Communications of the ACM*, Vol. 24, No. 11, p. 764-769
22. Monarchi, D. E. and Puhr, G. I. (1992) A Research Typology for Object Oriented Analysis and Design, *Communications of the ACM*, Vol. 35, No. 9, p. 35-47
23. Reenskaug, T. and Skaar, A. L. (1989) An Environment for Literate Smalltalk Programming, *OOPSLA '89*, p. 337-345
24. Rettig, M. (1992) Hat Racks for Understanding, *Communications of the ACM*, Vol. 35, No. 10, p. 21-24
25. Rosson, M. B. and Alpert, S. R. (1990) The Cognitive Consequences of Object-Oriented Design, *Human-Computer Interaction*, Vol. 5, p. 345-379
26. Rosson, M. B., Carroll, J. M., and Sweeney, C. (1991) A View Matcher for Reusing Smalltalk Classes, *CHI '91*, p. 227-283
27. Rumbaugh, J., *et al.*, (1991) *Object-Oriented Modeling and Design*, Prentice Hall
28. Shlaer, S. and Mellor, S. J. (1992) *Object Lifecycles: Modeling the World in States*, Prentice Hall.
29. Smith, D. B. and Oman, P. W. (1990) Software Tools in Context, *I.E.E.E. Software*, May 1990, p. 15-19
30. Sommerville, I., Haddley, N., Mariani, J. A., Thomson, R. (1990) The Designer's Notepad – a Hypertext System Tailored for Design, *Hypertext: State of the Art*, R. McAleese and C. Green (eds.), Ablex, p. 260-266
31. Wilde, N. and Huitt, R. (1991) Maintenance Support for Object Oriented Programs, *1991 I.E.E.E. Conference on Software Maintenance*, p. 162-170
32. Wirfs-Brock, R., Wilkerson, B., Wiener, L. (1990) *Designing Object-Oriented Software*, Prentice Hall

Towards a Cast Method for a Systematic Generation of Non-Orthogonal Complete Transformations

O.Bolivar Toledo, J.C.Quevedo Losada, R.Moreno Díaz jr,& S. Candela Solá.

Department of Computer Science and Systems
University of Las Palmas de Gran Canaria.
Las Palmas. Canary Island. 35017.
Spain

Introduction

Just as neurophysiologists worry about what natural neuron receptive fields are and what they are doing, people working in picture processing and in general, in artificial vision, worry about sizes of receptive fields and functions that should be performed on them by neuron- like elements.

A basic concept in Image Processing and Artificial Vision is that of a complete description or transform. A description is said to be complete if it contains all neccesary data and data properties to achieve some goal. From an analytical point of view, a complete description needs the preservation of the number of degrees of freedom or the number of independent properties of a visual field.

In non-structured artificial vision systems, what scientists and engineers seek, are kinds of complete transforms, which provide alternative descriptions of images that can be truncated for particular purposes, so that a much lower number of degrees of freedom are required to be handle. In this way, having appliyed an entire set of standard transformations to pictures to obtain descriptors we were aware that the search for orthogonality in the descriptor functions was really a handicap inherited from the times when global transforms (like Fourier, Hadamard, Haar) had to be made almost by hand. However with modern computer techniques it was not clear that there was any advantage except for the assurance of independence, that is, ensuring that the corresponding transform results were non redundant. We looked then for complete transforms, not necessarily orthonormal, but independent, so that the inverse transform will exist.

The first interesting result is that the partition of the receptive field is as important as the function performed on it [1][2]. Partition means that one is computing on the sensory data in parallel, although with overlapping.

The concept of Progressive Resolution Transform was introduced [3], in which it was emphasized, that no matter what function is performed, a vast class of partitions could be complete. That is, given an image (a "data field" in fact, because the image fills a data field in memory), then it is convenient to separate receptive fields from functions, so that a data field partition can be, by itself, complete.

The next logical problem was one of tradeoff, which is most probably what happens in the retina, since the resolution of fibres is not necessarily preserved, and there is a multiple meaning coding, which in this terminology will mean that each fibre of the optic nerve sends information pertaining to more than one operation on the receptive field. Here the question was faced by stating a theorem [4], that showed that for a family of L algebraic partitions of the same length, the computation of a number M=N/L functional coefficients, linearly indpendent in each partition, provides a complete description of the data field.

In this context, the next step is to consider the more general case, inspired by natural systems, in which the size of the receptive field is a variable.

In this work we will consider this new case in which the dimension of the partitions established is not the same and we have formulated a new theorem, based on the juxtaposition of linearly independent functionals in each partition, which provides a systematic way to obtain complete transformations.

We have formulated the theory for one dimension and considering the transformation to be separable, we have applied it to bidimentional images. At the same time we have investigated the existence of fast computational algorithms.

Nature does not follow theorems exactly, but they provide us with a clear way to approach nature. In any case, the parallel computing structures which result, are illuminating in order to understanding natural systems and to build artificial ones.

Receptive Field- Functional Transformation

Partitions of Constant Length

Let us consider an unidimentional data field D(N) with resolution R, addressed by index i (i=1...N), and consider also a set of L independent partitions of the i addresses. This set can be represented by a binary matrix.

A class of partitions used in previous works corresponded to one level of a foveal transformation. For a receptive field of N places, the number of partitions of this type generated would be L=N-d+1, d being the length of the partition.The dimension of the corresponding matrix would be N*(N-d+1).

These partitions are not, a priori, complete, due to the fact, that is, in general, $L < N$, although they are independent.

If we are looking for a complete representation, we have developed a theorem[5], which integrates the algebra and analysis of the data fields, stating that given an algebraic partition with certain restictions, the computation of a number M=N/L of analytical descriptors, independent in each partition, provides a complete and non redundant description of the data field.

The algebraic-analytical transformation, represented by a matrix N*N is obtained by the "application" of each functional vector to each partition.

$$\begin{bmatrix} F11 & . & . & . & F1d \\ F21 & . & . & . & F2d \\ . & & & & . \\ . & . & . & . & . \\ . & . & . & . & . \\ Fm1 & . & . & . & Fmd \end{bmatrix} x \begin{bmatrix} P11 & . & . & . & P11 \\ P21 & . & . & . & P21 \\ . & & & & . \\ . & . & . & . & . \\ . & . & . & . & . \\ Pn1 & . & . & . & Pn1 \end{bmatrix} = N \begin{bmatrix} & & & & & \\ & & & & & \\ & & & & & \\ & & & & & \end{bmatrix} \begin{matrix} \} & M_1 \\ \\ \} & M_2 \\ . \\ . \\ \} & M_k \end{matrix} = M$$

As we can see, groups of vectors M_1, M_2, ..., M_k have been generated, all of which have the same extension over the data field. In each group, the vectors are independent by construction. Likewise, the groups are independent among themselves since they correspond to independent partitions. Therefore, the N horizontal vectors are independent, thus the corresponding transformation is complete.

Partitions of Variable Length

We have now of final adequate conditions to be able to consider the most general case of variable length partitions, that is, those partitions in which the dimension of each subpartition is not the same. In this new case, for a data field of N dimension D(N), and a set of L partitions, we must compute a number $M=N/L$ of independent descriptors per partition, but now the independence of the functionals has to be guaranteed when they are applied to each of the columns, not only to just one.

For example, let us consider the partition:

$$\begin{bmatrix} 1 & 1 & 1 & 0 \\ 1 & 1 & 1 & 0 \\ 1 & 1 & 0 & 0 \\ 1 & 1 & 0 & 0 \\ 1 & 0 & 0 & 1 \\ 1 & 0 & 0 & 1 \\ 1 & 0 & 0 & 0 \\ 1 & 0 & 0 & 0 \end{bmatrix}$$

Two independent functionals for the first column could be of Haar type:

$$F_1 = 1/8 \ (1\ 1\ 1\ 1\ 1\ 1\ 1\ 1)$$
$$F_2 = 1/8 \ (1\ 1\ 1\ 1\ -1\ -1\ -1\ -1)$$

It is easy to see that when they are applied to the second column, the numbers obtained are not only non independent, but the same.

Nevertheless, focusing on the other limit case, if we select two independent functionals for the lower dimension column, for example those of weights:

$$(1,1)$$
$$(1,-1)$$

and two new functionals of the dimension corresponding to that of the larger column, are generated by juxtaposition, that is say:

$$F_1 = 1/8 \,(\, 1\ 1\ 1\ 1\ 1\ 1\ 1\ 1\,)$$
$$F_2 = 1/8 \,(\, 1\ -1\ 1\ -1\ 1\ -1\ 1\ -1\,)$$

it is easy to see that the generated numbers are now independent. The corresponding transformation matrix is:

$$
\begin{bmatrix}
1 & 1 & 1 & 1 & 1 & 1 & 1 & 1 \\
1 & -1 & 1 & -1 & 1 & -1 & 1 & -1 \\
1 & 1 & 1 & 1 & 0 & 0 & 0 & 0 \\
1 & -1 & 1 & -1 & 0 & 0 & 0 & 0 \\
1 & 1 & 0 & 0 & 0 & 0 & 0 & 0 \\
1 & -1 & 0 & 0 & 0 & 0 & 0 & 0 \\
0 & 0 & 0 & 0 & 1 & 1 & 0 & 0 \\
0 & 0 & 0 & 0 & 1 & -1 & 0 & 0
\end{bmatrix}
$$

This transformation is similar to the detector contrast progressive resolution transformation, introduced by Candela in 1988 [3].

The solution, found in this example, has been made possible because the average number of functionals to be calculated per column of the partition is equal or lower than the resolution of the column of lower dimension. So, obviously, it does not make sense to calculate three descriptors from the third and fourth columns, of the partition which only have two addresses.

The above leads us to formulate the following theorem:

THEOREM

Given a data field of N dimensions, and an arbitrary partition of L columns, such that d_m is the dimension of the lower dimension column, with $d_m \geq N/L$, then a set of N/L independent functionals of d_m length, juxtaposed up until the column of larger dimension, provide a complete transformation.

It is easy to demonstrate this theorem according to what has been presented in the example shown previusly. In effect, let us consider the transformation matrix:

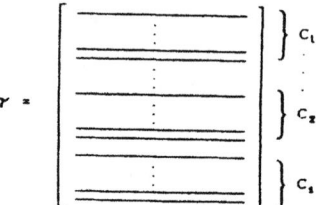

C_1 corresponds to the set of lower dimension, affected by the corresponding weights, which are independent by construction, while $C_2, C_3, \ldots C_l$. corresponds to the remaining columns, which will be independent by the additional condition of the independence of partitions, and/or by the extended functional used.

Limit Cases

In the transformation treatment developed previously, and from a completeness point of view, what remains to be considered are the situations where the dimension of the columns of lower dimension (d_m) is such that $d_m < N/L$. However, in this case, symmetry, which is one of the basic reasons for the separation of receptive field and function, is lost, in the sense that, depending on the partition length, we would need more or less vector coefficients. In this case, the utility of the separation carried out becomes questionable, instead of considering the transformation matrix directly, with the condition of independence of its vectors. In other words, because we need a number $N' > N/L$ of weight vectors, which form the functional, the product $N'L$ needed, exceeds the initial number of degrees of freedom and in order to avoid redundancies, the application of these vectors to columns must be carried out selectively.

However, some considerations can be carried out, which provide complete transform generation methods. According to what has been pointed out previously, the above situation correspond to partitions, where, on average, for reason of completness, it is essential to extract N/L independent descriptors, but certain corresponding columns can not provide then. This, could be due to the following:

1) Columns exist where it is possible to extract a number of descriptors higher than the average.

2) Addresses exist, which are lost from the data field, with data which cannot be retrieved by any function of the set of proccesors.

Case 1 corresponds to a partition of the type considered above (where each one of the addresses is considered once) and case 2 always corresponds to a situation of incompleteness. This last one, is theoreticaly interesting with regard to reliability or performance security, as far as it corresponds to the problem of "scotomas" (local disappearance of data), against the "disfunction problem" (local disappearance of processors), corresponding this, in a limit situation, with the typical method, used in artificial vision, of cutt the number of descriptors, which is enough in a practical situation.

It is possible to extend the systematic way developed to generate complete transformation to that corresponding to case 1.

As an illustration, let us consider the partition matrix:

$$\begin{bmatrix} 1 & 1 & 1 & 0 \\ 1 & 1 & 0 & 0 \\ 1 & 1 & 0 & 0 \\ 1 & 1 & 0 & 0 \\ 1 & 0 & 0 & 0 \\ 1 & 0 & 0 & 0 \\ 1 & 0 & 0 & 1 \\ 1 & 0 & 0 & 0 \end{bmatrix}$$

The number of descriptors per column must be $N/L = 8/4 = 2$, in average. But columns third and fourth cannnot provide it, therefore columns first and second must provide the additional number in order to keep the average. Particularly, the third and fourh columns can generate a unic descriptor each one, remaining 6 which must be distributed between columns one and two. Since the dimension of each one of then is equal or higher to the remainder average($6/3 = 2$), theorem proposed above, is again applicable to this columns.

From the point of view of modern retinal theory, the above solution, is not seem, a priori, to be sustained, since this would be equivalent to admit that ganglion cells, with extensive receptive fields, must carry out a proccess and codification function, greater than the carried out by cells with restricted receptive field. But, this would be so, only if completeness is required for the whole retina, which is a quostionable and even refutable argument, due to the fact that is the fovea (where the preccision is high) where are the cells of more restricted receptive fields.

Practical Illustration

As a practical illustration, consider the case of an image of 256*256.

1) In this case, we have considered a partition of L=128 columns and Hard Functionals. In the figure we show in a) the original image; in b) c) and d) the image transformed with ordered different sorting of the partitions matrix.

2) In this case, have considered a partition of L=64 columns and Hadamard Functional. In the figure we show in a) the original image; in b) the image transformed.

Figure a)

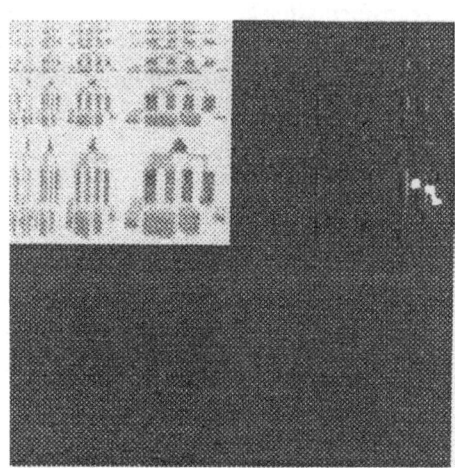

Figure b)

Figure c)

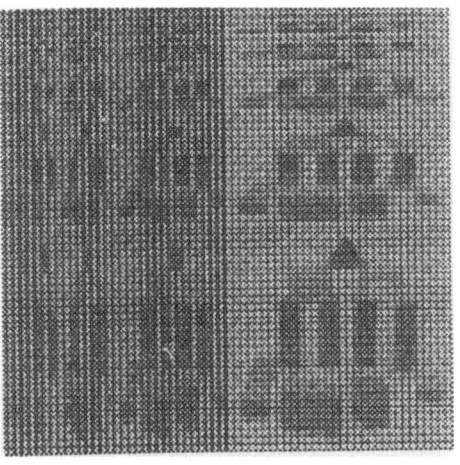

Figure d)

Figure a)

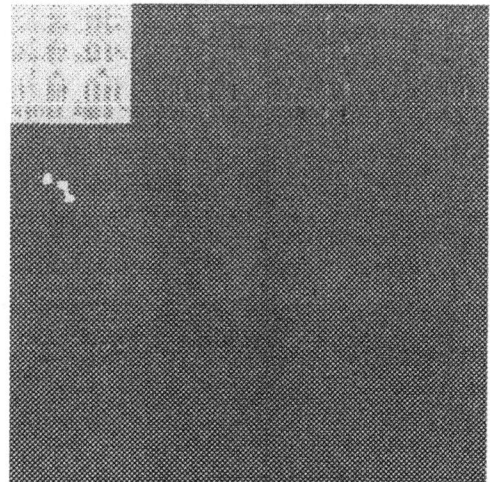

Figure b)

Conclusions

1) We have studied in depth the Theory of Receptive Field Functional Transformation, increasing their scope to situations of variable dimension partitions.

2) We have developed a theorem which allows the systematic generation of Complete Transform in the above case.

References

[1] J.A.Muñoz Blanco, J.C. Quevedo Losada, O. Bolivar Toledo. "The role of Partitions and Functionals in Descriptor Computation for Data Receptive Fields". Computer Aided Systems. Theory-Eurocast´93 (Las Palmas, Spain. 1.993) pp. 282-292.

[2] J.C Quevedo Losada, J.A. Muñoz Blanco, O. Bolivar Toledo. "Variable Receptive Field System Transformation and its Applications to Visual Field". United Kingdom Systems Society Conference on Systems Thinking in Europe (1.991 :Huddersfield. England) pp.225-230.

[3] S. Candela Solá. "Transformaciones de Campo Receptivo Variable en Proceso de Imágenes y Visión Artificial". Doctoral Thesis. 1988.

[4] O.Bolivar Toledo, S.Candela Solá & R.Moreno Díaz. "Complete Transforms and Their Incidence in Artificial Perception Systems Theory". Computer Aided Systems. Theory-Eurocast´91 (Krems, Austria, April 1.991) pp. 514-524.

[5] O. Bolivar Toledo. "Hacia una teoria de las Transformaciones en Campos Receptivos y Campos de Datos" Implicaciones en Teoría retinal y Proceso de Imágenes" Doctoral Thesis. 1990.

Tuning Fuzzy Logic Controllers by Classical Techniques

M. Santos*, S. Dormido**, A. P. de Madrid**, F. Morilla**, J. M. de la Cruz*

*Dpto. de Informática y Automática. Facultad de Físicas. (UCM.)
**Dpto. de Informática y Automática. Facultad de Ciencias. (UNED)
Ciudad Universitaria s/n. 28040-MADRID (Spain). FAX: (34)-1-3944687
e-mail: Matilde.Santos@eucmax.sim.ucm.es

Abstract. *Fuzzy Control* provides a good support to translate the knowledge of a skilled plant operator into rules, making *intelligent control* possible. But it is difficult to represent the expert's knowledge with no degradation, so a tuning phase is required. This is not an easy task, and there is not a general procedure for it. On the other hand, most of the control systems are still based on the conventional PID regulator. Aström has developed an empirical tool to predict the achievable performance of these controllers and to assess whether they are properly tuned. Based on Buckley's results, that have analytically proved the equivalence between one of the simplest fuzzy logic controller (FLC) and a PI, it is possible to apply Aström's tool to evaluate the performance of a FLC.

Key Words. Fuzzy control, Tuning fuzzy controllers, PID control, Self-tuning, Expert control, Adpative control.

1   Introduction

The conventional PID regulator is the most frequently used control element in the industrial world, because of its robustness and simplicity in the design and tuning of its parameters. There exists a clear relation between these parameters and the system response specifications, well known by the plant operators. To automate the adjustment procedure [8], there are many PID tuning techniques, elaborated during the last decades, that make the operator's task easier and reduce the cost.

Aström [1] has tried to characterize a class of systems where PID control is appropriate, defining some parameters to delimit them. He develops a formal tool to assess what can be achieved by PID control tuning with Ziegler-Nichols formulas [9]. By empirical results and approximate analytical reasoning, he defines two parameters, namely, the *normalized deadtime* and the *normalized process gain* to characterize the open-loop process dynamics. Two more parameters, the *peak load error* and the *normalized rise time* are also introduced to characterized the closed-loop response. Both of them can be used to predict the achievable performance of PID controllers and a measure of the difficulty of controlling a process, establishing whether the PID controller is properly tuned. So, he obtains an evaluation of PID regulators with Ziegler-Nichols tuning under certain conditions, characterized by one parameter of the open loop system. The resulting controller is not yet able to fit optimally the closed loop target but it will guide the process on a trajectory roughly close to the closed loop specifications.

But, as the involved processes are in general complex, time variant, with delays and non-linearities and, very often, with a poorly-defined dynamics, conventional methods are not able to guarantee the final control objectives. So, expert control strategies have been favored since they are based on the process operator's experience and do not need quantitative models. One of the most successful expert system techniques applied to a wide range of control applications has been fuzzy control, *"a rule-based control able to emulate the human expert's actions"*.

When the process is too complex to achieve a good physical description, the controller synthesis has to be based only on intuitions and heuristic knowledge. In this context, the fuzzy approach gives a good support to translate heuristic rules into numerical algorithms. A fuzzy controller is based on fuzzy information, set theory and fuzzy logic, for both the knowledge representation and inference.

It is difficult to represent perfectly the expert knowledge by linguistic control rules. Moreover, the fuzzy control system has many tuning parameters, and its performance depends on their value. One of the main problems of these fuzzy logic controllers is the selection and subsequent adjustment of their parameters. *There is no systematic procedure for tuning*; in practice, there is almost no other way but trial and error, a tedious and time-consuming task.

There have been several contributions to FLC tuning [10, 11, 12]. These methods give the best results when the system specifications cover, with adequate accuracy, all the system state space and when the fuzzy system is only asked to fit some examples, given by an expert. When the fuzzy system is designed to work in closed loop with a not well defined process, the previous methods show performance degradation. Furthermore, in many real processes the controller on-line tuning is very expensive from both the economic and time consumption point of view, or physically impossible. So, the problem is still open.

The aim of this paper is to establish a systematic procedure to tune the parameters of a FLC. Based on the equivalence of one of the simplest FLC and the PI regulator, analytically proved by Buckley [2, 3, 4], it is possible to describe the fuzzy controller, from an external point of view, as a PI controller. Then, we can apply the well-known PID tuning rules to tune the fuzzy controller, and also Aström's tool to evaluate the performance of the control system. Masking these parameters with the PI parameters, it is possible to establish a self-tuning method to respond when disturbances or changes in the process appear [5].

2   Intelligent PID Control

The aim of a controller is to reach or maintain a process in a specific state, by monitoring a set of variables, and selecting the adequate control actions. The classical block diagram of a feedback control system is shown in Figure 1.

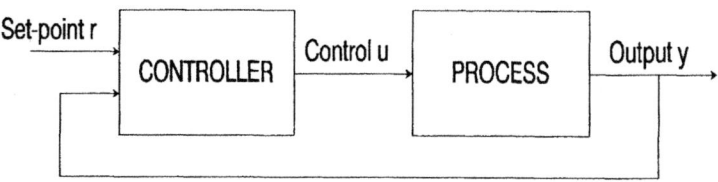

Fig. 1. Classical Control System

The main problem in the synthesis of a control system is the selection of a specific class of controller and the adjustment of its parameters, to verify certain specifications given for the controlled process. Most of the control systems are still based on the PID algorithm, but they incorporate some advanced capabilities of adaptability and/or self-tuning. It would be very desirable to incorporate some operator's knowledge, to help making decisions on how to choice the control algorithm, and provide diagnostics on the effectiveness of the control system.

A conventional PID controller is described by (1). To obtain the control, it incorporates three actions on the error signal: proportional, integral and derivative. Each of them is modified by a constant, which are respectively called proportional gain Kp, integral time Ti, and derivative time Td. Sometimes they are denoted as proportional, integral and derivative gains (Kp, $Ki = Kp/Ti$, and $Kd = Kp*Td$).

$$u(t) = Kp.\left[e(t) + \frac{1}{T_i} \int_0^t e(\tau)d\tau + T_d \frac{de(t)}{dt} \right] \tag{1}$$

The error signal $e(t)$ is defined as the difference between the process output $y(t)$ and the set-point or reference signal $r(t)$, and $u(t)$ is the absolute control action.

There are a large number of references about the selection of the type of controller: P, PI, PD or PID, and about the tuning of its parameters [6, 7]. Aström [1] has made an attempt to develop a tool to assess what can be achieved by PID control with a Ziegler-Nichols-like tuning formula, and whether it is properly tuned. Based on empirical studies and approximate analysis, two dimensionless numbers to characterize the open loop system dynamics,

- normalized deadtime θ
- normalized process gain κ,

and two dimensionless numbers to characterize the closed loop response have been defined.

- the peak load error λ
- the normalized rise time τ

The parameters in the time domain that characterize the stable process to be controlled are the static gain K, the apparent deadtime D, and the apparent time constant T. These parameters can be obtained from a step response experiment. In a similar way, a frequency domain characterization of the process dynamics can be introduced: the parameters are then the ultimate gain Ku and the ultimate period Tu.

For stable processes, Aström defines four new parameters. Analogous expressions can be introduced for processes with integral action.

Normalized deadtime: can be defined as the ratio of the apparent deadtime to the apparent time constant,

$$\theta = D / T \tag{2}$$

Normalized process gain: is defined as,

$$\kappa = K \cdot Ku \tag{3}$$

Peak load error: with l_{max} the maximum error due to the step disturbance, and l_o the amplitude of the step load disturbance,

$$\lambda = l_{max} / K \cdot l_o \tag{4}$$

Normalized rise time: is a measure of the response speed of the closed-loop system,

$$\tau = tr / D \tag{5}$$

where t_r is the closed-loop rise time.

Useful relations can be found by empirical experiments to analyze the controller behaviour:

- There exists a relation between the normalized deadtime θ and the normalized process gain κ. It means than they can be used interchangeably to assess the process dynamics.
- The product $\kappa.\lambda$ can be expected to be constant.
- The normalized rise τ time is approximately constant (and it is equal to 1 for stable process).

2.1 On-Line Assessment of PID Control Performance

The performance of the PID controller can be predicted from θ or κ. Based on one of them, it is easy to select the controller form, as it is summarized in Table 1.

Table 1. Choice of controller

	Tight control is not required	Tight control is required		
		High measurement noise	Low saturation limit	Low noise & high saturation
θ >1	I	I + B + C	PI + B + C	PI + B + D
0.6 < θ < 1	I or PI	I + A	PI + A	PI/PID+A+C
0.15 < θ < 0.6	PI	PI	PI or PID	PID
θ < 0.15	P or PI	PI	PI or PID	PI or PID

A = Feedforward compensation recommended, B = feedforward compensation essential, C = deadtime compensation recommended, D = deadtime compensation essential [1].

The rise time t_r can be measured when the setpoint changes. If the controller is properly tuned, then the normalized rise time τ should be equal to the apparent deadtime D. If the actual rise time is significantly different, it indicates that the loop is poorly tuned. Similarly, a maximum error that is significantly larger than predicted by the constant $\kappa.\lambda$ indicates that the loop is poorly tuned.

3 FLC-PI Equivalence

The fuzzy logic controller (FLC) design is based on the linguistic description of the control strategy that a skilled operator or expert should use in the manual control of the process. *A fuzzy regulator is a set of linguistic control rules, which have the possible values of the input variables as antecedents, and which also conclude the control action in linguistic terms.* This output is subsequently transformed into a deterministic value.

The input and output variables are identified with those of a conventional controller. Figure 2 shows the basic scheme. Error $e(t)$ and error change $ce(t)$ are the input variables, and absolute control action u or incremental control Δu is the FLC output. It is possible to have a FLC-PI controller with rules of one of the following types:

$$\text{If } e \text{ and } ie \text{ then } u$$
$$\text{If } ce \text{ and } e \text{ then } \Delta u$$

where ie is the integral of the error.

The parameters chosen to tune the FLC are the scale factors GE, GR and GU, gains which weight the input and output variables respectively. They are important tuning parameters since when they are varied the control action is adjusted within certain limits, without needing to change the definition of the rules or the linguistic terms.

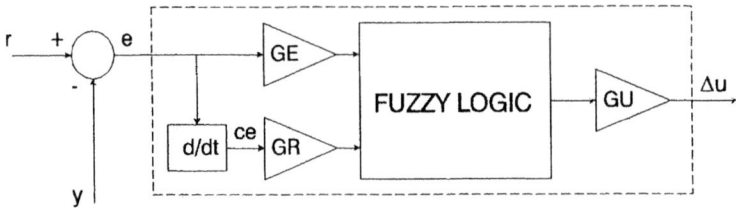

Fig. 2. FLC-PI Controller

Buckley's results have analytically proved the equivalence between one of the simplest FLC and a non-linear PI regulator [3]. The restrictions for the FLC are: two inputs and one output, with two and three linguistic labels respectively; triangular membership functions, which must be symmetrical from their center L, and center-of-area defuzzification method.

The formulas of the parameters of the discrete non-linear PI (Kp and Ki), being T the sampling period, as a function of the scale factors and the fuzzy controller variables, are given by the following expressions:

Zone 1: $GR \cdot |ce(nT)| \leq GE \cdot |e(nT)| \leq L$

$$Kp = \frac{0.5 * L * GU * GR}{2 * L - GE * |e(nT)|} \qquad Ki = \frac{0.5 * L * GU * GE}{2 * L - GE * |e(nT)|} \qquad (6)$$

Zone 2: $GE \cdot |e(nT)| \leq GR \cdot |ce(nT)| \leq L$

$$Kp = \frac{0.5 * L * GU * GR}{2 * L - GR * |ce(nT)|} \qquad Ki = \frac{0.5 * L * GU * GE}{2 * L - GR * |ce(nT)|} \qquad (7)$$

In the stationary case --when the error and error change tend to zero-- the static proportional gain Kp_s and the static integral gain Ki_s can be determined.

4   Tuning FLC

As the fuzzy controller is *overparametrized*, and some of these parameters are redundant, it is necessary to fix some of them to improve the system response. Their selection has a decisive influence on the FLC dynamics [10].

To be able to apply the widely studied tuning techniques of classical regulators to this type of fuzzy controller, some of the tuning parameters of the FLC can be reduced to parameters of a PI regulator. The scale factors *GE* and *GR* can be expressed as a function of the error, change of the error and a vector of parameters, which includes both the PI parameters and the rest of the FLC parameters. These gains are time-variant, as they depend on the error and the error change.

$$GR = gr\ (e(t),\ ce(t),\ GU,\ L,\ Kp,\ Ki) \qquad GE = ge\ (e(t),\ ce(t),\ GU,\ L,\ Kp,\ Ki)$$

$$GR = \frac{Kp*(2*L-f)}{0.5*L*GU} \qquad GE = \frac{Ki*(2*L-f)}{0.5*L*GU} \tag{8}$$

where $f = \max\ (GE*|e(nT)|\ ,\ GR*|ce(nT)|\ ) \le L$.

This kind of adjustment may be improved in order to include other heuristic aspects about the expert knowledge and to apply qualitative reasoning. The qualitative tuning of the scale factors seeks to improve the behaviour of the controller, based on the effects of the parameter changes on the system response. The general effects of varying the scale factors can be summarized as follows:

1. Increasing input gains implies a greater consideration of small input values. They have direct consequences on control: the response is faster and more oscillatory, reducing the stationary error. It thus improves the transient by reducing rise time and set up time, but it does increase the risk of instability with the overshoot increment. Increasing *GE* and *GR* makes the performance measure more sensitive around the set-point and less sensitive during rise time.

2. Reducing them produces the opposite effects; a rough control is achieved, which produces a slower response with less overshoot.

3. The variation in output gain has not yet been analyzed in depth. Its increase (to compensate by gain) improves the control by making the response more rapid, but it can also make the system less stable since it increases overshoot. Decreasing it can help to eliminate any initial instability of the closed loop system. It is the most destabilizing factor and significantly influences the convergence.

With this initial analytic adjustment, we obtain some parameters approaching the optimal ones. If the scale factors are fixed to the static values (9), the behaviour is good, but it improves notably when faced with sudden changes in the set-point if we make it *adaptive*, and the gains vary with time (8), modifying themselves to improve the response.

$$GR = Kp_s * 4, \qquad GE = Ki_s * 4, \qquad GU = 1 \tag{9}$$

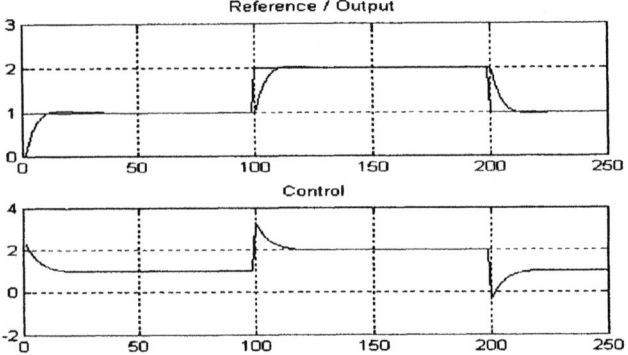

Fig. 3. Process output and control signal with a FLC-PI

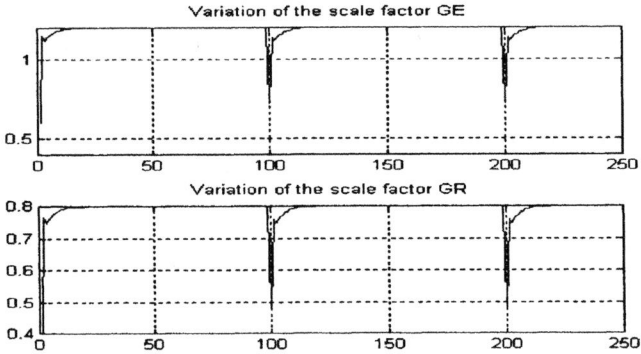

Fig. 4. Gain's variations

4.1 Self-Tuning Mode

With a supervisor module capable of monitoring changes in the process or disturbances, a self-tuning mode can be set up. Masking the FLC parameters (GE, GR) with the PI controller parameters (Kp, Ki), it is feasible to reformulate the first ones as a function of the PI ones (8), and then apply any classical PID self-tuning method to the new process, to estimate the new parameters of the FLC.

Figure 5 shows the response of a system with delay controlled by means of a fuzzy system. When the delay in the process varies suddenly, the process varies its behaviour, worsening its response. Estimating this new value and applying the Ziegler and Nichols tuning strategy, we obtain the new parameters of the conventional regulator, which are easily formulable for the fuzzy controller. The GE and GR gains have varied to adapt themselves to the changes in the dynamic of the system.

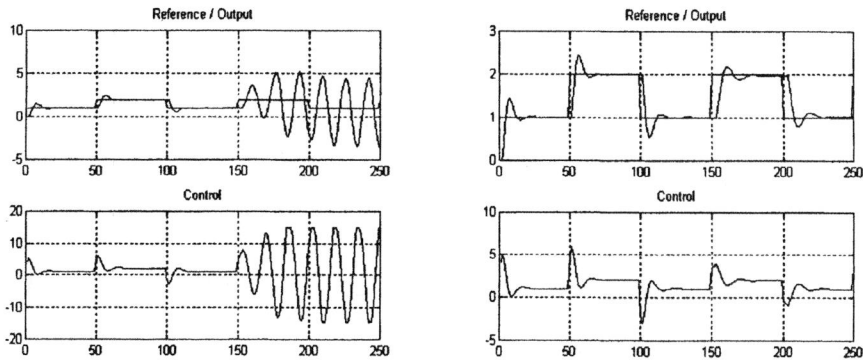

Fig. 5. Self-tuning of a FLC when the system delay changes

4.2 On-Line Assessment of FLC Control Performance

The empirical formulas obtained in [1] are reformulated, to be a function of the fuzzy controller parameters. That allows the establishment of a relation between them and the characteristic of the system, providing a tool to evaluate the performance of this type of controller, when it is applied to a restricted class of processes.

Once GE and GR are calculated for a given process, it is possible to determine the resulting normalized deadtime θ, which indicates if the required control is achieved, and so on the rest of the test characteristics.

In the stationary case, the scale factors are calculated by (9). For these values, the normalized deadtime θ is given by the expressions:

$$\theta = \frac{3.6}{K * GR * GU} \qquad \text{or} \qquad \theta = \frac{1.08}{K * D * GE * GU} \qquad (10)$$

a) For the systems shown in Figure 3, a first order process, the values of the FLC gains are:

$GE = 1.2$
$GR = 0.8$

With these values, the normalized deadtime is $\theta = 0.45$. The control obtained with the FLC is a tight control, although derivative action will improve its performance.

b) In a system with a small delay, as it is shown in the second graphic of Figure 5, the resulting values of the scale factors are:

$GE = 0.27$
$GR = 0.18$

When the delay changes, the self-tuning values are now:

$GE = 0.67$
$GR = 0.89$

The achieved control obtained with this FLC is similar to that described above, because $\theta = 0.2$ in the first case and $\theta = 0.40$ after the delay changes. Performance can be increased significantly by using derivative action. When the overshoot is excessive, setpoint weighting is recommended to obtain high quality FLC control.

c) For a new system, with a larger delay, the FLC does not control. The obtained scale factors are:

$GE = 0.075$
$GR = 0.3$

and the normalized deadtime is 1.2. As it is greater than 1, PID control based on Ziegler-Nichols or the equivalent FLC is not recommended.

5 Conclusions

The first conclusion that comes to mind is the lack of a systematized tuning procedure for Fuzzy Logic Controllers. Stability, efficiency, performance, etc., depend on how the controller parameters are selected, but there is not a unique solution to this problem: some of the parameters are redundant, the controller is non linear, etc., so tuning is a difficult and time-consuming task. However, great efforts have been carried out in this sense, but the proposed solutions are only valid in specific situations, and the problem is still open.

More general solutions are necessary, but the plant operator must feel comfortable with the new proposed procedures. Most industrial controllers are PID controllers. In a conventional PID controller, only three parameters are needed, and the plant operators know the direct relation between them and the response objectives as their characteristics. So, new tuning methods should try to keep the PID look-and-feel if we want self-tuned fuzzy controllers move from Academia to industry.

Acknowledgment: This work has been partially supported by the Spanish CICYT under project TAP 122/92.

References

1. K. J. Aström, C. C. Hang, P. Person, and W. K. Ho: Towards intelligent PID control. Automatica 28, 1, 1-9 (1992).

2. J. J. Buckley: Universal fuzzy controllers. Automatica 28, 6, 1245-1248 (1992).

3. H. Ying, W. Siler, J. J. Buckley: Fuzzy control theory: a nonlinear case. Automatica 26, 3, 513-520 (1990).

4. H. Ying, J. J. Buckley: Fuzzy ontroller theory: Limit theorems for linear fuzzy control rules. Automatica 25, 3, 469-472 (1989).

5. S. Dormido, M. Santos, A. P. de Madrid, F. Morilla: Autosintonía de controladores borrosos utilizando técnicas clásicas basadas en reguladores PID. Proc. of III FLAT, España, 1993, pp. 217-225.

6. F. Morilla, S. Dormido, J.L. Fernández, M.A. Canto: A systematic study of PID controllers tuning methods. Proc. IASTED Int. Symp. Modeling Identification and Control, pp. 383-386, 1989.

7. F. Morilla: Controladores PID: ajuste de parámetros. Automática e Instrumentación, 207, 155-160 (1990).

8. K.J. Aström: Automatic tuning of PID Controllers. Instrument Society of America, 1988.

9. J.G. Ziegler, N.B. Nichols: Optimun setting for automatic controllers. Trans. ASME 64, 759-768 (1942).

10. M. Santos: Contribución a las técnicas de sintonía de los controladores basados en la Lógica Borrosa. PhD. Dissertation, 1994.

11. C.C. Lee: A self-learning rule-based controller employing approximate reasoning and neural net concepts. Int. Intelligent Systems, 6, 1, 71-92 (1991).

12. F. Herrera, M. Lozano, J.L. Verdegay: Un algotimo genético para el ajuste de controladores difusos. III FLAT, España, 1993, 251-258.

Computer Aided Design of Protocol Converters

Hasan Ural and Huaqing Zeng

Department of Computer Science
University of Ottawa
Ottawa, Ontario, K1N 6N5, Canada
ural@csi.uottawa.ca

Abstract. A protocol converter enables the users in different networks to communicate with each other. An approach for the automatic construction of a protocol converter is proposed. This approach first constructs the service specification of the conversion system, then constructs an adapter from the service specification and finally, produces a suitable converter from the adapter.

1 Introduction

Currently, there are thousands of networks operating. Many of these networks use different protocol architectures such as SNA, DNA and OSI. To provide universal services, it is very important to interconnect these networks. Protocol conversion is a procedure to resolve the mismatch between computer network protocols so as to enable the users on different networks to communicate with each other.

If we have two mismatched protocols P and Q, as shown in Fig. 1a and 1b, the users $U_{E0P}(U_{E0Q})$ and $U_{E1P}(U_{E1Q})$ of protocol P(Q) can communicate with each other through protocol entities $E0_P(E0_Q)$ and $E1_P(E1_Q)$. If we wish U_{E0P} to communicate with U_{E1Q}, we have two possible approaches. The first approach is the *adapter approach* where an adapter D, as shown in Fig. 1c, is placed at the same layer of U_{E0P} and U_{E1Q}. Any message sent among these two users is relayed by the adapter D. The second approach is the *converter approach* where a converter C is placed at the same layer of protocol entities, $E0_P$ and $E1_Q$, as shown in Fig. 1d. The second approach is more popular since with the first approach the adapter may not be transparent.

In 1986, Green considered the general problem of protocol conversion and examined many of its practical aspects in his pioneering work [8]. He pointed out that no general solution methodology is known and suggested that the formal methods used in specification and verification of protocols might form the basis for a deeper and more systematic conversion. Since then, some approaches based on formal methods have been proposed in [1-7], [10-13], [15-16], [18-20]. However, none of the previous work suggested fully automatic generation of a converter as they all assumed that the service specification of the conversion system (or the message mapping function between two protocols) is given or derived manually [11].

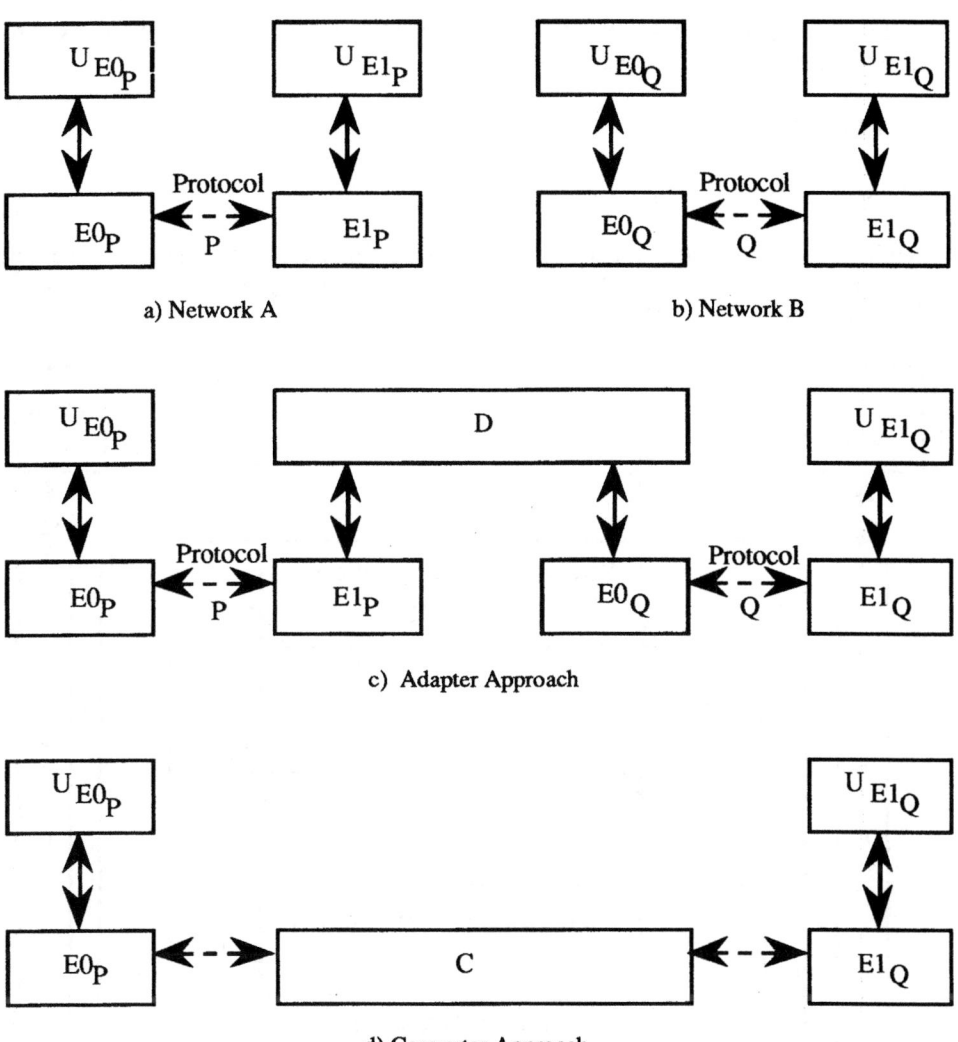

Fig. 1. Protocol Conversion Problem

In this paper, we propose an approach to construct a protocol converter automatically. Given specifications of two different protocols and specifications of services required and supplied by both protocols together with suitable interface adapters, the major steps of the proposed approach are:
a) generate the service specification of the conversion system
b) construct an adapter for the conversion system
c) generate a converter from the adapter for the conversion system
d) find a subsystem of the converter derived in c) which has no deadlocks and no unspecified receptions.

Section 2 gives the terminology in a layered protocol architecture and the definition of the protocol conversion problem. Section 3 prescribes the model and formalization used in this paper and describes the proposed approach. Section 4 presents an example. Section 5 gives the concluding remarks.

2 Preliminaries

2.1 Architecture

In a layered protocol architecture, e.g., ISO/OSI reference model, there are three kinds of protocol interactions involving a given layer N, namely, Peer Protocols, Control Protocols, and Interface Protocols.

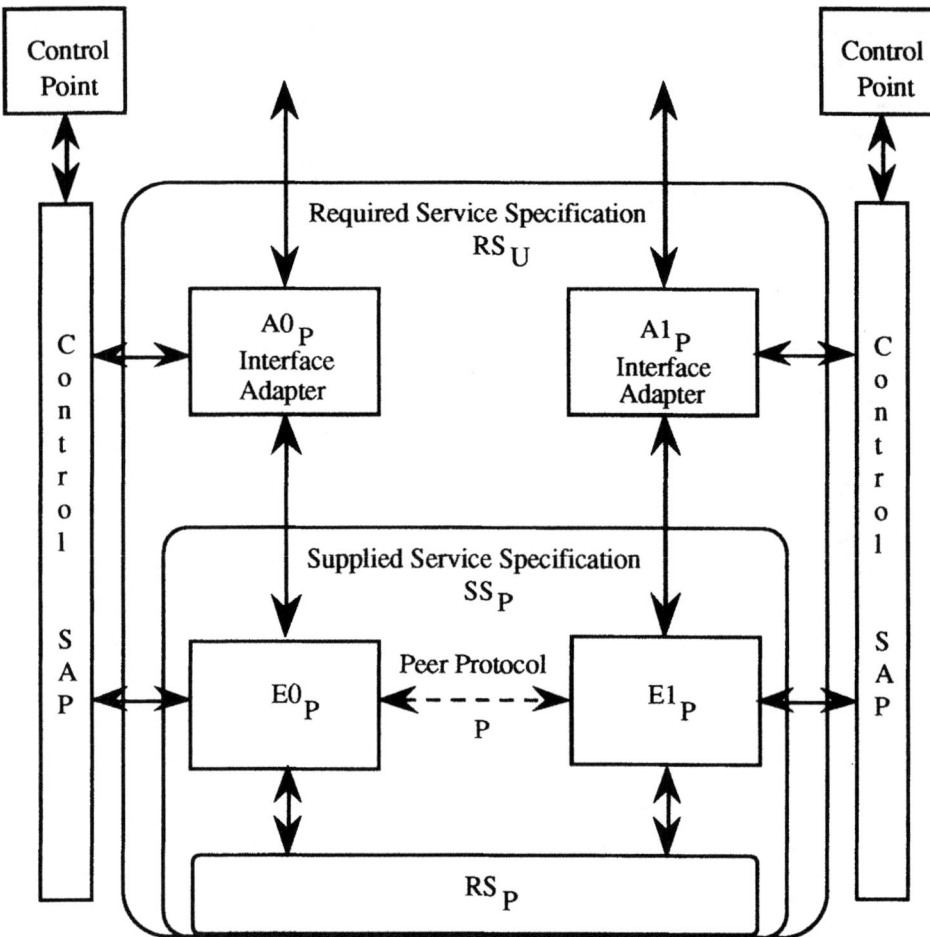

Fig. 2. A Layered Protocol Architecture

The peer protocols are for the message delivery between communicating entities in the same layer. For a peer protocol of layer N, the communicating entities are located in separate nodes of the network and the exchange of messages takes place by means of the service supplied by the lower layer. Fig. 2, shows an example for peer protocols where two protocol entities E0p and E1p utilize the *required service* RSp and provide the *supplied service* SSp to their upper layer. Usually, given a network architecture, the peer protocols of all layers are fully documented as standards, such as X.25, TCP/IP and so on.

The control protocols are those that command the instantiation, initialization, activation, deactivation, test and recovery of protocol entities in all layers. They also set up the parameters and tables in various layers such as X.25 user facility parameter and TCP/IP address translation. In each node of a network, much of this control function is collected into a single point. The control point interacts with other control points distributed in networks or may be controlled by Network Control Center (NCC), if it exists. Since the control protocols are usually poorly and inconsistently documented, it is very difficult to find a formal method to solve the conversion problem between control protocols. Usually, the control points, which are distributed over all the nodes in the network, exchange the network condition information with each other and control the protocol entities and adapters in the node by changing the configuration tables and some special signaling commands such as RESET and RESTART, according to the network condition information they receive. In this paper, we will not discuss how to design the protocol converters for control protocols.

The interface protocols are those that are between protocol entities in two consecutive layers. When we observe the existing networks, we can find many examples of different protocols running in the same layer, e.g., HDLC and SDLC, two different Data Link Layer protocols. When we design a protocol entity, we not only design the interactions between this entity and its peer but also consider the default required service for the protocol entity, e.g., X.25 packet layer protocol is designed for running on top of HDLC. But in some networks, X.25 protocol may run on top of a lower layer protocol which is not HDLC but say, SDLC, so there is a gap between the protocol's required service and the service supplied by lower layer. The interface protocols fill this gap. As shown Fig. 2, there are interface protocols over peer protocol entities E0p and E1p, which are called interface adapters, A0p and A1p.

2.2 Protocol Conversion Problem

The formulation of a protocol conversion problem between Network A and Network B, is based on the following assumptions:
a) Layer M protocol (called protocol P) of Network A and Layer N protocol (called protocol Q) of Network B are two different protocols.
b) There is a common peer protocol U which is Layer M+1 protocol of Network A and Layer N+1 protocol of Network B.
c) The required service RS_U for the common peer protocol U is defined.

In Network A, as shown in Fig. 3a, E0p and E1p are two peer protocol entities, and together with the required service RSp of protocol P, they can provide the supplied service SSp of protocol P. Since in the general case, there may be a gap between SSp and RS_U, there may be a need for two interface adapters A0p and A1p. Based on RSp, protocol entities E0p and E1p together with interface adapters A0p and A1p can supply RS_U. Similarly, in Network B, as shown in Fig. 3b, based on RS_Q, protocol entities $E0_Q$ and $E1_Q$ together with interface adapters $A0_Q$ and $A1_Q$ can supply RS_U.

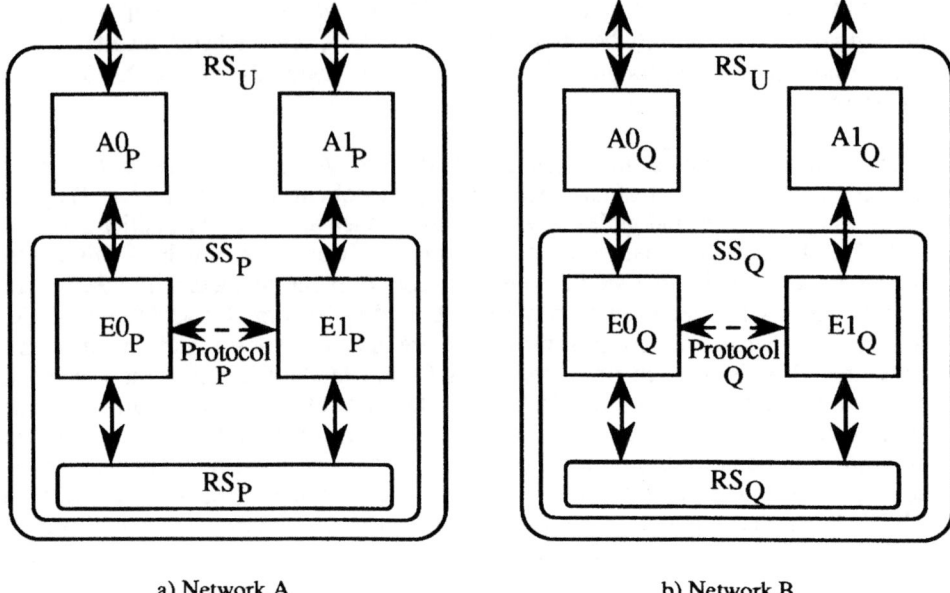

a) Network A   b) Network B

Fig. 3. Two Network Architectures

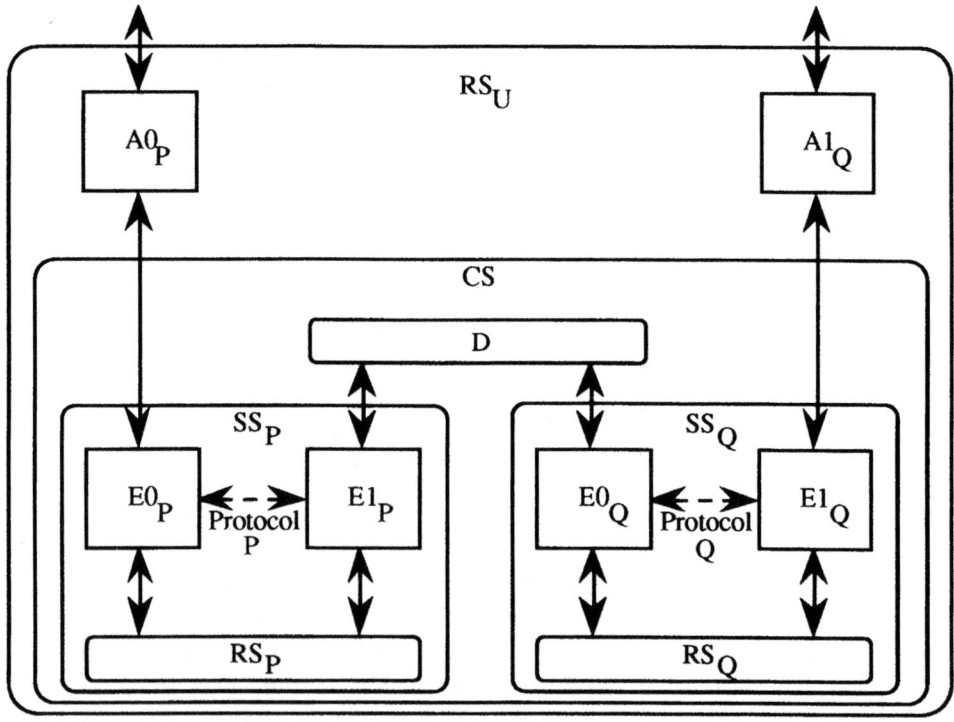

Fig. 4. The Adapter Approach

As stated earlier, to solve the protocol conversion problem between Network A and Network B, there are two approaches. In the adapter approach, we must construct an adapter D (as shown in Fig. 4) which together with SS_P, SS_Q, $A0_P$ and $A1_Q$ also supplies RS_U. This means that any entity in Network A using RS_U service can communicate with any entity in Network B using RS_U service. Here the service supplied by SS_P, D and SS_Q is called the *service specification of the conversion system, CS*. In the converter approach, we must construct a converter C (as shown in Fig. 5) which together with $E0_P$, $E1_Q$, $A0_P$, $A1_Q$, RS_P and RS_Q also supplies RS_U. This means that any entity in Network A using RS_U service can also communicate with any entity in Network B using RS_U service. The service supplied by $E0_P$, C and $E1_Q$ based on RS_P and RS_Q is the same as CS, that is, $E0_P$, C and $E1_Q$ together compose a new protocol whose service specification is CS. Clearly, observed from $E0_P$, the converter C acts as $E1_P$ and observed from $E1_Q$, the converter C acts as $E0_Q$. For the converter to be used in networks, it should be free from deadlocks and unspecified receptions which are undesirable properties of communication systems.

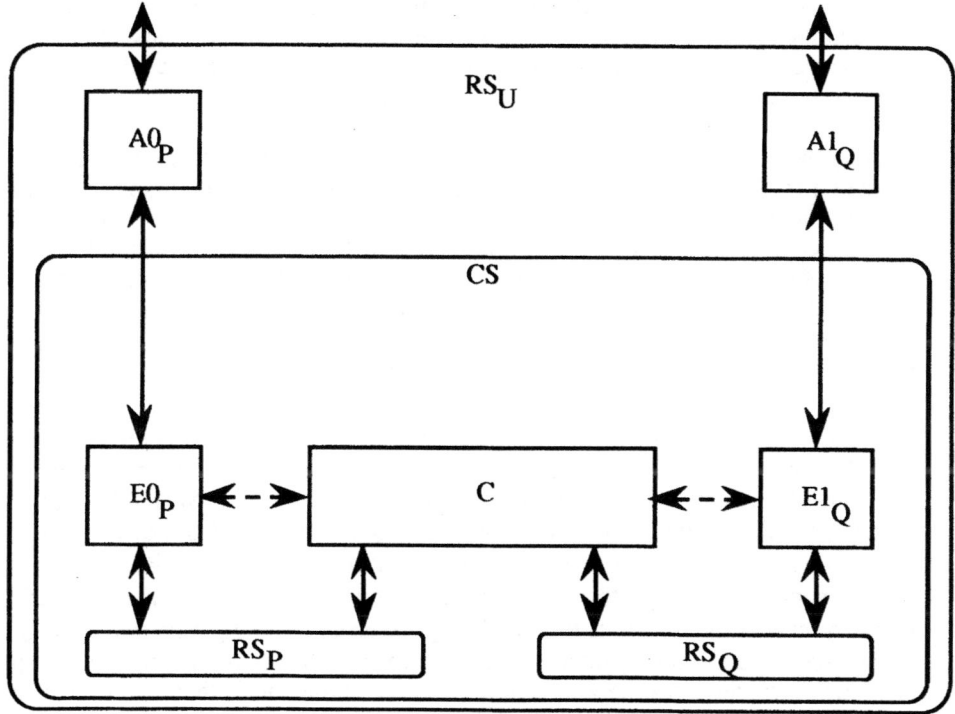

Fig. 5. The Converter Approach

3   Protocol Converter Construction

In this paper, a communication system is modeled as a collection of communicating finite state machines (CFSMs) and thus each protocol entity, interface adapter, service specification is represented by a CFSM.

A CFSM X is defined as a quadruple $X = (S, M, \delta, q_0)$ where

1. S is a finite, non-empty set of states;
2. M is a finite set of messages which consists of the *send message set* M^- and the *receive message set* M^+. $M = M^- \cup M^+$ and $M^- \cap M^+ = \varnothing$. Based on the components of M, a set V is defined as the *event set* of X, where

   $V = \{e \mid\ e = -m$ if $m \in M^-$ i.e., e is the transmission of message m or

   $e = +m$ if $m \in M^+$ i.e., e is the reception of message m$\}$
3. $\delta: S \times V \to S$ is a state transition function that defines the next state of X, $\delta(s, e)$, when event $e \in V$ occurs at the current state $s \in S$;
4. state $q_0 \in S$ is the initial state of X.

A CFSM $X = (S, M, \delta, q_0)$ can be represented by a directed graph where the set of nodes corresponds to S and the set of edges corresponds to δ, i.e., the labels of edges correspond to the events of X.

A composition operator $\otimes$ over two CFSMs F1 and F2 defines a CFSM $F = F1 \otimes F2$ in a manner similar to that of CSP [9]. We only consider the case that two CFSMs will not send the same message or will not receive the same message. Message m sent by F1 and received by F2 is called a *matching message* and the event related to a matching message is called a *matching event*, as in CSP [9].

Using the composition operator, one can build a procedure, called *Cons*, that can be utilized for the solution of the following subproblem (Submodule Construction Problem) which is used in the solution of the protocol conversion problem [14]:

Given a specification Z of a module which consists of K submodules
$X_1, X_2, ..., X_K$ $(K \geq 2)$ and given the specifications of submodules
$X_1, X_2, ..., X_{K-1}$, construct the specification of the submodule X_K.

The specification of the submodule X_K can be constructed by using the procedure Cons as follows: $X_K = Cons(X_1 \otimes X_2 ... \otimes X_{K-1}, Z)$.

In this paper, we suppose that the required service of a protocol is directly coupled with the protocol entities which use the service. Also, each protocol entity is assumed to be directly coupled with the interface adapter directly above it. With this assumption in mind, consider two different protocols P and Q. In protocol P, protocol entities $E0_P$ and $E1_P$ as well as the required and supplied service specifications RS_P and SS_P are given. In protocol Q, protocol entities $E0_Q$ and $E1_Q$ as well as the required and supplied service specifications RS_Q and SS_Q are given. Let RS_U represent the required service specification for a protocol U. For protocol P, there are two interface adapters $A0_P$ and $A1_P$ which use SS_P to supply RS_U. For protocol Q, there are two interface adapters $A0_Q$ and $A1_Q$ which use SS_Q to supply RS_U. Assume that the suitable interface adapters $A0_P$, $A1_P$, $A0_Q$ and $A1_Q$ exist. Then, the steps of the proposed approach are:

Step 1: generate the service specification CS of the conversion system
In order to construct the conversion system, we must define the service specification of the conversion system. As shown in Fig. 4, the interface adapters $A0_P$ and $A1_Q$ can supply RS_U based on the service specification CS. In this paper, we discuss only one way traffic for ease of presentation since the traffic in the other direction can be carried

232

out in the same way. The service specification of the conversion system is a semantic regulation between the event sets of two protocols [15]. Since RS_U, $A0p$, $A1_Q$ are given, the service specification CS of the conversion system can be obtained as CS = $Cons(A0p \otimes A1_Q, RS_U)$. If CS = $\emptyset$, this means that we can not find the service specification of the conversion system of protocols P and Q for RS_U.

Step 2: construct a suitable adapter for the conversion system
Let CS be the service specification of the conversion system and let SSp and SS_Q be the specifications of supplied service of protocols P and Q, respectively. Then, an adapter D for the conversion system can be obtained as D = $Cons(SSp \otimes SS_Q, CS)$. If D = $\emptyset$, protocol conversion by adapter is not possible.

Step 3 : construct a converter C from the adapter for the conversion system
The adapter D constructed in Step 2 can be used for the conversion system through the adapter approach. If we can use an adapter D for the conversion system, then we can construct a converter C to solve the conversion problem through the converter approach as C = $E1p \otimes D \otimes E0_Q$.

Step 4. find a suitable sub-CFSM C' of the converter derived in step 3
CFSM C derived in step 3 is a converter for the conversion system. But it can not be guaranteed that C will be deadlock free and unspecified reception free when it is put together with $E0p$ and $E1_Q$ to compose a new protocol [14]. Okumura proved that any sub-CFSM of C is also a converter [16]. We can apply Okumura's theorem and algorithm to find a sub-CFSM of C which together with $E0p$ and $E1_Q$ will be a deadlock free and unspecified reception free protocol.

4 Example

Suppose that we have two different protocols, P and Q, in two networks where a common required service specification RS_U shown in Fig. 6 exists. According to this specification, every received data request "+Dreq" sent by user in Network A, will be responded by a data indication "-Dind" sent by user in Network B.

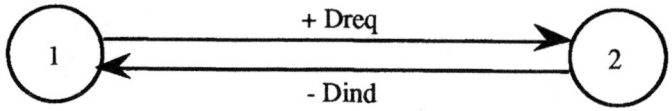

Fig. 6. Common Required Service Specification RS_U

Protocol P is the classical unidirectional Alternating Bit Protocol (ABP) which performs error recovery. The two peer ABP protocol entities are shown in Fig. 7a, 7b, where service primitives are given by upper-case letters. Two entities are based on an unreliable medium, shown as Fig. 7c which loses messages (data or ack). The service specification SSp of the ABP supplied by the entities is shown in Fig. 7d). Since there is a gap between SSp and RS_U, we assume the existence of two interface adapters $A0p$ and $A1p$ as shown in Fig. 7e and Fig. 7f. Obviously, $A0p \otimes SSp \otimes A1p = RS_U$ and $E0p \otimes Medium \otimes E1p = SSp$.

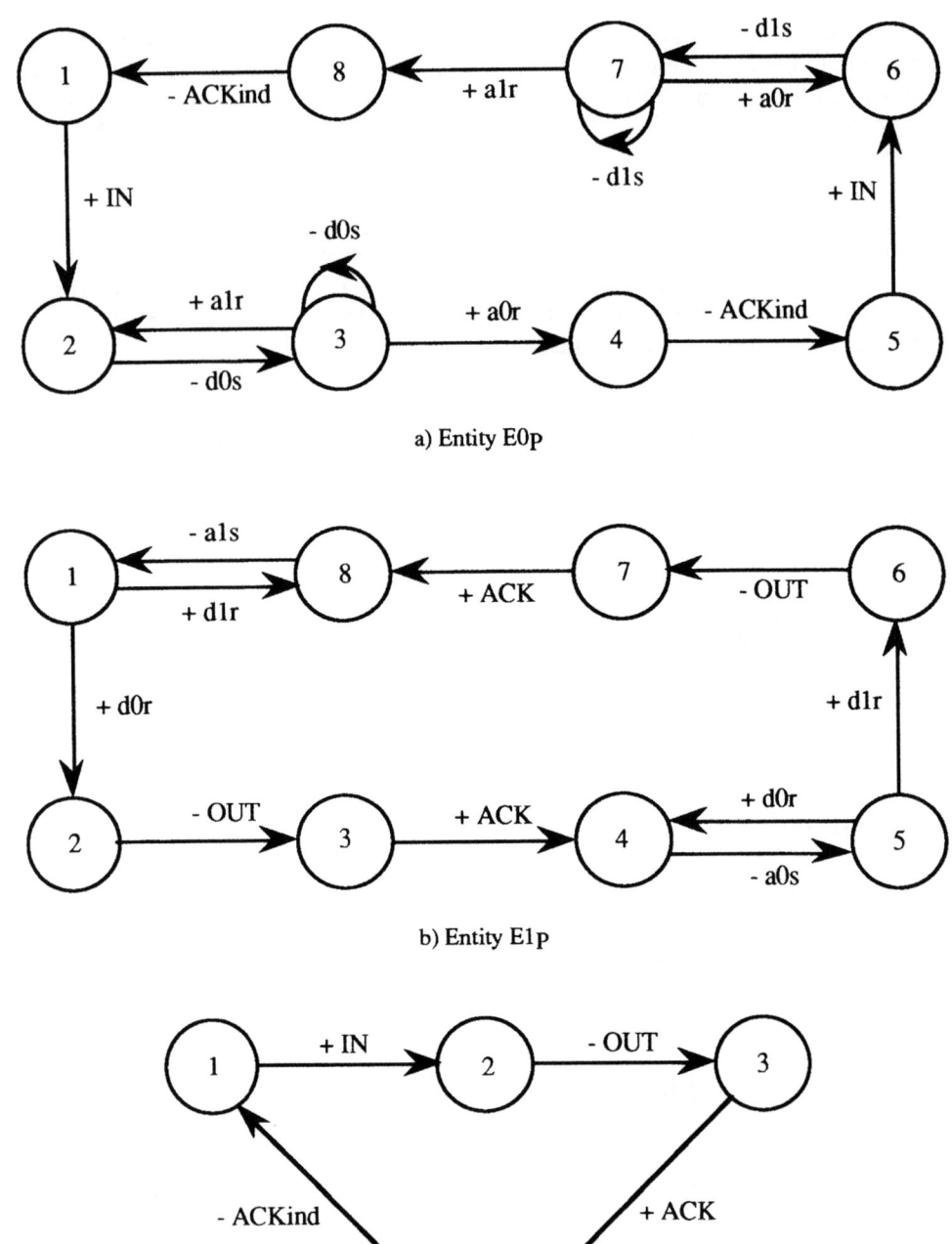

a) Entity E0p

b) Entity E1p

d) Specification of Supplied Service SSp

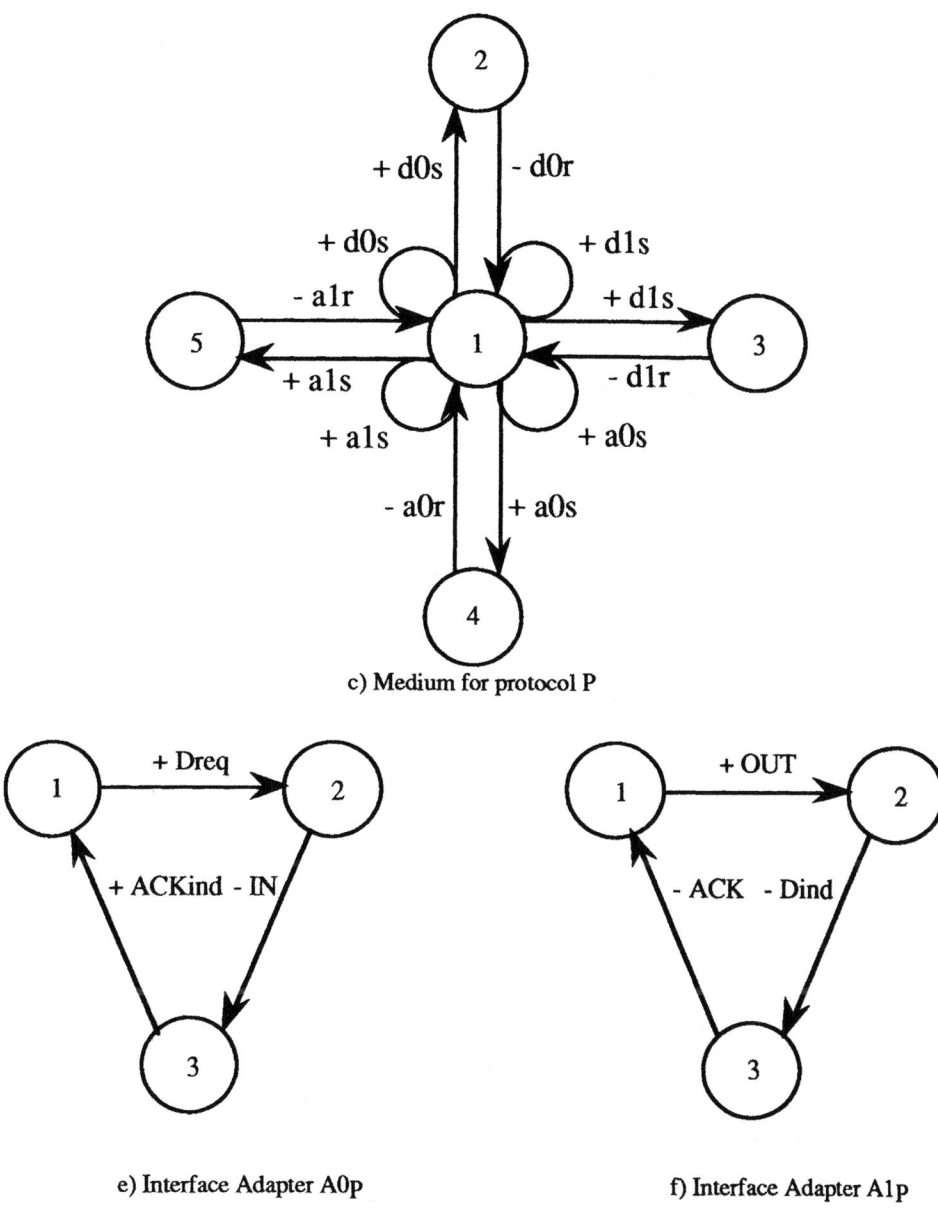

c) Medium for protocol P

e) Interface Adapter A0p   f) Interface Adapter A1p

Fig. 7. Protocol P : Alternating Bit Protocol

On the other hand, protocol Q is a simple data transfer protocol with empty medium, that is, two entities $E0_Q$ and $E1_Q$ are directly coupled. Consequently, no error recovery is performed in this protocol. The two peer entities of protocol Q are shown in Fig. 8a and Fig. 8b. The service specification SS_Q is shown in Fig. 8c. The interface adapters $A0_Q$ and $A1_Q$ for the gap between RS_U and SS_Q are shown in Fig. 8d and 8e. One can verify that $A0_Q \otimes SS_Q \otimes A1_Q = RS_U$ and $E0_Q \otimes E1_Q = SS_Q$.

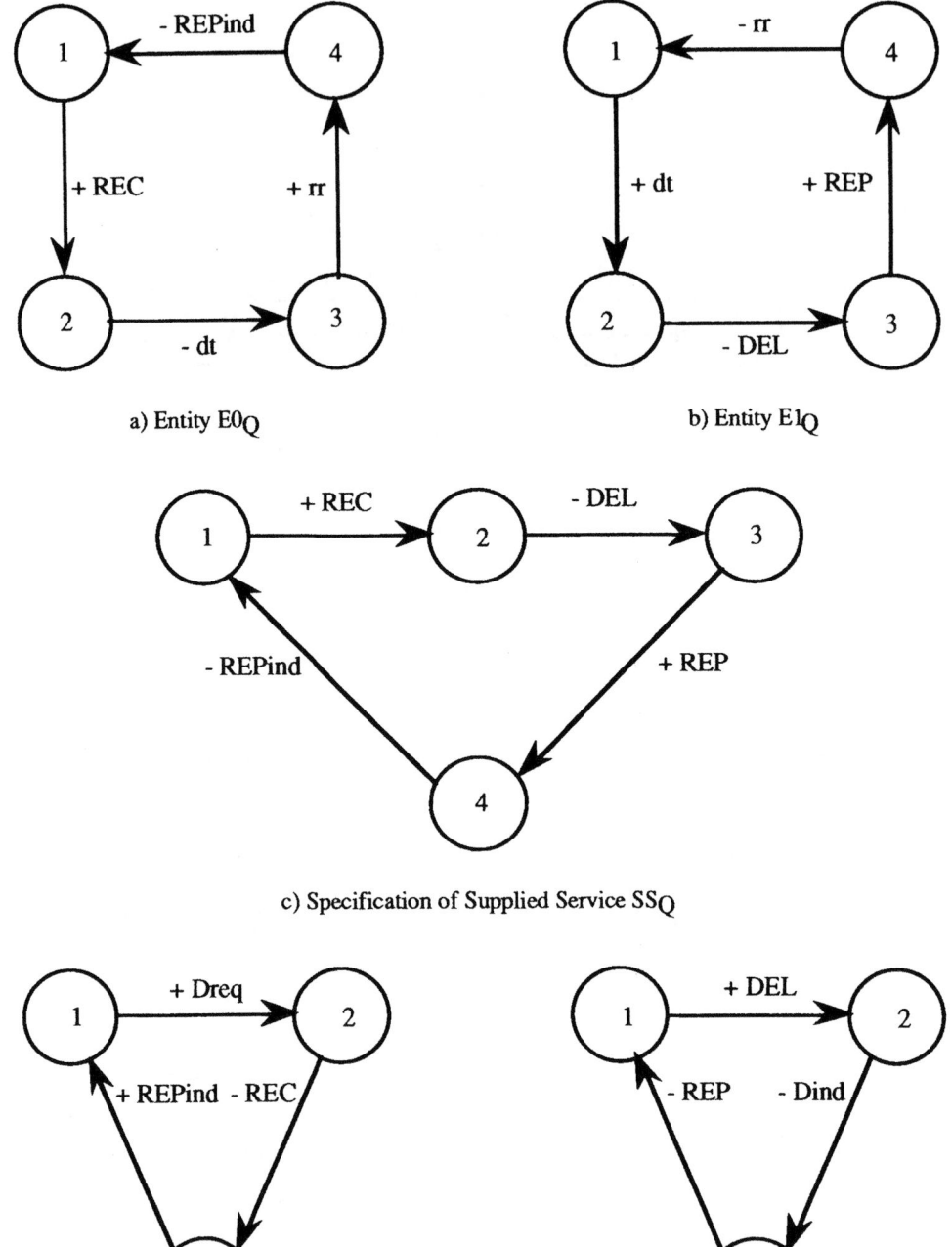

a) Entity E0Q

b) Entity E1Q

c) Specification of Supplied Service SSQ

d) Interface Adapter A0Q

e) Interface Adapter A1Q

Fig. 8. Protocol Q : Simple Protocol

Now we apply Step 1 of our approach, CS = Cons(A0p ⊗ A1Q , RSU), and get CS as shown in Fig. 9. CS gives the relation between the service primitives "+IN" and "-ACKind" of Protocol P and the service primitives "-DEL" and "+REP" of Protocol Q. One can verify that A0p ⊗ CS ⊗ A1Q=RSU, so CS is a suitable service specification of the conversion system between protocols P and Q for RSU.

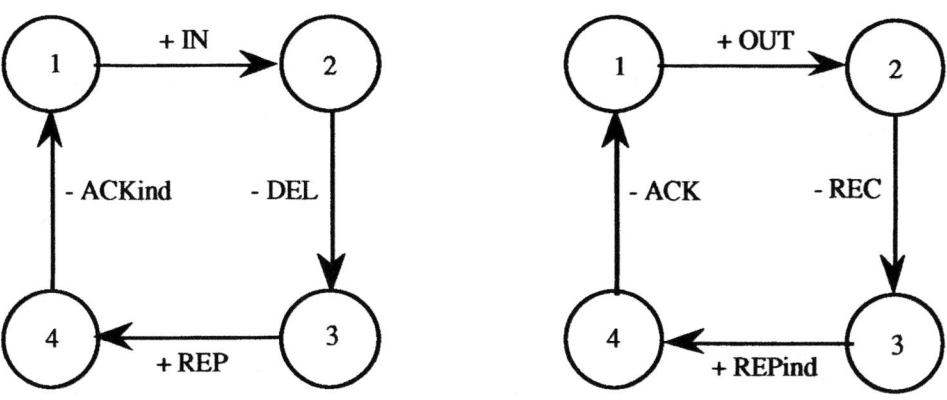

Fig. 9. Service Specification   **Fig. 10.** Adapter D

Then, we continue with Step 2 and construct the adapter D by D = Cons(SSP ⊗ SSQ, CS). CFSM D is shown in Fig. 10. One can verify that SSP ⊗ D ⊗ SSQ = CS. This means that D is an adapter for the conversion system for protocols P and Q.

Now, we can use protocol entities of protocols P and Q to generate the protocol converter C as C = E1p ⊗ D ⊗ E0Q as in Step 3. The converter C constructed by this approach is shown in Fig. 11. Obviously, together with E0p and E1Q, C can form a new protocol whose service specification is CS. C together with A0p, E0p, E1Q and A1Q can supply the required service specification RSU. One can verify that C already has the deadlock freedom and unspecified reception freedom properties when put together with E0p and E1Q. So, we need not perform Step 4 and C is a suitable protocol converter for protocols P and Q.

5   Conclusions

We have proposed an approach to solve the protocol conversion problem automatically. The converter derived by this approach may also be used in a different conversion system for protocols P and Q if we select a suitable common required service specification RSU. We used the submodule construction method in this approach. This method has exponential time complexity with respect to the size of the input module and submodules in the worst case [4][14][16]. Thus, our approach has also exponential time complexity in the worst case. However, we only use submodule construction method in Step 1 and Step 2. In Step 1, we construct CS from RSU, A0p and A1Q, in Step.2, we construct D from CS, SSP and SSQ.

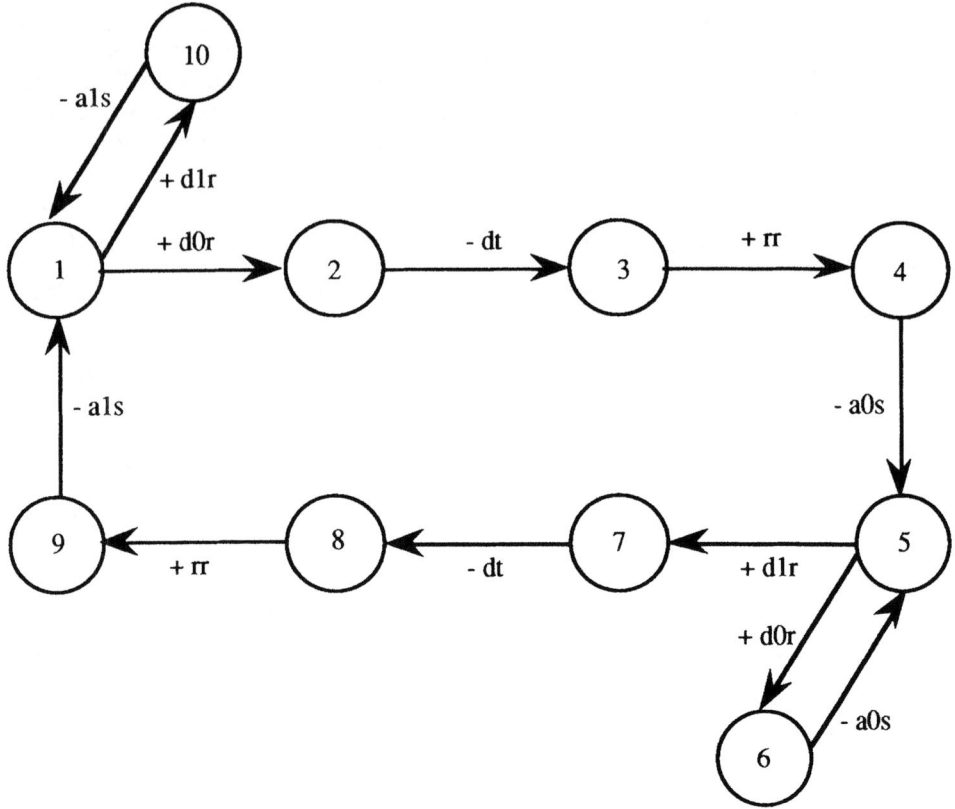

Fig. 11. Converter C for Protocols P and Q

Usually, the interface adapters and service specifications are simple, so the time consumed in these two steps is reasonable in most cases. It is of current interest to find a procedure that will not yield the exponential time complexity.

References

[1]   G.v. Bochmann and P. Mondain Monval, "Design Principles for Communication Gateways", IEEE Journal on Selected Areas of Communications, Vol.8, No.1, pp. 12-21, 1990.

[2]   G.v. Bochmann, "Deriving Protocol Converter for Communications Gateways", IEEE Trans. on Communications, Vol.38, No.9, pp. 1298-1300, 1990.

[3]   K.L. Calvert and S.S. Lam, "Adapters for Protocol Conversion", Proc. IEEE INFOCOM'90, pp. 552-560.

[4]   K.L. Calvert and S.S. Lam, "Formal Methods for Protocol Conversion", IEEE Journal on Selected Area in Communications, Vol.8, No.1, pp. 127-142, 1990.

[5]   K.L. Calvert and S.S. Lam, "Deriving a Protocol Converter: a Top-Down Method", Proc. ACM SIGCOMM'89, pp. 247-258.

[6]   K.L. Calvert and S.S. Lam, "An Exercise in Deriving a Protocol Converter", Proc. ACM SIGCOMM'87, pp. 151-160.

[7]   J. Chang and M. T. Liu, "Using Protocol Validation Technique to Solve Protocol Conversion Problems", Proc. IEEE IPCCC'90, pp. 539-546.

[8]   Paul E. Green,Jr, "Protocol Conversion", IEEE Trans. on Communications, Vol.34, No.3, pp. 257-268, 1986.

[9]   C.A.R. Hoare, *Communicating Sequential Process*, Prentice-Hall,1985.

[10] S.G. Kelekar and G.W. Hart , "Synthesis of Protocols and Protocol Converters Using the Submodule Construction App. roach", Proc. IFIP Protocol Specification, Testing and Verification, XIII, pp. F1-1-16, 1993.

[11] D. M. Kristol, D. Lee, A. N. Netravali and K. K. Sabnani, "Efficient Gateway Synthesis from Formal Specification", Proc. ACM SIGCOMM'91, pp. 89-97.

[12] S.S. Lam, " Protocol Conversion", IEEE Trans. on Software Engineering, Vol.14, No.3, pp. 353-362, 1988.

[13] S.S. Lam "Protocol Conversion- Correctness Problem", Proc. ACM SIGCOMM'86, pp. 19-28.

[14] P. Merlin and G.v. Bochmann, "On the Construction of Submodule Specification and Communication Protocols", ACM Trans. on Programming Languages and Systems, Vol.5, No.1, pp. 1-25, 1983.

[15] K. Okumura, "A Formal Protocol Conversion Method", Proc. ACM SIGCOMM'86, pp. 30-38.

[16] K. Okumura, "Generation of Proper Adapters and Converter from A Formal Service Specification", Proc. ACM SIGCOMM'90, pp. 564-571.

[17] J.C Shu and M.T. Liu, "Approach to Indirect Protocol Conversion", Computer Networks and ISDN System, Vol.21, No.2, pp. 93-108, 1991.

[18] Y.W. Yao, W.S. Chen and M.T. Liu, " A Modular Approach to Constructing Protocol Converters", Proc. IEEE INFOCOM90, pp. 572-579.

[19] Y.W. Yao and M.T. Liu, " Constructing Protocol Converter with Guaranteed Service", Proc. IEEE INFOCOM91, pp. 960-969.

[20] Y.W. Yao, W.S. Chen and M. T. Liu, "A Parallel Model for Constructing Protocol Converters", Proc. IEEE GLOBECOM'90, pp. 1902-1907.

3   CAST Tools and Environments

Modelling, Analysis and Evaluation
of Systems Architectures

Ernst Schmitter
Siemens AG Corporate Research and Development
ZFE ST SN 1, 81730 Munich, Germany

Abstract. Systems architectures get normally more and more complexity be increasing functionality, by implementing additional features, by using microelectronic and digital components, by dominant processing of signals and information, by software-based applications, by optimizing the price/performance ratio etc. The realization of a lot of industrial systems is based on the use of processors, memories, I/O components, communication nets, programs, and application workloads. This implies a common system view with comparable structures, similar modelling approaches, and computer aided system evaluation. The paper discusses an analytical solution in assistance of systems architectures design and represents practical experiences using stochastic Petri nets realized by a tool TOMSPIN (TOol for Modelling based on Stochastic PetrI Nets). Motivated to introduce a powerful methodology in CAST the presentation will give an overview of the concept and its use.

1. Introduction

The development of systems architectures has become very complex and difficult due to the demand of today´s applications for extensive functionality, for high quality, for high performance and reliability as well as for the application in quite different areas. In order to deal with this increasing complexity of the resulting systems the developer needs assistance in designing the system structure and its architecture, in implementing the adequate reliability and necessary dependability, in analyzing the system environment and its specified application, in establishing the quality of the overall system architecture, and in optimizing the relations within the system architecture. Complying with such requirements stochastic Petri nets as a representation of graphical modelling formalism for performance and dependability analysis of complex systems deal with the architecture design especially on the system level.

It is obvious that the development of innovative systems has to be improved by new methods and powerful tools, which support system modelling, system analysis, and system evaluation during its whole life cycle from the specification to the product use. Additionally these methods and tools should support the reducing of the system complexity and the improving of CAST. In order to develop systems by methods of modelling and by use of tools first of all you have to specify a concept of an adequate design and development process which enables you to get a global view of the overall system architecture including hardware components, software modules, environment and application.

Figure 1 shows such an approach demonstrating this process: based on different system requirements and on application areas the system designer specifies the system functionality and its realization, defines a expected system behavior, and gets a first system design by using known design methods and architecture measures like fault tolerance, through put, performance/cost relation. Based on given boards, subsystems, components, communication relations and medium etc. you describe a system architecture by specifying the system structure and behavior but also a scenario of its realization. For a real solution you also need a description of the system environment defined by I/O periphery, special components or processors, system conditions and restrictions etc. Optimizing the system architecture and its design the knowledge of system workload defined by the application is also necessary which can be estimated or supported by analysis of measurements monitoring the system. In the same way system monitoring of real systems is also a basis of performance analysis tuning different system criteria. With input data you can model the system on an abstract level depending on the questioning of the development process and system requirements. Using adequate methods and tools the system evaluation is started to determine the system criteria

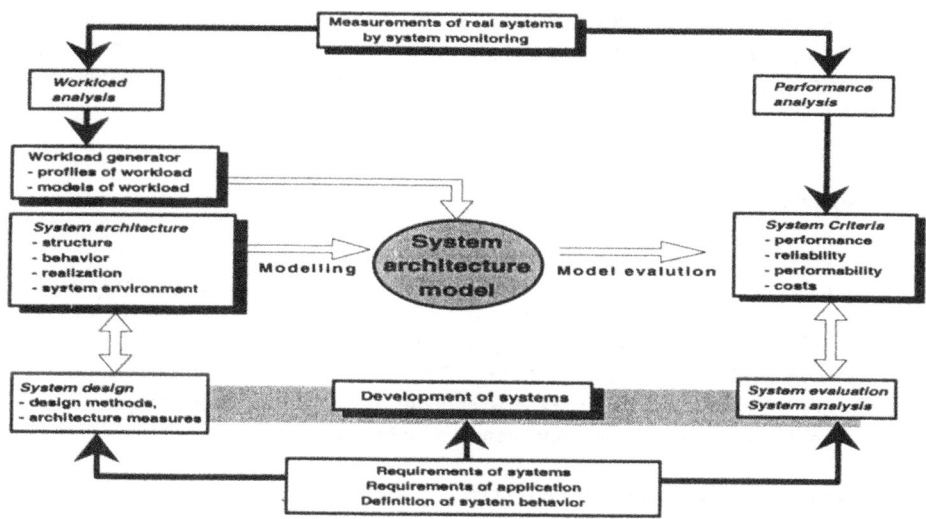

FIGURE 1: ITERATIVE SYSTEMS ARCHITECTURES DESIGN

During system development and its different design steps system developers have to analyse alternative architectures in order to make sound design decisions with respect to given system requirements (reliability, performance) and the real application. Adequate models allow the user to make meaningful conclusions about the overall behavior. The results of system evaluation give evidence about the consequences of architectural decisions and, thus, support the correctness and the optimization of the system design without expensive synthesis/analysis iterations. Using these models during further phases of the system life cycle especially system evaluation allows to get insight in the actual system behavior and to identify system weak points and bottlenecks that are possible places for architectural evolutions /8/.

2. Analytical methods within the system technology

2.1 Stochastic Petri nets

In the history of system modelling you can find different approaches. Besides the traditional simulation analytical methods become more and more important especially stochastic Petri nets. Originally invented in the context of communication systems they allow the modelling and analysis of systems which comprise properties such as synchronization and concurrency. Today, the theory of Petri nets is well developed and many qualitative properties like possibilties of deadlocks or reachability of system states can detected, and various extension have been proposed for quantitative performance analysis and dependability prediction /3/.

A stochastic Petri net provides a graphical representation of a discret event dynamic system. In order to obtain these quantitative results, the underlying stochastic process has to be analysed by numerical methods. The class of probability distribution functions and the structure of the stochastic Petri nets determine the structure of the underlying stochastic process and therefore the applicability of certain numerical methods, e.g. the solution of the matrix integral equations. Discussing the state of the art, advantages and disadvantages can be listed:

- Advantages of stochastic Petri nets

  - good description of all states of the systems;
  - method to analyse and to evaluate parallel activities including synchronization and parallelism;
  - identification of deadlocks and blocking states of systems;
  - numerical model solution with high flexibility;
  - consideration of different experiments during one model evaluation;
  - analysis of both transient and steady state behavior.

- Disadvantages of stochastic Petri nets

  - restrictions in describing time delays modelling deterministic system states;
  - use of probability distribution functions with median instead of exact values;
  - restriction regarding the possible structure of the net;
  - vast amount of system states occurring with a real system (state space explosion);

2.2 System evaluation by analytical methods

The computer aided system evaluation is a complex process of system modelling and model analysis as part of the iterative system design, but nevertheless a more and more important phase of system architecture development. Analytical modelling and the use of stochastic Petri nets is only one method of the methodology (Figure 2). Solving the problems of system evaluation you have to find out evaluation measures as a function of architecture parameters and (measured or estimated) workload parameters. Based on our different experiences ideas for potential evolution of current methodology can be descibed:

- Model construction methods use system structure, modularization, hierarchical modelling and combination of paradigms to handle increasing complexity of systems and their models.

- The optimization of systems evaluation is based on approaches using parallel simulation to speed up the solution, improving numerical solvers by taking into account machine architectures, extending the amount of product form stochastic Petri nets, and using approximate solutions and performance bounds.

- The management of the system architecture complexity needs the definition of the evaluation problem by abstraction which is basis for an effective model generation. New techniquies, structuring and library support improve this model generation.

- Finally the overall view of different criteria, efficient algorithms and hierarchical analysis optimize the model evaluation.

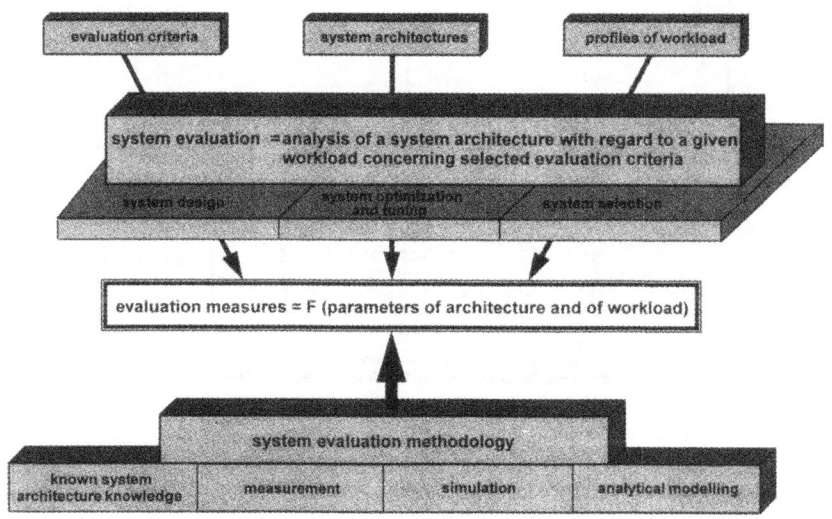

FIGURE 2: BASIS RELATIONS OF SYSTEM EVALUATION

Depending on the actual questioning the different modelling approaches have to be considered with respect to the system design and development, the adequacy of the decision level and the realization expenditure. That is followed by requirements of available tools and method features like diversity, extent and depth, control, capability for the user and realistic time constraints. Crucial aspects are the integration of tools in engineering environments and their suitability for the solution of sophisticated evaluation related questions of practical significance.

The use of stochastic Petri nets generates the system behavior by the description of all system states which

- optimizes system architecture design on the upper system levels
- supports the design and analysis of alternative architectures
- assists architectural decisons and their consequences on the system architecture
- adds to the traditional simulation approach
- extends the issues of computer aided systems technology

Motivated by the possibilities of stochastic Petri nets in solving relevant industrial problems we considered GSPN /7/ as an adequate approach some years ago, developed some additional features in the tool TOMSPIN /4/ and used it for solving concrete questionings of industrial systems architectures. Results of these analytical methods are decision bases for many developments till today.

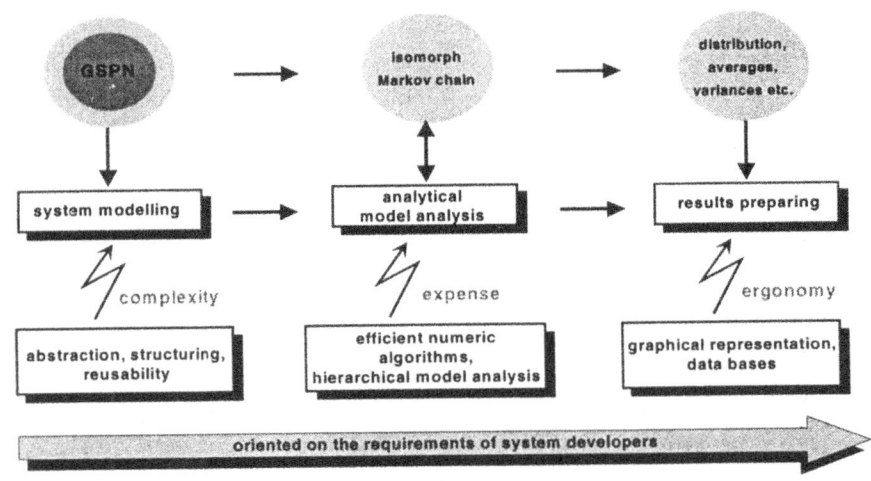

FIGURE 3: SYSTEM EVALUATION PROCESS

Summarizing our ideas (Figure 3) the evaluation process is characterized

- by using GSPN tools (e.g. TOMSPIN) for system modelling which implies the challenge to solve the complexity reduction by abstraction, structuring, and reusability;

- by a transformation of the Petri net model in isomorph Markov chain which is basis of the analytical model. Efficient numeric algorithms and hierarchical model analysis reduce its expension;

- by an adequate ergonomy preparing the results of distribution, averages, variances etc in graphical representation and data bases.

3. TOMSPIN, example of an analytical method and tool

3.1 Modelling using TOMSPIN

TOMSPIN (TOol for Modelling with Stochastic PetrI Nets) has been designed as an easy-to-use, graphical software package. Being aware of the features and intentions of related tools in the GSPN domain the essential features of TOMSPIN can be pointed out:

- enhancement of the descriptive and modelling power of stochastic Petri nets by adding several language extensions to standard GSPNs.
- development of efficient modelling techniques in order to limit or to handle the increasing model complexity.
- effective solution of GSPNs with respect to both their steady state and transient behavior.

Fehler! Textmarke nicht definiert. implementation of adequate algorithms for solution of "stiff" Markov chains.

- extensive support for cost-effective system engineering be providing a user friendly modelling environment that meets the demands of system developers.

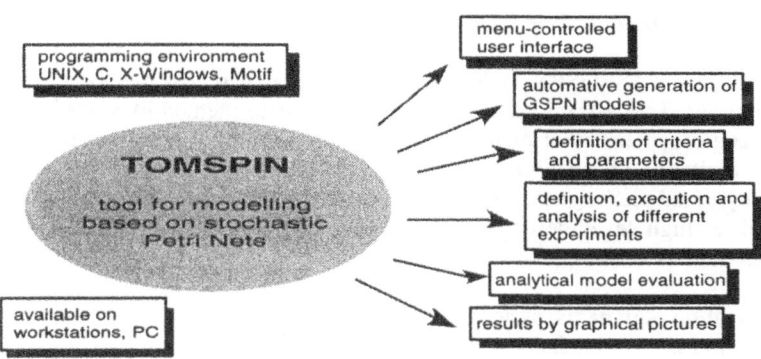

FIGURE 4: TOMSPIN FEATURES

TOMSPIN is tightly structured in a hierarchical and modular manner so that it can easily be upgraded adding new functional components (Figure 4). Its programming environment are UNIX, C, X-Windows and Motif, and as the tool is available on workstations and PCs. At the moment the following features are implemented:

- menu-based graphical user interface
- automatic generation of GSPN models
- definition of criteria and parameters
- definition, execution, and analysis of different experiments
- analytical model evaluation
- results supported by graphical pictures

These features support an efficient modelling of systems architectures during all modelling phases descibed by

1. Specification of an evaluation problem
2. Structured model generation based on a library
3. Transformation on mathematical basis model
4. Hierarchical analytical model analysis
5. Problem solving

Including data bases of methods and models and the modelling rules TOMSPIN is a powerful tool regarding systems architectures modelling be one method. It implements the well-known GSPN-methodology, but at the moment TOMSPIN does not have an own graphical Petri net editor as almost all other tools. However, we use a commercial-editor (SIGRAPH) to draw Petri nets and we translate the nets to the TOMSPIN input language. Reasons for that are the requirements of the real industrial applications which demand optimized algorithms for the analyser and the possibilities of hierarchical system modelling. Therefore today TOMSPIN is able to analyse more than 2 Million system states during a normal evaluation session.

3.2 Model-based analysis of TOMSPIN

Supporting extended GSPN TOMSPIN offers also an efficient method of dependability modelling /6/. The restrictions of the classical fault-tree method like independence of the failure and repair behavior of the system components, like static system structures and like the limitation of reliability analysis are solved by the analysing features and the tool flexibility based on the generated system architecture model. Additionally TOMSPIN can be used as high level description of Markov models. A user-friendly generation of compact models allows

- evaluation of dynamic system structures,
- description of redundancies and concurrent activities,
- sychronization of signals and processes,
- definition of failure series and failures following an other failure,
- analysis of multiple faults,
- specification of agreeable repair and maintenance strategies,
- definition of optional characteristic value for reliability and availability.

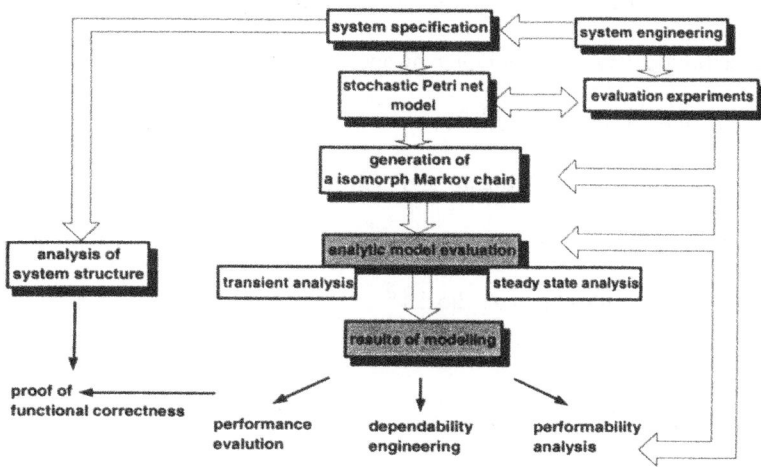

FIGURE 5: MODEL-BASED ANALYSIS OF TOMSPIN

At the moment TOMSPIN realizes a powerful model-based analysis described by Figure 5 which is part of our efficient methodology to evaluate a wide-range area of systems architectures. Looking at advanced requirements for modelling tools a transient analysis of models with hundred thousands of states /1/, sensitivity analysis /2/, experimental support, data interfaces to business graphics tools (e.g. EXEL), animation, and model debugging tools characterize the improvement of development methodology.

3.3 Model evaluation by TOMSPIN

For the system development and its different design steps the developers have to make meaningful conclusions about the modelling and analysing results and about the overall behavior. These results give evidence about the consequences of architectural decisions and, thus, support the correctness and the optimization of the system design without expensive synthesis / analysis iterations. Depending on the system criteria and on the required evaluation experiments the computation of the Petri net and its isomorph Markov chain and the transient / steady state analysis result in measures of performance, of dependability engineering, and of performability /6/.

Features of TOMSPIN and its flexibility offer the developer a lot of different model evaluation aspects as last step of the powerful system evaluation:

- selecting of the criteria and their combinations

- comparing of alternatives

- determing parameters and variants of the evaluation experiments

- descibing of environment inputs by adequate parameters

- defining of optimization criteria

- interpreting of the experiments and their results

- computing of concrete measurements

- finding out architectures deadlocks

- putting out performance bottlenecks

- preparing results by graphics

- etc.

4. Evaluation methodology by CAST

Discussing an approach of a methodology (Figure 6) many relations and dependencies can be found in order to support and to improve system architecture design and its evaluation:

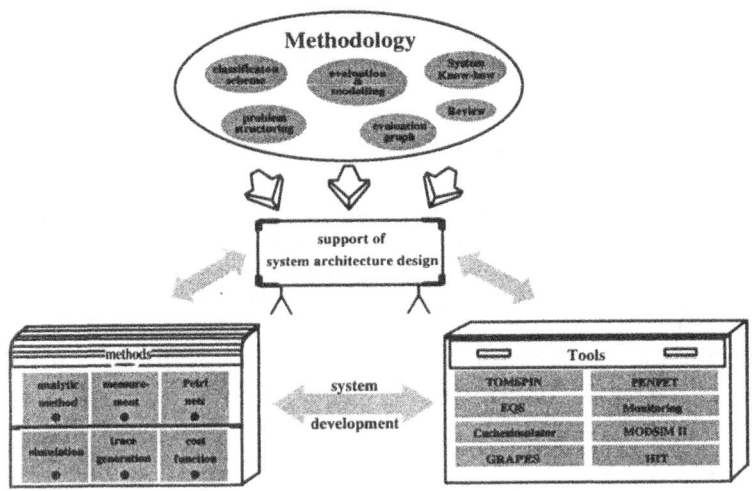

FIGURE 6: METHODOLOGY FOR SYSTEM ARCHITECTURE DESIGN

- Each single methodology influences the process of support and improvement: classical approaches like problem structuring and classification schemes reduce the complexity; system know-how, system engineering and review can validate the design, modelling, analysis and evaluation help to find an optimal system architecture etc.

- The realization of a methodology is given by different methods: system designers and developers have the choice to select simulation, analytical methods like Petri nets, trace generation, cost function or measurement, as adequate and cost-effective design support.

- In order to deal with the increasing complexity of the resulting systems powerful tools are required which implement the enumerated methods and which could be integrated in engineering environments, e.g. TOMSPIN, Cachesimulator, Monitoring, HIT, MODSIM II, GRAPES, PENPET etc. /5,6/

Experiences with industrial systems give us an idea how system complexity can be managed: the definition of the evaluation problem by abstraction is basis for an effective model generation. New techniques, structuring and library support improve this model generation. Finally the overall view of different criteria, efficient algorithms and hierarchical analysis optimize the model evaluation /5/.

Methods of modelling and evaluating systems architectures are only one part of system technology /8/, but the described approach based on stochastic Petri net and its realization by the tool TOMSPIN give us an usable basis of computer aided system evaluation and system optimization.

5. Conclusion

The paper presents and discusses a methodology of modelling, analysis and evaluation by using stochastic Petri nets implemented in a tool. The experiences with any method and tool discover todays limitations too, e.g. increasing model complexity like stochastic Petri nets with some million states, inflexible simulation programs, small use of any timing or dynamic system state etc. We believe that the combination of methodologies, e.g. problem structuring, and modularization for hierarchical solution, or combination of different methods, e.g. simulation and stochastic Petri nets for hybrid modelling will help to relax these limitations.

References

/1/ H. Choi, V.G. Kulkarni, K.S. Trevedi "Transient Analysis of Deterministic and Stochastic Petri Nets", in Proceedings of 14th Int. Conference on Application and Theory of Petri nets, Chicago, Illinois, USA, June 1993, pp 166-185

/2/ H. Choi, V. Mainkar, K.S.Trevedi "Sensitivity Analysis of Deterministic and Stochastic Petri Nets", in Proceedings of MASCOTS '93, Int. Workshop on Modeling, Analysis and Simulation of Computer and Telecommmunication Systems, San Diego, USA, Jan. 193, pp. 271-276

/3/ R. German "Analysis of Stochastic Petri Nets with Non-Exponentially Distributed Firing Times", Ph.D. Thesis, Technical University of Berlin, Germany, 1994

/4/ G. Klas, R. Lepold "TOMSPIN, a Tool for Modeling with Stochastic Petri Nets", CompEuro 92, The Hague, May 4-8, 1992, pp. 618-623

/5/ G. Klas "Hierarchical Evaluation of Generalized Stochastic Petri Nets", Ph.D. Thesis, Technical University of Munich, Germany, 1993

/6/ R. Lepold "Performability Evaluation of Degradable Computer Systems based on Stochastic Petri Nets", Ph.D. Thesis, University of Mulhouse, France, 1992

/7/ M.A. Marsan, G. Balbo, G. Conte "A Class of Generalized Stochastic Petri Nets for Performance Analysis of Multiprocessor Systems", ACM Transactions on Computer Systems, Vol. 2, No. 2, May 1984, pp. 93-122

/8/ E. Schmitter "The Implications of Industrial Systems' Complexity on Methodology for System Design and Evaluation", in Proceedings of 5th Int. Workshop on Petri Nets and Performance Models, Toulouse, France, Oct. 19-22, 1993, pp 192-201

Systematic Strategy for Performance Prediction in Improvement of Parallel Programs *

Roman Blasko

Department of Software Technology and Parallel Systems
University of Vienna, Liechtensteinstrasse 22, A-1090 Vienna, Austria
e-mail: roman@par.univie.ac.at

Abstract. Second-generation supercompilers should comprise performance analysis tools for creating an information feed-back to improve parallel programs. We have developed a technique and performance analysis tool, PEPSY, embedded in our supercompiler, which is based on automatic generation of the abstract model of the parallel program and performance analysis using a simulation technique. Real application programs might consist of a number of procedures calling one another and creating a complex call structure. We have designed a systematic strategy for analysis of the call structure, which significantly reduces the total time necessary for the performance prediction of the whole program.

1 Introduction

Second-generation supercompilers, such as VFCS (Vienna Fortran Compilation System), should comprise performance prediction as an indispensable phase of the program development process. A Fortran program is parallelized by VFCS [15] into an explicit parallel program with message-passing statements. VFCS is based on a *Single-Program-Multiple-Data (SPMD) computation model*, where each processor executes the same program, however for a different data segment. It is assumed, that at most one process is runing on each processor of the target computer, and thus each process may be associated with a distinct private memory [8]. An initial version of the program is written in Fortran'77, or Vienna Fortran programming language [15]. A considered target computer is being the distributed memory computer system Intel iPSC/860.

Performance analysis for VFCS is supported by the tools such as Weight Finder [6], PPPT [7] and PEPSY [4] developed by our team and MEDEA [5] and PARMON [10] developed independently. A common problem, for performance evaluation of large real programs, is the time necessary for analysis and computation of performance characteristics. There are several abstract description tools (e.g. Petri nets based) for parallel systems, where a common problem is the

---

* The work described in this paper is being carried out as a part of the research project " Performance Prediction and Expert Adviser for Parallel Programming Environment", funded by the FWF, Vienna, Austria, under the grant number P9205-PHY.

size of the model, determined by their application to real systems. Model-based tools use the model created by a user. This is not a feasible approach, in our case, where we have a program with hundreds of lines of the code, being changed by VFCS during an iterative improvement process. Performance evaluation tools based on the measurements cannot be used before a real run of the parallel program. We have tried to avoid these disadvantages by the systematic strategy outlined below.

The tool, PEPSY, is designed for performance prediction of parallel programs, to be done before generating the final code for the target compiler of the parallel machine, i.e., before a real run on the machine. Performance results should be used for comparison of several variants of the parallel program, so as to find an optimal parallelization strategy for VFCS. This implies that our focus is to develop techniques suitable for evaluation of relative differences between several program versions, identify the most time consuming parts of the program, and not to concentrate on absolute values, e.g., the precise run time values usually measured on real machines. This paper presents the systematic strategy for performance prediction of parallel programs with a complex procedure call structure, and evaluates effectiveness of the designed strategy, using PEPSY.

Performance prediction is carried out in the VFCS programming environment and a system overview of the information flow, during the program improvement process, is described in Sec.2. The systematic performance analysis, Sec.3, is presented by a control flow and modeling of parallel programs on the statement level and procedure level, using a bottom-up approach for analysis and top-down approach for global performance prediction. Effectiveness of the designed strategy is evaluated by effectiveness parameters defined in Sec.4. Experiments illustrating the effectiveness of the strategy and indicating the most time consuming parts of the parallel program are in Sec.5.

2 System Overview

The basic functional blocks of VFCS are the *front-end* (FE), *transformation system* (TS) and *back-end* (BE). A general block schema, with illustration of the main information flows for the program improvement process, is in Fig.1. After processing of the initial sequential program (SP') by the front-end, we get a *sequential program* (SP), as an input to TS. TS carries out transformations of the program on its internal representations and produces an intermediate *parallel program* (PP) containing message-passing statements. If the PP properties are not satisfactory and could be improved by the next transformations, then the program is sent again to TS, for the next iteration of the improvement process. The final product of the parallelization process is an optimized PP, to be processed by the back-end generating the final parallel program (PP'). Then, PP' is processed by the *target computer* (TC), representing the target compiler and machine, in this context.

For more effective programming and improvement of parallel programs, we

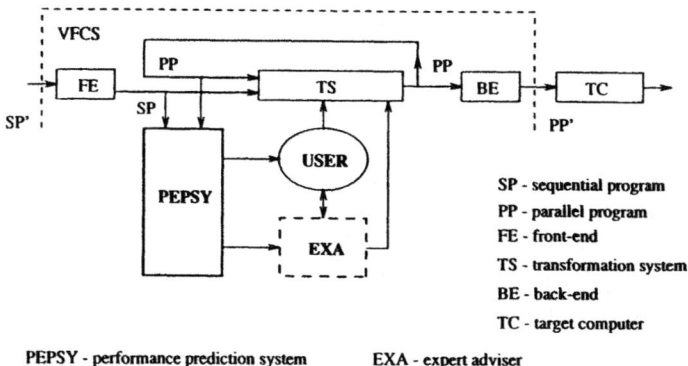

Fig. 1. System view of the performance prediction

have developed the tool, **PEPSY** (PErformance Prediction SYstem) [4], that analyzes the parallel program or program unit before/after processing by TS, during the compilation process. The obtained performance data enable to recognize the most critical parts the program, to be improved by TS, in the next iteration of the preparation. The performance model of the parallel program is generated automatically, in the *Process Graph* form [3], and the performance analysis is based on the discrete-event simulation. PEPSY evaluates the detailed behavior of the modeled parallel processes, and performance indices defined for parallel systems such as the activity or utilization of the parallel processes, parallelism degree, communication and waiting times of parallel processes, etc. The performance characteristics can be evaluated for steady or transient states of the behavior, with an optional evaluation and sampling interval.

PEPSY creates a feed-back analyzing the input to TS or/and product of TS. However, this solution is still a semi-automatic or *off-line* parallelization system. The second subsystem will be an *EXpert Adviser* (EXA) [2, 9], to be developed for automation of the compilation process, also in an *on-line* mode. Development of all the mentioned functional blocks leads to the creation of an automatic parallelization system. This new environment can be characterized as a *simulative expert system*, which derives new knowledge from the performance prediction based on simulation [11].

3 Systematic Performance Analysis

PEPSY provides the user with a set of performance data on detailed and global behavior of each procedure, parallelized into a set of parallel processes with complex interactions. If the parallel program has a complex call structure of the called procedures, it is not so easy to find the proper strategy for the most effective program improvement process. We utilize a call graph of the parallel program, for the systematic performance analysis. The present version of our strategy is restricted to an acyclic call graph, which is the most topical case in the considered compiler.

3.1 Behavioral Features of the Program Parallelized by VFCS

A general structure of the parallelized program, in the framework of the SPMD model, consists of the host module creating a host process (HP) and the node module, to be replicated for the specified set of node processes (NP) with own segments of distributed data; see Fig.2. Considering the general structure of the program, we distinguish two stages of model building:

- *intraprocess* modeling, where various program constructs and their combinations in one node process are described by model attributes, and
- *interprocess* modeling, where process interactions and work distribution are described by model parameters derived from the communication and mask statements.

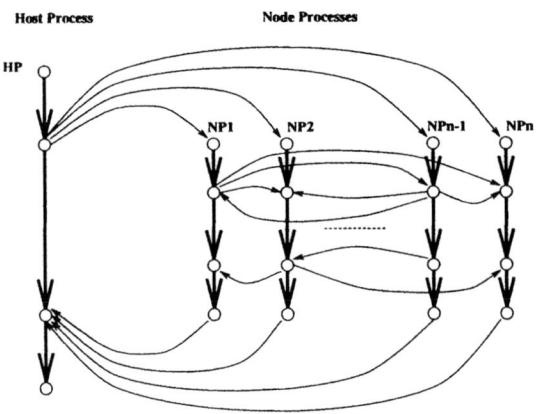

Fig. 2. General structure of the parallel program

The host module and the set of node modules may be considered at the statement or procedure level. The node module is a key component, which may be parallelized in different ways. Because of this, the considered and analyzed program part, in the rest of the paper, is the node module parallelized into the set of the parallel node processes.

A program P is a set of $n >= 1$ declared procedures. The procedures correspond to Fortran program units; they are *non-recursive* and their *declarations cannot be nested.* The procedures are called *functions* (FUNCTION) or "proper" *procedures* (SUBROUTINE), depending on whether or not their execution yields a result. The "proper" procedure has one input and one output point.

3.2 Control Flow at the Statement Level

A basic structure of the model is derived from data produced by a control flow analysis. The control flow analysis, on the intraprocedural level, generates the control flow graph [14].

Definition 1. A *control flow graph* G is a triple $G = (N, E, s)$, where (N, E) is a directed graph, $s \in N$ is the *initial node* and there is a path from s to every node of G.

The control flow graph is to be used to model the control structure of programs, at the desired level of abstraction where *nodes* represent: single Fortran statement, intermediate language statements (such as triple or quadruple notation) or a sequence of statements. The *edges* represent control transfers between the nodes.

3.3 Modeling of the Parallel Program at the Statement Level

The structure of the control flow graph, at the statement level, is used for generating the basic structure of the model. This structure represents one program unit, i.e., the procedure in one sequential process. The basic structure of the model is attributed by the quantitative parameters characterizing the language or machine features stored in the database and in some cases also by parameters evaluated by Weight Finder [6]. This structure creates a template for generating the complete model of the parallel program, consisting of several parallel processes. The parallel model is generated after parallelization of the program and is completely represented by the process graph [3].

Definition 2. A *Process Graph (PG)* is a directed graph, $PG = (N, A)$, where N is a set of nodes and A is a set of directed arcs, $A \subset N \times N$. A node represents a statement with defined input/output behavior specified by a set of node attributes and a directed arc represents a control flow between nodes.

We have defined the set of PG node types covering the program constructs in Fortran'77 and Message Passing Fortran [1]. If the parallel program is structured into several procedures creating independent program units, then there is a problem of how to proceed with the analysis at the level of procedures.

3.4 Control Flow at the Procedure Level

A product of the control flow analysis at the *interprocedural level* is the call graph representing a calling relationships of procedures [14].

Definition 3. Let P denote a program. The *call graph* G of P is a directed graph $G = (N, E)$, where there is a one-to-one correspondence between N and P, and $(p, q) \in E$ iff procedure p contains a call whose execution may result in the direct activation of procedure q. One or more calls may be mapped to the same edge of the call graph.

Even if a program does not execute recursive calls, the process of determining the call graph may lead to a cyclic graph when conditional calls occur. This case is called *static recursion* and it is excluded by the algorithm implemented in

VFCS [8]. Throughout the following, we assume that the call graph is an *acyclic graph*. The call graph generated by the front-end of VFCS, can be used as a navigating structure for the systematic performance analysis of the program at the issue.

3.5   Modeling of the Parallel Program at the Procedure Level

The SPMD model and mapping technique of VFCS imply that only one procedure is running on the node processors simultaneously. This enables the use of a *partial evaluation strategy* for the parallel program, when the behavior of each procedure (program unit), parallelized into several processes, is time independent on the behavior of others. The *systematic modeling and analysis strategy* for the parallel program, consisting of the large call structure, is as follows.

The procedure call statement is modeled as one PG node with one input and one output. All procedures which contain no procedure call, are evaluated in the first stage. Their processing times are used as delay times in the nodes representing the calls, in their callers. The second stage of the performance analysis, concerning the callers, is much less time consuming as it uses the delay times instead of the repeated analysis of the called procedures, especially when these procedures are in the loop bodies. In the next stage, the analysis proceeds by evaluation of the "next levels" of the procedure callers, in the call graph structure, and continues till the top of the call hierarchy. In a simple case, the call graph is the directed tree [13] and the analysis proceeds from the procedures represented by leaves, up to the program unit represented by the root of the call graph, usually the main unit. This technique is used systematically in a *bottom-up* manner, making the global model of the program more simple, and reducing the performance analysis time, necessary for the processing of the total program model.

3.6   Performance Prediction of the Parallel Program

The performance results provided by PEPSY clearly indicate the most time consuming nodes (statements) in the program, at each level of the call structure. One of the most important goals of the performance prediction, and interpretation of performance results, is to identify the most critical parts of the program which should be modified for better performance. After the mentioned systematic performance analysis, in bottom-up manner, we have obtained the temporal behavior of the parallel program at every procedure call level and also at the top level, i.e., the main program level. Then, we proceed in *top-down* manner, from the main program to the most expensive procedure calls or parts of the program, indicated by the performance parameters. A relative time significance or "cost" of the program parts navigates the process of modifications and improvement of the whole program. This navigation information, obtained as one of the main results of the performance prediction, is utilized by the user directly or by the expert adviser, to be developed for the parallelization treatment.

4 Effectiveness of the Strategy

Evaluation of the effectiveness, of the designed systematic strategy, enables us to characterize its practical usefulness and importance for treating the real applications. We need to define some terms and effectiveness indices for a comparison of this systematic and usual flat performance prediction, with respect to the processing time needed.

4.1 Basic Terms

We distinguish the following time parameters for evaluation of the effectiveness.

Definition 4. *Run time* of the real system, t_r, is a time period necessary for execution of the specified system activity.

Definition 5. *Simulation time* of the model, t_s, is a time period simulated by a model of the real system, and determined by the specified system activity.

Definition 6. *Processing time* of the model, t_p, is a time period necessary for processing of the simulation model, for the specified simulation time, t_s.

In our case, the real system is the parallel program with a measured run time and the analyzed system activity is a complete execution of the program, on the parallel computer. The simulation time is a simulated run time. Usually, the run and simulation times are used for the basic validation of the model, comparing the modeled and real behavior. The processing time of the model determines the effectiveness of the designed performance prediction technique. The computational work, necessary for processing of the simulation (performance) model, can be described by another model, called a *processing model*. The processing model of the simulation model is a *metamodel* of the analyzed system, concerning the required processing time.

Using the systematic approach for the analysis of the parallel program behavior, every procedure is evaluated only once, analyzing its performance model. This fact allows to define a *reduced processing model* of the whole program, as a set of the simulation models for all the procedures, characterized only by their processing times. Then, the *reduced processing time*, t_{pr}, necessary for the analysis of the whole simulation model, is described as

$$t_{pr} = \sum_{i=1}^{n} t_{pr}(i), \tag{1}$$

where $t_{pr}(i)$ is the reduced processing time of the i-th procedure simulation model, and n is a number of different procedures called in the program.

Evaluating the effectiveness of the systematic strategy, the processing requirements of the reduced processing model should be compared with the processing requirements of the usual (non-systematic) flat evaluation. The *non-reduced processing model* is derived from the simulation model of the program, directly. The total *non-reduced processing time*, t_{pn}, represents the time requirements for processing of the whole simulation model. Such model consists of the nodes representing the procedure calls, and control branching nodes. A node, representing the procedure call, has a delay time corresponding to t_{pn} of the procedure, and a node representing the control branching, e.g., conditional-GOTO or DO-loop head statement, has the delay time $dt = 0$. However, the procedure or program at the issue should also be included in the non-reduced processing model. This is done attaching one node, sequentially at the beginning or at the end of the model, with the delay time equal to the processing time, t_p $(t_p = t_{pr})$, of its simulated model.

The total non-reduced processing time, t_{pn}, of the whole simulation model can be computed *analytically* in a bottom-up way, in simple cases, or can be evaluated by *simulation*, using the mentioned processing model. The non-reduced processing model can be described, obviously, by the Process Graph, and evaluated by PEPSY, as well.

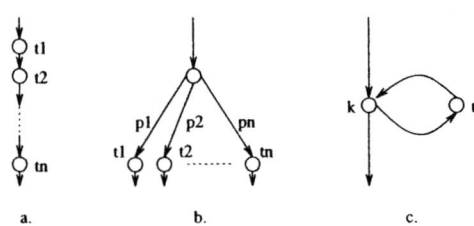

Fig. 3. Basic interconnection structures

The analytical evaluation of the non-reduced processing time is based on the composition of the following *basic interconnection structures* among the nodes of the processing model. A total (non-reduced) processing time, t_p, for a *sequential connection*, see Fig.3.a, is described by Formula (2), and for a repeated activation of the structure by Formula (3),

$$t_p = \sum_{i=1}^{n} t_p(i) \qquad (2) \qquad\qquad t_p = k * \sum_{i=1}^{n} t_p(i) = k * t_p' \qquad (3)$$

where $t_p(i)$ is a delay time in the i-th node, n is a number of the nodes interconnected sequentially, k is a number of repeated activations of the structure, and t_p' is the repeated processing time. The t_p of a *parallel connection*, see Fig.3.b, is described by Formula (4), and for a repeated activation of the structure by Formula (5),

$$t_p = \sum_{i=1}^{n} p_i * t_p(i) \qquad (4) \qquad\qquad t_p = k * \sum_{i=1}^{n} p_i * t_p(i) = k * t_p' \qquad (5)$$

where p_i is an activation probability of the i-th branch in the structure. The t_p of a *DO-loop structure*, see Fig.3.c, is described as $t_p = k * t_p'$, where k is a number of iterations, and t_p' is the processing time of the loop body.

4.2 Effectiveness Indices

For evaluation of the effectiveness, we have defined the following indices.

Definition 7. *Reduction factor*, r, of the systematic strategy is defined as a ratio of the non-reduced processing time, t_{pn}, and reduced processing time, t_{pr},

$$r = t_{pn}/t_{pr}. \tag{6}$$

Usually, $r > 1.0$, and no reduction is achieved when $r = 1.0$. This value indicates that there is no repeated call of the procedure in the analyzed parallel program, when no reduction of the processing time is possible. The reduction might be considered effective with $r > 10.0$.

Definition 8. *Efficiency of simulation*, e, is defined as a ratio of the run time, t_r, and reduced processing time, t_{pr},

$$e = t_r/t_{pr}. \tag{7}$$

Definition 9. *Overhead of simulation*, o, is defined as a ratio of the reduced processing time, t_{pr}, and run time, t_r, i.e., an inverse of e,

$$o = t_{pr}/t_r = 1/e. \tag{8}$$

Usually $e < 1.0$, because the simulation takes more time than the real execution, and $o >> 1.0$ indicates how many times the simulation is slower than the real execution.

In many cases, simulation is criticized as too slow a technique. But for complete evaluation and comparison of it with the measurement technique, we should consider also the time necessary for preparation the final code of the parallel program by the back-end, instrumentation for measurements, compilation by the target compiler, execution with measuring and processing of the measured data from an experiment. All these negative features of measurements promote advantages of the modeling and simulation approach. In any case, when the target parallel computer is not available, or a considered parallel configuration is larger than the available one, then the simulation, as in other application fields, is still the only one way feasible for the behavioral performance analysis.

5 An Example

To illustrate the practical effectiveness of the designed evaluation strategy, we use the *Jacobi Relaxation Iterative Method* [12], which can be used to approximate the solution of a partial differential equation discretized on a grid. At each step, it updates the current approximation at a grid point by computing a weighted average of the values at the neighboring grid points. An excerpt from the Jacobi relaxation code, after parallelization by VFCS, is shown in Fig.4. We distribute arrays $F, U, UHELP$, and RES block-wise into the $(2*2)$ processor array. Owing to this, every processor contains a local segment with $(64 * 64)$ data elements for every distributed data array. For those array elements, which are local to a specific processor, all operations are executed on the local memory. Non-local data elements can only be accessed via the message-passing statements, $EXSR$, which have a major effect on the overall program performance.

We see all procedure calls, used in the program, in the source code of the $MAIN$ program unit and the procedure $RESIDU$. The whole program consists of four procedures, i.e., $JACOBI, INIT, RESIDU, NORM$, and the program unit $MAIN$. The call graph, generated by VFCS, is in Fig.5. The *non-optimized*

```
   1:        TASK PROGRAM MAIN
   .....
   4:        INTEGER I,ITER
   5:        REAL F(128,128),U(128,128),UHELP(128,128),RES(128,128)
   .....
   9:        AINIT = MASTER()
  10:        N = 128
  11:        ITER = 50
  12:        OMEGA = 0.66
  13:        CALL INIT(F,U,128)
  14:        CALL RESIDU(0,RNORMO,RES,F,U,128)
  15:        DO 10 I=1,50
  16:           CALL JACOBI(OMEGA,F,U,UHELP,128)
  17:           CALL RESIDU(I1D,RNORM,RES,F,U,128)
  18: 10     CONTINUE
  19:        STOP
  20:        END
```

```
   1:        SUBROUTINE RESIDU(ITER,RNORM,RES,F,U,N)
   2:        INTEGER J,I
   3:        INTEGER ITER
   4:        REAL RNORM,RES(N,N),F(N,N),U(N,N)

   5:        DO 11 J1D=2,128-1
   6:           DO 10 I2D=2,128-1
   7:              EXSR U(I,J+1) {[0/0,0/1]}
   8:              EXSR U(I,J-1) {[0/0,1/0]}
   9:              EXSR U(I+1,J) {[0/1,0/0]}
  10:              EXSR U(I-1,J) {[1/0,0/0]}
  11:              OWNED(RES(I,J))-> RES(I,J) = F(I,J) - 4 * U(I,J)
       *                    + U(I-1,J) + U(I+1,J) + U(I,J-1)   + U(I,J+1)
  12: 10        CONTINUE
  13: 11     CONTINUE
  14:        CALL NORM(ITER,RNORM,RES,128)
  15:        END
```

Fig. 4. Source code for the MAIN unit and procedure RESIDU

version of the source code is obtained after parallelization, according to the specified data distribution, done by VFCS; see Fig.4. The *optimized version* is created by the communication and masking optimization transformations [8], implemented in VFCS. In the following experiments, we consider both these versions of the program as two variants.

5.1 Evaluation of the Effectiveness

The goal of this section is to evaluate and illustrate the effectiveness of the designed systematic strategy, by practical evaluation and comparison of the run time and processing time necessary for evaluation of the concrete program.

Firstly, we evaluate the *run time*, t_r, for all procedures and the main unit, separately. The values are in Tab.1, in the first two columns, in *msec*, for the non-optimized and optimized version of the program. The run time for the whole program is 58.83*sec* and 0.66*sec* for the two versions.

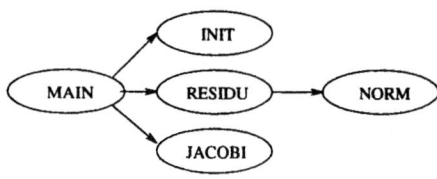

Fig. 5. Call graph

Secondly, we proceed in the *bottom-up* way, using the *systematic strategy*, and evaluate the *simulation time*, t_s, and *processing time*, t_p, for all program units. The call graph of the program is a tree with three leaves, Fig.5. The leaves should be analyzed in the first phase, for evaluation of both the simulation and the processing time.

In the next part of the text, we concentrate on evaluation of the *processing time* only, unless mentioned otherwise. The values of the processing time, for the reduced processing model, are in Tab.1, in the two middle columns, in *sec*, for the non-optimized and optimized version of the program. After evaluation of the $INIT, NORM$, and $JACOBI$, we evaluate the procedure $RESIDU$, using the simulation time of the $NORM$ procedure, as the delay time in the node representing its call in the performance model of $RESIDU$. The processing time of $MAIN$ is evaluated in the third phase, after $RESIDU$, using the simulation times for all called procedures. The whole *reduced processing time*, t_{pr}, necessary for simulation using the systematic strategy, is computed by Formula (1), i.e. 57.93*sec* and 15.02*sec* for the non-optimized and optimized version, respectively. All partial reduced processing times, see Tab.1, are measured during simulation by PEPSY.

procedure /	RUN TIME		RED. PROC. TIME		PROC. TIME	
prog.unit	non-optim.	optim.	non-optim.	optim.	non-optim.	optim.
JACOBI	634.04	7.25	23.997	2.425	- " -	- " -
INIT	21.85	21.87	10.033	9.996	- " -	- " -
NORM	9.18	2.74	4.754	1.327	- " -	- " -
RESIDU	657.39	13.61	19.063	1.198	24.53	2.53
MAIN	58 831.97	661.86	0.078	0.076	2 461.02	260.37

Table 1. Measured time and processing time

To evaluate the effectiveness of the systematic strategy, we need also to evaluate the *non-reduced processing time, t_{pn}*, which would be necessary for simulation, using the total performance model of the program. Such a model consists of submodels of all mentioned program units, where each procedure call is substituted by the corresponding performance submodel of the procedure. The non-reduced processing time can be evaluated analytically, in this simple case, using the processing model. The *non-reduced processing models* for $MAIN$ and $RESIDU$ are in Fig.6.a and Fig.6.b, respectively. The non-reduced processing times are in the last two columns of Tab.1, for both the versions.

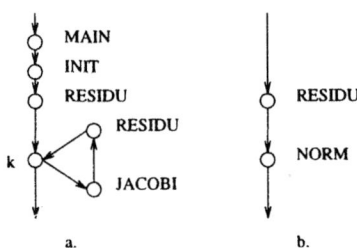

Fig. 6. Processing model

The leaves of the tree have the same values, because those were not reduced. The processing time for $RESIDU$, evaluated by the processing model in Fig.6.b and Formula (2), is 24.53*sec* and 2.53*sec* for both the versions. The non-reduced processing time for $MAIN$, evaluated by the processing model in Fig.6.a, and Formulas (2) and (3), is 2461.02*sec* and 260.37*sec* for both the versions.

After evaluation of all three times, i.e., t_r, t_{pr}, and t_{pn}, we evaluate other effectiveness parameters of the strategy. The most explanatory one, is the *reduction factor, r*, given by Formula (6). The reduced processing time, 57.93*sec*, is necessary for evaluation of all non-optimized program units, modeled and simulated by PEPSY using the systematic strategy. The non-reduced processing time, for the $MAIN$ unit, is 2461.02*sec* = 41.02*min*. Comparing both the times by Formula (6), we get the reduction factor $r = 42.48$ for the non-optimized version of the program. The reduction factor, for the optimized version of the program with the reduced processing time 15.02*sec* and the non-reduced processing time 260.37*sec*, is $r = 17.33$. The reduction of the processing time is *more than one magnitude*, in this simple case, for both versions of the program. This implies the usefulness and importance of the designed systematic strategy for more complex cases, when the non-reduced processing time would be hardly feasible. The extreme situation is on the top level of the call graph, i.e. the $MAIN$ unit, where the evaluation, of this level's behavior only, needs 0.078*sec* and 0.076*sec* of the reduced processing time, for the two versions, respectively; see Tab.1. Comparing those times with the non-reduced ones, we obtain the reduction $r = 31551.54$ and $r = 3425.92$ for the non-optimized and optimized version of the parallel program. These values are extremely high and for some experiments very useful.

The *efficiency* and *overhead* of simulation are the parameters comparing the simulation with the real processing. Using Formula (7) for efficiency, we have obtained the values $e = 1.02$ and 0.044, for the non-optimized and optimized version, respectively. The value $e > 1.0$ indicates that the simulation is faster than the real processing. This is very useful in cases when the real parallel computer is not available; even so the behavioral characteristics of the parallel program can be obtained in a feasible time. The overhead of simulation, given by Formula (8), is $o = 0.98$ and $o = 22.69$ for the two versions.

5.2 Performance Analysis

This section is devoted to the recognition of the most time-consuming parts of the parallel program using PEPSY and the designed systematic strategy. The performance prediction and interpretation of performance results should be carried out in the *top-down* manner. We introduce only a few results, just for illustration.

The processing of the non-optimized version of the $MAIN$ unit, see the source code in Fig.4, is characterized by values in Fig. 7. In the upper portion of Fig.7, we see the *activity for groups of nodes* in the performance model. Every group represents one process, or virtual processor, of the modeled parallel program. The activity can be interpreted as utilization of the processes or processors. All four processes are active during the full time of the processing, (100 %), which means that all four processors are 100 % utilized, equally balanced, when a parallelism degree for four processors is 4.0. After this global view of the processes, we can look at more detailed characteristics of any process. The lower portion of Fig.7. illustrates the *activity of all nodes* in the selected Process $No.1$. Nodes $No.16$ and 17 take 50.1 % and 48.9 % of the process activity, respectively. These nodes, which represent the procedure call for $JACOBI$ and $RESIDU$, respectively, are the most "expensive" statements in the $MAIN$ program unit, see Fig.4.

The processing of the non-optimized version of $RESIDU$, with the source code in Fig.4, is characterized by the performance in Fig.8. All four processes are utilized almost equally, with the mean value 97.2 %, see the upper portion of Fig.8. Additional characteristics for all processes are obtained by evaluation of the *communication and waiting*, illustrated by "c" and "w" pseudographically, in the middle portion of Fig.8. About 93.4 % of every process activity is devoted to communication and 2.8 % of the communication time is spent in waiting for synchronization among parallel processes. The degree of parallelism is 3.89. An evaluation of the internal activity of the selected Process $No.1$, is illustrated by the *activity of all nodes*; see the lower portion of Fig.8. Nodes $(7 - 10)$, representing the communication statements in $RESIDU$, Fig.4, are the most expensive ones in this procedure. The processing of the $NORM$ procedure, Node $No.14$, is relatively minor, taking 1.77 % of the process activity only. This is the

265

Fig. 7. Evaluation of the MAIN unit

brief illustration, how to find out the most critical parts of the parallel program, which are worth of further modifications and improvement.

procedure	JACOBI	INIT	NORM	RESIDU	MAIN
RUN TIME	634.04	21.85	9.18	657.39	58 831.97
SIM. TIME	775.01	39.87	13.38	756.39	77 366.88

Table 2. Run time and simulation time

The simulation time is one of the most sensitive parameters for verification of the model. In Sec.1, we mentioned that evaluation of the absolute values of time is not the goal of our prediction, because it needs a precise model of the target computer architecture. Despite of this, for rough verification of the performance models, we present a comparison of the run times, t_r, and simulation times, t_s, for all procedures in the non-optimized version of the parallel program; see Tab.2, in *msec*.

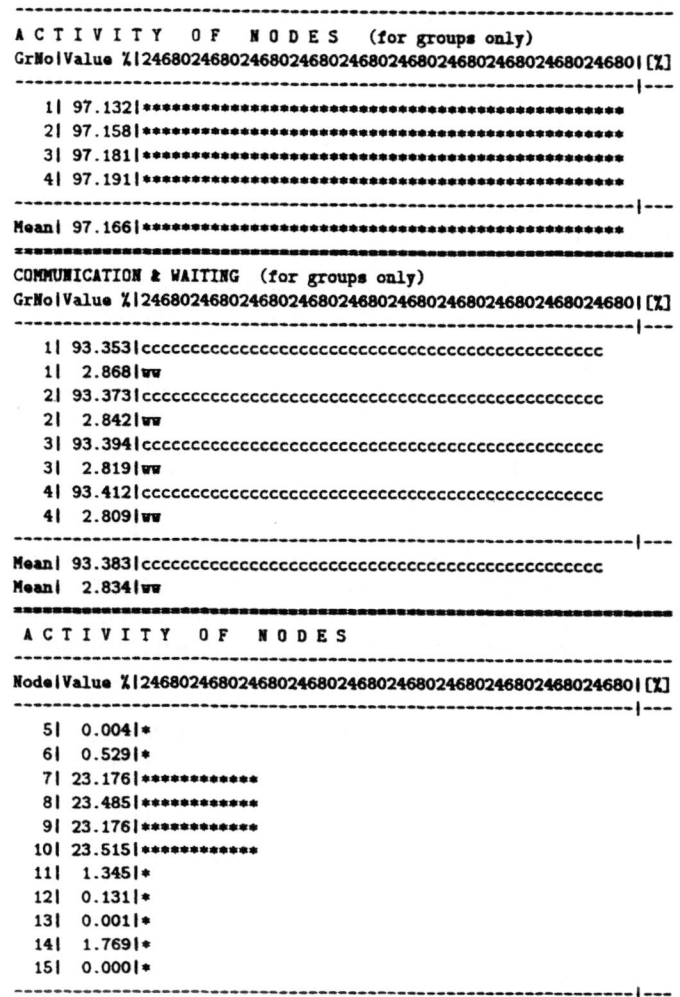

Fig. 8. Evaluation of the RESIDU procedure

6 Conclusions

The designed performance prediction technique, for improvement of the parallel program, is based on the automatic generation of the abstract model of the parallel program, and on the analysis of its dynamic behavior by the simulation. The systematic evaluation of the procedure call structure, in the bottom-up manner, reduces the analysis time. The clear indication of the most time consuming parts of the program, in the top-down way, enables one to concentrate on the most critical parts of the program. This strategy guides the user or the future expert adviser for the most effective technique of the improvement process. Our future research will be devoted to the automation of this approach and its integration with the expert adviser.

References

1. Blasko R.: "Parameterization and Abstract Representation of Parallel Fortran Programs for Performance Analysis", Proc. of the AICA'93 Conference, Gallipoli (Italy), Sept. 1993, pp.75-91.
2. Blasko R.: "Performance Prediction and Expert Adviser for Automatic Parallelization of Fortran Programs", Journal "Computers and Artificial Intelligence", 22 p., (in print).
3. Blasko R.: "Process Graph and Tool for Performance Analysis of Parallel Processes", Troch I., Breitenecker F.(eds.): Proc. of the IMACS Symposium on Mathematical Modelling (1.MATHMOD), Vienna, February 1994, pp.60-63.
4. Blasko R.: "Automatic Modeling and Performance Analysis of Parallel Processes by PEPSY", Fortschritte in der Simulationstechnik, Band 9, (eds. Kampe G., Zeitz M.), ASIM'94 Symposium, Stuttgart, October 1994, Vieweg, pp.241-246.
5. Calzarossa M., Massari L., Merlo A., Pantano M., Rossaro P.: "Techniques and Tool for the Analysis of Parallel Programs", Rech. Rep. R3/98, Progetto Finalizzato C.N.R. "Sistemi Informatici e Calcolo Parallelo", October 1992, (in Italian).
6. Fahringer T., Huber C.: "Der Weight Finder. Ein Profiler Fuer sequentielle Fortran'77 Programme", Rech. Report, Inst. for Statistics and Computer Science, University of Vienna, Sept. 1992.
7. Fahringer T., Zima H.P.: "A Static Parameter based Performance Prediction Tool for Parallel Programs", Proc. of the 7-th ACM Int. Conf. on Supercomputing 1993, Tokyo, July 1993.
8. Gerndt H.M.: "Automatic Parallelization for Distributed-Memory Multiprocessing Systems", PhD. Thesis, Rheinischen Friedrich-Wilhelms-Universitat, Bonn 1989, 170p.
9. Hulman,J., Andel,S., Chapman,B.M., Zima,H.P.: "Intelligent Parallelization Within the Vienna Fortran Compilation System", Proc. of the Fourth Workshop on Compilers for Parallel Computers, Delft, Netherlands, December 13-16, 1993.
10. Lenzi P., Serazzi G.: "PARMON: Parallel Monitor", Tech. Rep. R3/95, Progetto Finalizzato C.N.R. "Sistemi Informatici e Calcolo Parallelo", October 1992, (in Italian).
11. Oren T.: "Simulation Environments: Challenges for Advancement", Chinese J. of Systems Engineering and Electronics, Vol.4, No.4, 1993, pp.12-24.
12. Press W.H., Flannery B.P., Teukolsky S.A., Vetterling W.T.: "Relaxation Methods for Boundary Value Problems", Cambridge University Press, 1991, pp. 674-676.
13. Tremblay J.P., Manohar R.: "Discrete Mathematical Structures with Applications to Computer Science", McGraw-Hill, 1975.
14. Zima H.P., Chapman B.M.: "Supercompilers for Parallel and Vector Computers", Addison-Wesley Publ.Comp., ACM Press, 1990, 376p.
15. Zima H.P. et al.: "Vienna Fortran - A Language Specification, Version 1.1", NASA Contract Report, ICASE Langley Research Center, Hampton, Virginia, March 1992, 84p.

Design and Implementation of Multimedia Environment for Simulation

Francisco Bustío, Pedro Corcuera, Eduardo Mora

Department of Applied Mathematics and Computer Sciences
University of Cantabria
Avda. de los Castros s/n. 39005 Santander. SPAIN

Abstract. The importance of simulation in all fields is widely recognized due to the impact it has on design, production and engineering. This paper presents a design methodology aimed at developing multimedia interfaces for the simulation of industrial processes, using an object oriented point of view. A division of its composing elements into three functional blocks is proposed. The design methodology is based on the carrying out of a series of interrelated tasks, which from a software engineering perspective, gives maximum performance, both for the development process and its subsequent updating.

The methodology was applied to the simulation of a nuclear power plant with success. In this case, two operational methods (batch mode and interactive mode) are proposed, determined by the simulation modules used. In each case, the operating mode is described and analyzed from the programmers's and the user's point of view.

1 Introduction

The complexity of the problems associated with nuclear plants, as well as other important factors, means that is essential for the former to have at their immediate disposal the capacity to simulate their procedures to a high degree of accuracy. Currently computer technology offers rapid solutions and graphics facilities which contribute to the development and improvement of huge simulation programs and their user interfaces based on multimedia features. Multimedia, by means of hypertext, sounds, images and video, gives the user both, a faster and more efficient analysis and a realistic feeling of the processes being simulated.

This paper presents a design methodology aimed at developing graphic interfaces for the simulation and control of industrial processes. There are three basic aspects to consider in this kind of developments: the adjustment of the simulation programme to the procedures to be simulated, the design of the graphic interface and the connection between both.

In the next two sections, the structure and design of a multimedia interface is analyzed, and three type of connections between the simulation programme and the interface are proposed for the development of the application. The problem of the adjustment of the simulated programme is not the concern of this present paper, although some comments are made regarding this subject.

Finally, the paper presents two developments designed for use in the Santa María de Garoña Nuclear Plant, in Burgos, Spain, in which two operational modes are proposed (batch mode and interactive mode), analyzing the objectives attainable in each case and studying the operation scheme for both alternatives of connection between the graphics interface and the simulation programme.

2 The Structure of Multimedia Interfaces

The concept of the multimedia interface arises from the need to facilitate communication between the user and the computer, so that the former can use the computer as intuitively as possible [1, 2].

The modules to consider in the development of a multimedia simulation application are: the simulation programme, the graphic interface and the connection between both.

The simulation programme deals with the task of making all of the calculations corresponding to the phenomenology to be represented. For this purpose, commercial programs, whose results have been well tested, are generally used.

The multimedia interface is made up of a series of elements at the user disposal, by means of which interaction with the simulation programme is established.

Regarding the composition of the modules, the possible coupling models [3] are: programme driven (the data representing events in the programme controls the interface) or interface driven (the user events control the application), or a mixture of the two.

2.1 Functional blocks in a Multimedia Interface

In order to facilitate the design of a multimedia interface and to make subsequent adaptations and simple systematic improvement easier, a division of the elements of which it is composed needs to be considered. This division can be made according to the function of each of the elements, in the following ways:

User Input. This can be said to be the set of all the objects which are permanently in attendance, ready to act on the actions of the user, in response to which they can generate messages to be processed by the corresponding control model. The most important of these are the input keys and fields, with all their variants: menus, windows, scroll bars, icons, etc.

Data Display. This group consists of all the multimedia representations needed by the user. They are the means by which he can visualize the state of the information which he is dealing with. Normally, objects belonging to this block do not allow any direct interaction with them, though the case may arise where this might prove necessary. If so, the object would then need to assume characteristics of the first category, the user input block.

The most important objects in this block are the output fields, graphs and images, which represent the reality to be simulated.

Control Model. This block incorporates the functional part of the application. In the area of simulation work, it permits a greater degree of accuracy and is capable of discerning between application data, application control and control flow.

Application data consist of the information shared by the interface and the simulation programme, which can be modified by either of them during the running of the programme.

Application control is the part of the interface which makes the application run in such a way that the users´s commands are attended to. These can change the course of the simulation or change the information to be presented to the user.

The control flow block is responsible for coordinating the internal flow of information among the objects, altering some of their attributes and guaranteeing coherence among the different elements. Its behaviour and structure must be designed according to the requirements of the user, in order to be able to model the problem in terms of objects. This allows the communication between the objects to be produced automatically and consequently also allows the programmer to forget about the problem of updating, since the objects themselves process the information supplied by the control flow block (see Figure 1).

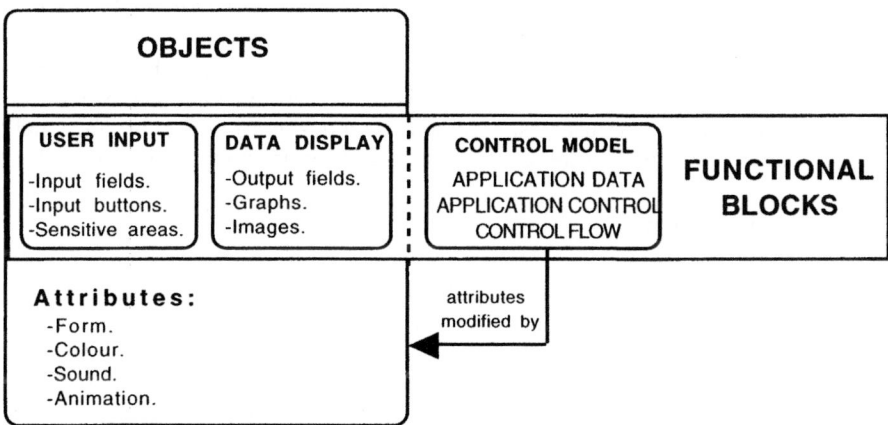

Figure 1. Object´s layout, functional blocks and interaction.

3 Design Methodology

During the design it is very important to consider the inherited and reuse properties of objects modeled and the minimal changes criteria for maintenance. Moreover, this step can be assessed as the realization of a series of interrelated tasks. These are:

Design and analysis of the data sources for the interface. These can be: files, simulation programme variables and/or variables of the interface itself. It will also be necessary to consider the kind of information, origins, format, etc. contained within these sources. The data sources also influence the structuring of the functional block Application Data. The resulting model is built using a standard method, such as one of those dealing with data flow or data structuring.

Screens Design. This operation should be carried out in consultation with the final user, in order to satisfy the desired functional requirements. In this stage, the capacity of the multimedia tools being used must be borne in mind.

This task corresponds to the identification of kinds of objects in OOP (Object Oriented Programming) using either bottom-up methodology or top-down or a mixture of both [4]. Part of this task consists of assigning the attributes which correspond to each of the objects present on each screen, such as colour, shape, size, movement, sound, etc.

Similarly, it is necessary to identify the actions which can be carried out on each object. As the range of possibilities is vast, only some of them will be used.

Materialization of the previously defined specifications, using the kinds of objects which the multimedia tool allows one to create and manage.

The elements which the screens are composed of can be defined by means of a specific metalanguage. Normally, to make this operation easier, a purpose-built tool is used, allowing one to work directly with the objects to be incorporated and generating its specification in this specific metalanguaje.

It is also necessary to establish the protocol for communication between objects, which allows the planned actions to be put into effect. This normally consists of a data structure which is accessible at every moment by any object and can inform about the state of each of the others.

4) Communication between the interface and the programme. This is established by means of a module whose function is to create and update the connection between the elements (variables, parameters, values, etc.) of the interface and the programme, so that a two directional exchange of information can be attained.

4 Implementation

This design methodology has been put into effect for the development of two applications of process simulation of a nuclear plant.

The development has been carried out using DataView (tool for multimedia and OOP developing) and with the simulation codes TRAC (Transient Reactor Analysis Code) and MAAP (Modular Accident Analysis Program) which simulates the phenomenology of the former. These codes are written in Fortran and they are widely used for nuclear plants.

Each application corresponds to two operational methods proposed: the batch mode and the interactive mode. The implementation of these operation modes is described in the following sections.

Initially the proposed objectives were the continuous assessment of the models by means of comparison with real sequences, in order to attain the adjustment of the parameters so that they adequately simulate the systems considered, as well as the compilation of successive simulations to be studied later. Working in the Batch mode these objectives were successfully achieved.

The next step consisted of the building of an application allowing one to work interactively with MAAP code. As a consequence of this, a graphic simulator has been developed for the training of operators and other staff. This interactive simulator also allows one to carry out studies of the behaviour of systems and to assess the operating procedures.

4.1 The Batch Mode

This operating mode is the one used normally when the aim is to analyze the results of the simulations in a friendly way.

With the development of this application one objective was the continuous assessment of the models by means of comparison with real sequences to attain the adjustment of the parameters, so that they should adequately simulate the systems considered. Another objective was to obtain a compilation of successive simulations for subsequent studies, such as risk analysis, study of the behavior of the system, and assessment of operating procedures (Figure 2).

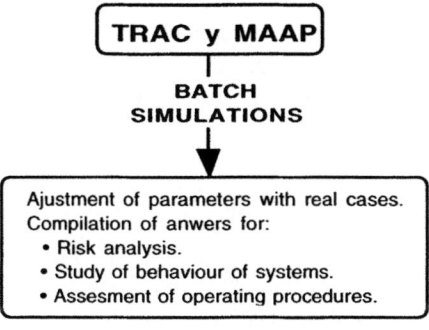

Figure 2. Batch mode objectives.

In practice, the application follows the steps indicated in Figure 3. This operating mode corresponds with the interface driven model. The process is set into motion by the data files which take in both the current situation and the operations and incidents which will take place during the simulation. The results of the simulation are stored in output files, this being the end of the batch process. Subsequently, these files receive a multimedia representation treatment.

For the development and subsequent implementation of this application, the steps described in section 3, have been followed, that is:

An extensive analysis of the structure and content of the simulation programmes result files was made. Using this information, the data was selected and brought together to be connected with the environment´s variables.

The multimedia representation was defined according to the information to be visualized, the objects and their organization by means of the various screens. The graphic editor DV-draw was used for the materialization of the screens.

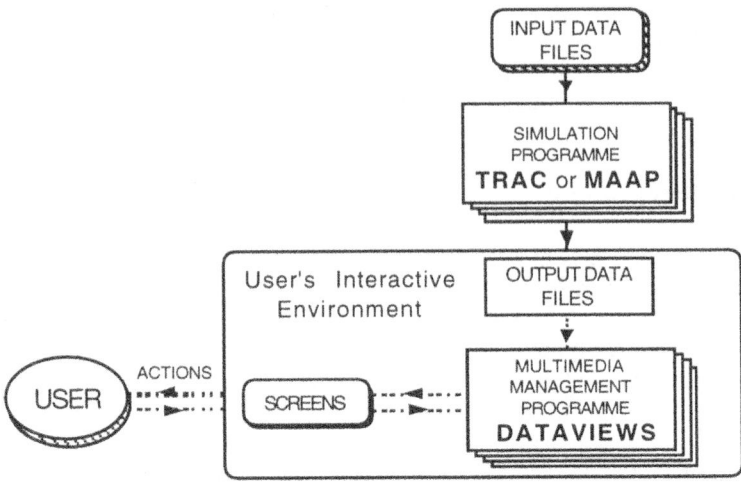

Figure 3. The process in batch and subsequent visualization

The information flow control was established, both inside the interface and in communication with the output files, the results of the simulation. The DV-tools library was used for this operation. One of the application´s screens is shown in Figure 4.

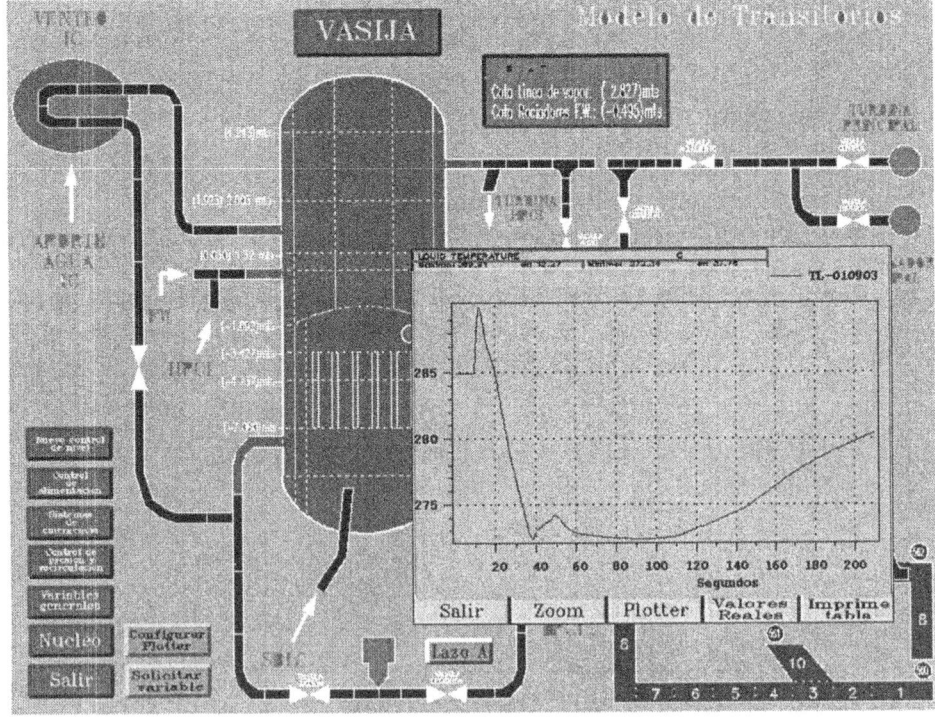

Figure 4. Example of screen for Batch Mode.

In this mode, the two above mentioned coupling models are mixed together, although the programme driven model is the predominant. The interaction process is carried out through the screens which consist of objects whose properties are associated with variables of the MAAP code.

4.2. The Interactive Mode

The next stage consisted of the building of an application which would enable work to be carried out interactively with the MAAP code. As a consequence of this, a graphic simulator, called TS-G, for the training of operators and other staff has been developed. This interactive simulator also allows one to undertake studies and analyses of any incidents which might have taken place. Figure 5 shows a schema of these objectives.

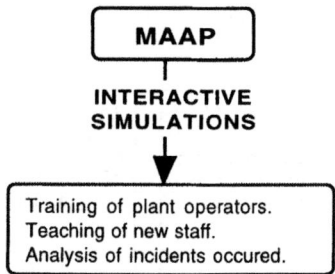

Figure 5. Interactive mode objectives.

The main difference between this mode and the batch mode is found in the data source which, in this case, is made up of the states of the systems (valves, pumps, etc.) determined by the simulation process. A schema of the operational method in the interactive mode can be seen in Figure 6.

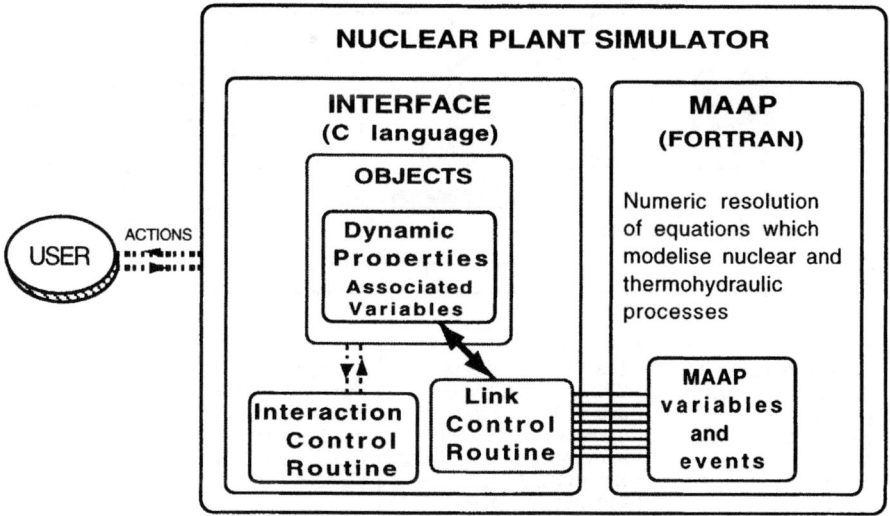

Figure 6. Interactivity with MAAP

This applications screens represent systems by means of objects whose dynamics properties are associated to the systems´ variables. A routine of creation of connections, which is established when starting the programme, make these properties link up with the variables and events of the MAAP code, thus establishing interactive communication. The application must also dispose of the corresponding event control routines which enable connection with the user.

The displays of the simulator are a replica of the ones being used in the SPDS (Safety Parameters Display System) at the control room. TS-G operating method was conceived in such a way that several users can interact simultaneously on the process. This allows a workstation to be assigned to an instructor who can provoke various malfunctions which the operators must correct from the other consoles. For this purpose the multistation capacities offered by the graphic standard X11 [5] have been used. Figure 7 shows the final layout.

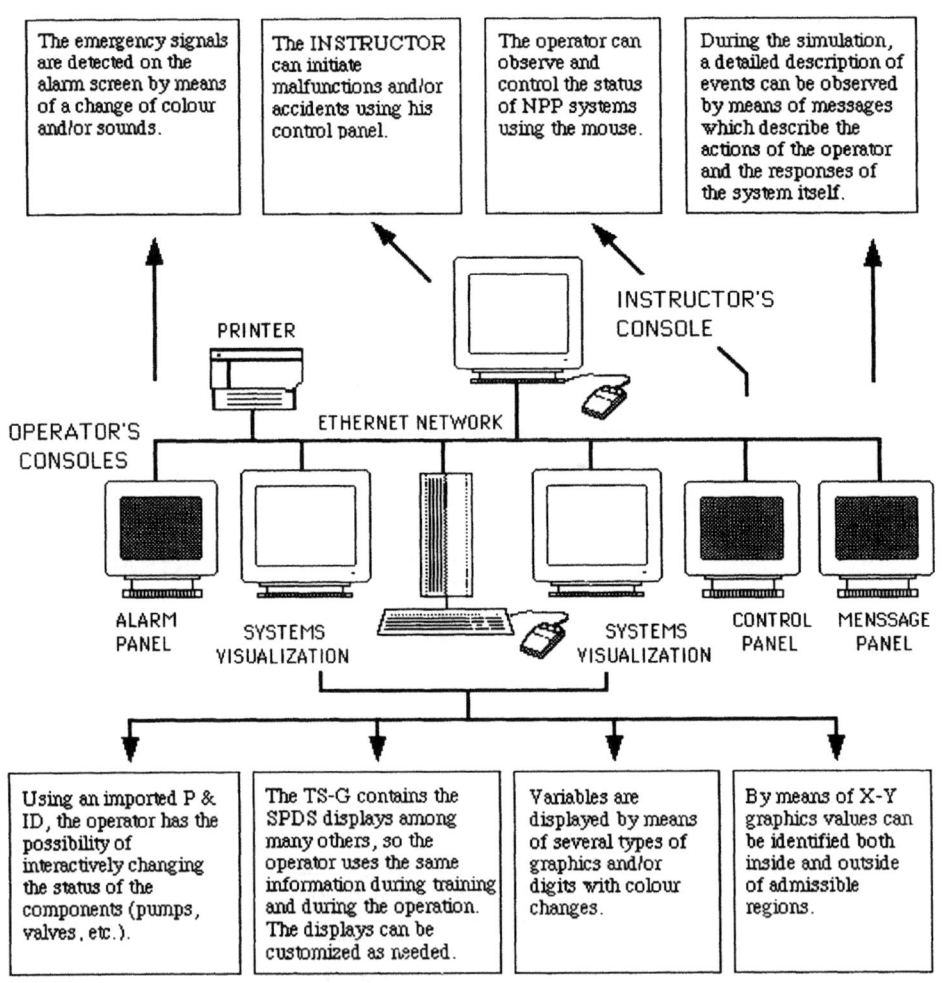

Figure 7. Final layout and features of Interactive Mode.

Another possibility which can be incorporated into the simulator is that of visualizing the plant´s data in real time, by making use of the data acquisition system already installed in it. In this process a microwave radio-link is used to overcome the problem of distance (the University of Cantabria is 140 km. from the Nuclear Power Plant).

The following hardware has been used in the development of this application:

• Personal Iris 3D, Silicon Graphics Workstation where the application is put into effect. This is also the station which the instructor work from.

• Five X11 terminals, dedicated to the visualization of different systems, one of which is used as the operator station.

• A printer to print graphics of the evolution of selected variables.

5 Conclusions

In this paper, a design methodology for multimedia environments for the simulation of industrial processes is presented. After a brief analysis of the structure of multimedia interfaces for readily available programmes, the following procedure is proposed.

1) Design and analysis of the data sources. With regards to this stage, it should be noted that it is essential to make a detailed study of the simulation programme, especially in the area of structuring and data input/output.

2) Screens Design. This task can be performed jointly with the final user and, for this purpose, an Object Oriented Analysis method is recommended.

3) Materialization of the specifications. For this stage, it is extremely important to select the appropriate software multimedia to be used in the development, in order to be sure that it can incorporate the previously defined requirements.

4) Communication interface-programme. Normally a routine deals with establishing this connection, defining common memory zones, accessible from the interface and the programme, so that a two directional exchange of information can be attained.

This methodology has been successfully applied in the simulation of the operating of a nuclear plant in two operating modes (batch and interactive modes). In the description of this process, some methodological proposals are also made.

Currently we are working on a further development, the incorporation of an expert system to advise the operator on operations to be executed in certain emergency situations.

Acknowledgments

This work has been supported by Nuclenor S. A.. We would like to acknowledge Nuclenor's Staff, especially J. V. López and J. González for their advice.

References

1. Dumas, J.S. Designing User Interfaces for Software. Prentice-Hall, 1988.

2. Jonassen, D., Mandl, H. Designing Hypermedia for Learning. Sringer-Verlag. 1990.

3. Bustío, F., P. Corcuera, J. García and E. Mora: Multimedia environments designs for simulation. Cybernetics and Systems Volume 25, Number 1. January-February 1994. Pages 63-71.

4. Meyer, B. Object Oriented Software Construction. Prentice-Hall, 1988.

5. Quercia, Valerie, and Tim O´Reilly. X Window System User´s Guide vol.3. Sebastopol, CA: O´Reilly & Associates, 1988.

TurboBrain: A Neural Network with Direct Learning Based on Linear or Non-Linear Threshold Logics

Daniel M. Dubois [1], Germano Resconi [2], Alessandro Raymondi [2]

1: Institut de Mathématique, Université de Liège
Avenue des Tilleuls 15, B-4000 Liège, Belgium

2: Istituto di Matematica, Universita Cattolica
Via Trieste 17, I-25121 Brescia, Italy

Abstract. This paper deals with a significant extension of the neural Threshold Logic pioneered by McCulloch and Pitts. The output of their formal neuron is given by the Heaviside function with an argument depending on a linear weighted sum of the inputs and a threshold parameter. All Boolean Tables cannot be represented by such a formal neuron. For example the exclusive OR and the Parity Problem need hidden neurons to be resolved. A few years ago, Dubois proposed a non-linear fractal neuron to resolve the exclusive OR problem with only one single neuron. Then Dubois and Resconi introduce the Non-linear Threshold Logic, that is to say a Heaviside Function with a non-linear sum of the inputs which can represent any Boolean Tables with only one neuron where the Dubois' non-linear neuron model is a Heaviside Fixed Function. In this framework the Supervised Learning is Direct, that is to say without recursive algorithms for computing the weights and threshold, related to the new foundation of the Threshold Logic by Resconi and Raymondi. This paper will review the main aspects of the linear and non-linear threshold logic with direct learning and applications in pattern recognition with the software TurboBrain. This constitutes a new tool in the framework of neural CAST, Computer Aided Systems Theory and Technology, proposed by F. Pichler.

1 Introduction

This paper deals with a significant extension of the McCulloch and Pitts formal neuron [10,12,14] in the framework of the Threshold Logic [1,2]. The models of the Dubois' Fractal Neural Network [3,4,5] and the Fukushima et al [9] Neocognitron give the original framework of the new Non-linear Threshold Logic.

This Non-linear Threshold Logic avoids some important critics on this Threshold Logic collected in a book by Minsky and Papert [11]. A difficult problem is described in Rumelhart et al [15,16]: "One of the problem given a good deal of discussion by Minsky and Papert is the parity problem, in which the output required is 1 if the input pattern contains an odd number of 1s and 0 otherwise. This is a very difficult problem because the most similar patterns (those which differ by a single bit) require different answers. The XOR problem is a parity problem with input patterns of size two.".

A few years ago, Dubois [3,4,5] proposed the following Quadratic Activation Function to resolve the XOR problem with only one single neuron without hidden neuron

(1) $$y = \alpha + 4\mu(s_1 x_1 + s_2 x_2)(1 - \beta(s_1 x_1 + s_2 x_2))$$

where $x_1 \in \{0,1\}, x_2 \in \{0,1\}, s_1 \in [-1, +1], s_2 \in [-1, +1], \mu \in [0,1], \alpha \in [0,1], \beta \in [-1, +1]$. The 16 Boolean rules with two inputs and the 4 Boolean rules with one input are well obtained from this equation (1) with the values of the parameters given in Dubois [5]. Dubois and Resconi [6] validated completely this proposition by the demonstration that Dubois' Quadratic Activation Function is a Heaviside fixed function solution. For example, a single non-linear neuron with two inputs resolved the Boolean Table for exclusive OR, XOR, which is the following XOR Heaviside fixed activation function:

(2) $$y = 2.(x_1 + x_2).(1 - (x_1 + x_2)/2) = \Gamma (2x_1 + 2x_2 - (x_1^2 + x_2^2 + 2x_1 x_2))$$

Thus the classical XOR problem considered until now as impossible to realise with only one single neuron is so completely resolved with a Non-linear Threshold Logic. Moreover, the Non-linear Threshold Logic resolved all the functions with one single neuron with every number of inputs. It is shown that the very difficult parity problem has also a vary elegant solution with one single non-linear neuron. The Non-linear Threshold Logic will permit to diminish a great quantity of hidden neurons in the neural networks.

2. The Non-linear Threshold Logic

Let us start with the following Non-linear Threshold Function for one single neuron with n inputs introduced by Dubois and Resconi [6]

(3) $$y = \Gamma(g(x)) = \Gamma\left(a_{00} + \sum a_{i0} x_i + \sum a_{ij} x_i x_j + f(x)\right)$$

where a_{00}, a_{i0}, a_{ij} are real numbers and f is a non-linear function, beyond the degree two, of the variables $x_1, x_2, \dots, x_n$, with $x_{ij} \in \{0,1\}$, $i = 1,2, \dots, n$. The parameter a_{00} is similar to the threshold in the McCulloch and Pitts model and the linear term $\sum a_{i0} x_i$ is similar to the weighted inputs in the McCulloch and Pitts model, and $\Gamma(g(x))$ is the Heaviside function. Recall that $\Gamma(g(x)) = 1$ if and only if $g(x) > 0$, otherwise $\Gamma(g(x)) = 0$. In the case of a single neuron with two inputs (without loss of generality for the method), the equation (3) is written as

(4) $$y = \Gamma(g(x)) = \Gamma\left(w_1 x_1 + w_2 x_2 + (a_{11} x_1^2 + a_{22} x_2^2 + 2a_{12} x_1 x_2) - \theta\right)$$

where x_i are the 0 or 1 inputs of a single neuron, w_i the weights on the corresponding inputs, a_{ij} are the new coefficients of the quadratic extension of the linear argument function of the McCulloch-Pitts model, θ is a threshold parameter and

(5)
$$g(x) = w_1 x_1 + w_2 x_2 + \left(a_{11} x_1^2 + a_{22} x_2^2 + 2a_{12} x_1 x_2\right) - \theta$$

is the argument of the Heaviside function $\Gamma(g(x))$. The general mathematical formalism of this new Non-linear Threshold Logic is described in Dubois and Resconi [6]: it is demonstrated that all the Boolean functions can be realised with one single neuron.

The argument function $g(x)$, given by equation (5), can be a "*Heaviside Fixed Function*" denoted $h(x)$ by Dubois and Resconi [6], when $g(x)$ satisfies to the following identity:

(6)
$$\Gamma(h(x)) = h(x) = y$$

Let us explicitly demonstrate now that such a general "Heaviside Fixed Function" $h(x)$ exists for the Boolean logic of one neuron with two inputs. The function $h(x)$ will be shown to be a non-linear continuous function and it is then possible to replace the non-derivable Heaviside function by this non-linear Heaviside fixed function, the derivatuve of which being possible. So, in that case, an output y is equal to the $h(x)$ argument of the Heaviside function, $y = h(x)$ where $h(x)$ is a derivable continuous function, for which, from the output value y, a finite number of values x can be obtained.

The 16 Boolean tables with 2 inputs x_1 and x_2, given in TABLE 1, can be defined by

x_1 x_2	y
0 0	y_1
1 0	y_2
0 1	y_3
1 1	y_4

where the output y can take the values y_i which are 0 or 1. From relation (6) and the equations (4) and (5), we deduced the general conditions (see Dubois and Resconi [6])

(7)
$$\theta = -y_1$$
$$w_1 = y_2 - a_{11} + \theta$$
$$w_2 = y_3 - a_{22} + \theta$$
$$a_{12} = (-y_2 - y_3 + y_4 - \theta)/2$$

for the parameters of the following quadratic activation function (equation (5) where g is now a Heaviside fixed function h):

(8)
$$g(x) = w_1 x_1 + w_2 x_2 + \left(a_{11} x_1^2 + a_{22} x_2^2 + 2a_{12} x_1 x_2\right) - \theta$$

The 4 eqs. (7) define 4 parameters, thus the values of two parameters are to be determined, a_{11} and a_{22} for example. These parameters are similar to the linear weights w because in Boolean logic, $x^2 = x$.

TABLES 2 and 3 give the parameters for eqs. (1) and (8).

TABLE 1: Classification of the 16 Boolean rules for 2 inputs and 4 rules for 1 input

R1			R3			R5			R7		
x_1	x_2	y	x_1	x_2	y	x_1	x_2	y	x_1	x_2	y
0	0	0	0	0	0	0	0	1	0	0	1
0	1	1	0	1	1	1	1	1	0	1	1
1	0	1	1	0	1	1	0	1	1	0	1
1	1	1	1	1	0	1	1	0	1	1	1

R2			R4			R6			R8		
x_1	x_2	y	x_1	x_2	y	x_1	x_2	y	x_1	x_2	y
0	0	0	0	0	1	0	0	1	0	0	0
0	1	0	0	1	0	0	1	0	0	1	0
1	0	0	1	0	0	1	0	0	1	0	0
1	1	1	1	1	1	1	1	0	1	1	0

R9			R11			R13			R15		
x_1	x_2	y	x_1	x_2	y	x_1	x_2	y	x_1	x_2	y
0	0	0	0	0	1	0	0	0	0	0	1
0	1	0	0	1	0	0	1	0	0	1	0
1	0	1	1	0	1	1	0	1	1	0	1
1	1	0	1	1	1	1	1	1	1	1	0

R10			R12			R14			R16		
x_1	x_2	y	x_1	x_2	y	x_1	x_2	y	x_1	x_2	y
0	0	1	0	0	0	0	0	1	0	0	0
0	1	1	0	1	1	0	0	1	0	1	1
1	0	0	1	0	0	1	0	0	1	0	0
1	1	1	1	1	0	1	1	0	1	1	1

R17		R18		R19		R20	
x	y	x	y	x	y	x	y
0	1	0	0	0	1	0	0
1	0	1	1	1	1	1	0

TABLE 2: Parameters of eq. (1) for the Boolean rules given in Table 1 (Dubois [5])

R1	R3	R5	R7
$s_1 = +1$ $s_2 = +1$ $\mu = 3/8$ $\alpha = 0$ $\beta = +1/3$	$s_1 = +1$ $s_2 = +1$ $\mu = 1/2$ $\alpha = 0$ $\beta = +1/2$	$s_1 = +1$ $s_2 = +1$ $\mu = 1/8$ $\alpha = +1$ $\beta = +1$	$s_1 = +1$ $s_2 = 0$ $\mu = 1/4$ $\alpha = +1$ $\beta = +1$
R2	**R4**	**R6**	**R8**
$s_1 = -1$ $s_2 = -1$ $\mu = 1/8$ $\alpha = 0$ $\beta = -1$	$s_1 = -1$ $s_2 = -1$ $\mu = 1/2$ $\alpha = +1$ $\beta = -1/2$	$s_1 = -1$ $s_2 = -1$ $\mu = 3/8$ $\alpha = +1$ $\beta = -1/3$	$s_1 = -1$ $s_2 = 0$ $\mu = 1/4$ $\alpha = 0$ $\beta = -1$
R9	**R11**	**R13**	**R15**
$s_1 = +1$ $s_2 = -1$ $\mu = 1/8$ $\alpha = 0$ $\beta = -1$	$s_1 = +1$ $s_2 = -1$ $\mu = 1/8$ $\alpha = +1$ $\beta = +1$	$s_1 = +1$ $s_2 = 0$ $\mu = 1/4$ $\alpha = 0$ $\beta = 0$	$s_1 = 0$ $s_2 = -1$ $\mu = 1/4$ $\alpha = +1$ $\beta = 0$
R10	**R12**	**R14**	**R16**
$s_1 = -1$ $s_2 = +1$ $\mu = 1/8$ $\alpha = +1$ $\beta = +1$	$s_1 = -1$ $s_2 = +1$ $\mu = 1/8$ $\alpha = 0$ $\beta = -1$	$s_1 = -1$ $s_2 = 0$ $\mu = 1/4$ $\alpha = +1$ $\beta = 0$	$s_1 = 0$ $s_2 = +1$ $\mu = 1/4$ $\alpha = 0$ $\beta = 0$
R17	**R18**	**R19**	**R20**
$s_1 = -1$ $s_2 = 0$ $\mu = 1/2$ $\alpha = +1$ $\beta = -1/2$	$s_1 = +1$ $s_2 = 0$ $\mu = 1/2$ $\alpha = 0$ $\beta = +1/2$	$s_1 = +1$ $s_2 = 0$ $\mu = 1/2$ $\alpha = +1$ $\beta = +1$	$s_1 = -1$ $s_2 = 0$ $\mu = 1/2$ $\alpha = 0$ $\beta = -1$

TABLE 3: Parameters of eq. (8) for the Boolean rules in Table 1 (Dubois & Resconi [6])

R1	R3	R5	R7
$\Theta = 0$	$\Theta = 0$	$\Theta = -1$	$\Theta = -1$
$w_1 = +3/2$	$w_1 = +2$	$w_1 = +1/2$	$w_1 = +1$
$w_2 = +3/2$	$w_2 = +2$	$w_2 = +1/2$	$w_2 = 0$
$a_{11} = -1/2$	$a_{11} = -1$	$a_{11} = -1/2$	$a_{11} = -1$
$a_{22} = -1/2$	$a_{22} = -1$	$a_{22} = -1/2$	$a_{22} = 0$
$a_{12} = -1/2$	$a_{12} = -1$	$a_{12} = -1/2$	$a_{12} = 0$

R2	R4	R6	R8
$\Theta = 0$	$\Theta = -1$	$\Theta = -1$	$\Theta = 0$
$w_1 = -1/2$	$w_1 = -2$	$w_1 = -3/2$	$w_1 = -1$
$w_2 = -1/2$	$w_2 = -2$	$w_2 = -3/2$	$w_2 = 0$
$a_{11} = +1/2$	$a_{11} = +1$	$a_{11} = +1/2$	$a_{11} = +1$
$a_{22} = +1/2$	$a_{22} = +1$	$a_{22} = +1/2$	$a_{22} = 0$
$a_{12} = +1/2$	$a_{12} = +1$	$a_{12} = +1/2$	$a_{12} = 0$

R9	R11	R13	R15
$\Theta = 0$	$\Theta = -1$	$\Theta = 0$	$\Theta = -1$
$w_1 = +1/2$	$w_1 = +1/2$	$w_1 = +1$	$w_1 = 0$
$w_2 = -1/2$	$w_2 = -1/2$	$w_2 = 0$	$w_2 = -1$
$a_{11} = +1/2$	$a_{11} = +1/2$	$a_{11} = 0$	$a_{11} = 0$
$a_{22} = +1/2$	$a_{22} = +1/2$	$a_{22} = 0$	$a_{22} = 0$
$a_{12} = -1/2$	$a_{12} = -1/2$	$a_{12} = 0$	$a_{12} = 0$

R10	R12	R14	R16
$\Theta = -1$	$\Theta = 0$	$\Theta = -1$	$\Theta = 0$
$w_1 = -1/2$	$w_1 = -1/2$	$w_1 = -1$	$w_1 = 0$
$w_2 = +1/2$	$w_2 = +1/2$	$w_2 = 0$	$w_2 = +1$
$a_{11} = +1/2$	$a_{11} = +1/2$	$a_{11} = 0$	$a_{11} = 0$
$a_{22} = +1/2$	$a_{22} = +1/2$	$a_{22} = 0$	$a_{22} = 0$
$a_{12} = -1/2$	$a_{12} = -1/2$	$a_{12} = 0$	$a_{12} = 0$

R17	R18	R19	R20
$\Theta = -1$	$\Theta = 0$	$\Theta = -1$	$\Theta = 0$
$w_1 = -2$	$w_1 = +2$	$w_1 = +2$	$w_1 = -2$
$w_2 = 0$	$w_2 = 0$	$w_2 = 0$	$w_2 = 0$
$a_{11} = +1$	$a_{11} = -1$	$a_{11} = -2$	$a_{11} = +2$
$a_{22} = 0$	$a_{22} = 0$	$a_{22} = 0$	$a_{22} = 0$
$a_{12} = 0$	$a_{12} = 0$	$a_{12} = 0$	$a_{12} = 0$

The proposition to consider a non-linear threshold function which is a Heaviside Fixed Function has some similarities with the model of the following activation function of Fukushima et al [8],

(9) $\qquad \phi(g(x)) = \max(g(x),0)$

where ϕ is equal to the greatest value between $g(x)$ and 0, so that the function ϕ is equal to its positive argument g which is a non-linear function.

2.1 Exclusive OR and Parity Problems

Let us demonstrate that our Non-linear Threshold Logic permits to give a general Heaviside fixed function which resolves the parity problem with any input patterns. Let us start with the parity problem with two inputs from the exclusive OR (XOR) only with one single neuron. For XOR, the outputs are given by $y = (0,1,1,0)$, so from the first equation of (7), we obtain the numerical value of the threshold $\theta = 0$, and, in choosing the numerical values of the free parameters as $a_{11} = 0$ and $a_{22} = 0$, the three other parameters are then given by $w_1 = 1$, $w_2 = 1$, $a_{12} = -1$.
Putting these values in the general Heaviside Fixed Function (8), we obtain a Heaviside fixed activation function for the XOR, which is the parity problem for two inputs

(10) $\qquad y = x_1 + x_2 - 2x_1 x_2$

which can be written as

(11) $\qquad y = (1 - (2_1 x - 1)(2x_2 - 1))/2$

and we can verify that it gives the correct numerical values of the output y in function of the different inputs x_1 and x_2 of the table XOR.
Equations (10) and (11) can be generalised with every number of inputs n for resolving the Parity problem by the following equation (12)

$$y = \sum x_i - 2 \sum_{1 \neq 2} x_{i1} x_{i2} + (-2)^2 \sum_{1 \neq 2 \neq 3} x_{i1} x_{i2} x_{i3} + \dots + (-2)^{n-1} \sum_{1 \neq 2 \neq \dots \neq n} x_{i1} x_{i2} \dots x_{in}$$

which can be written in the following compact form

(13) $\qquad y = (1 - (-1)^n (2x_1 - 1)(2x_2 - 1) \dots (2x_n - 1))/2$

Indeed, the general Heaviside Fixed Function in our Non-linear threshold logic for every number n of input variables x_n, is capable of resolving the Parity problem.

The Threshold Logic, using linear argument of the Heaviside function, is not capable of resolving all Boolean functions with only one neuron. As pointed out by Rumelhart et al [14], "it requires at least N hidden units to solve parity with patterns of length N" ... "there is always a recoding (i.e. an internal representation) of the input patterns in in the hidden units in which the similarity of the patterns among the hidden units can support every required mapping from the input to the output units".

Our Non-linear Threshold Logic needs only one single non-linear neuron with N inputs with the equations (12-13). The non-linear part gives an internal representation inside the single neuron similar to the internal representation by the hidden neurons.

2.2 Unifying Linear and Non-linear Logics

In this section we want to realise Boolean functions by the Non-linear Threshold Logic and to obtain the *formal solutions* both for the linear and non-linear weights. In fact by a particular operation the input function that we want to realise with a non-linear neuron is expanded so that it holds also the non-linear terms: *realising with a linear neuron the expanded function we will find both the linear and non-linear coefficients of the input function.* The main result of this section is that a non-linear realisation *is equivalent* to a linear realisation.

General Theorem [7]: Any Boolean function F of dimension n, can be realised by a single non-linear neuron with integer parameters $w_1,..., w_n , a_{11},..., a_{nn} ,..., a_{12...n} , \theta$ such that the following equation (14)

$$g(x_1,...,x_n) = w_1 x_1 + w_2 x_2 + ... + w_n x_n + a_{11} x_1 + a_{12} x_1 x_2 + ... + a_{1n} x_1 x_n + ... + a_{123n} x_1 x_2 x_3 x_n - \theta$$

gives $g(x_1,...,x_n) = 1$ if $F(x_1,...,x_n) = 1$ and $g(x_1,...,x_n) = -1$ if $F(x_1,...,x_n) = 0$

The proof is given in Dubois and Resconi [7, p.53-54].

Remark 1 [7]: The parameters w_i , a_{ijn} , θ are related to the parameters w'_i , a'_{ijn} , θ' of the Heaviside Fixed Function: $w_i = 2w'_i, a_{ijn} = 2a'_{ijn}, \theta = 2\theta' - 1$.

Remark 2 [7]: Since the weights realising a Boolean function F by the Heaviside Fixed Function are unique, the preceding remark shows that the weights obtained by the theorem are also unique.

A simple algorithm for direct learning can be obtained for computing the linear and non-linear weights and threshold of any Boolean function from this theorem. It is only necessary to test once each bit of the function and to add at each step the necessary linear or non-linear term. Indeed, from a Boolean function $F(x_1,..., x_n)$, *ordered as follows* $F(0,0,...,0) = y_0$, $F(1,0...,0) = y_1$, ... , $F(1,1,...,1) = y_{2n-1}$, a function $g(x_1, ... ,x_n)$, initially equal to zero, is iteratively built in the following order

1. If $y_0 = 0$ then $g(x_1, ... ,x_n) = -1$ *so that* $\theta = -1$
   If $y_0 = 1$ then $g(x_1, ... ,x_n) = 1$ *so that* $\theta = 1$
2. If $y_1 = 0$ and $g(x_1, ... ,x_n) = 1$ then $g(x_1, ... ,x_n) = 1 - 2x_1$ *so that* $w_1 = -2$
   If $y_1 = 1$ and $g(x_1, ... ,x_n) = -1$ then $g(x_1, ... ,x_n) = -1 + 2x_1$ *so that* $w_1 = +2$
   Otherwise $g(x_1, ... ,x_n)$ remains unchanged.

In practice, the general rule consists to check each output bit $y_i = F(x_1, \ldots, x_n)$ with $0 \leq i \leq 2^n-1$, in computing the following error function

(15) $\qquad e_i = 2y_i - 1 - g(x_{i1}, \ldots, x_{in})$

and then correcting this error in adding to $g(x1, \ldots, xn)$ a term given by

(16) $\qquad e_i . x_{a1} x_{a2} \ldots x_{ak}$ with $1 \leq k \leq n$ and $a_j \in \{h \mid x_{ih} = 1\}$

that is to say to add a term which is the product of the error by all the input variables equal to 1. This procedure builds a function which obeys the preceding theorem.

For supervised learning and pattern recognition, only Partial Boolean Tables dealing with samples to be learnt are to be considered.

Definition of partial functions [7]: F is a partial Boolean function of dimension n if

(17) $\qquad y_i = F(x_{i1}, x_{i2}, \ldots, x_{in}) \quad i \in S$ where $S \subset \{0, \ldots, 2^n - 1\}$

x_{ij} are some of all the possible values 1,0 of the independent variables $x_1, \ldots, x_n$, y_i are the values 0,1 of the Boolean function F associated to the inputs x_{ij} . This definition means that the Boolean *function F isn't defined in all the input space but only in a subset of it* (indicated with S).
Now when we realise a partial Boolean function F in the form (17) we have

(18) $\qquad y_i = \Gamma \left[\sum_{j=1}^{n} w_j x_{ij} - \theta \right]$

where x_{ij} are the input values and y_i are the output values of the neuron, only if $i \in S$.

Definition of new partial functions [7]: Given a Boolean function F of dimension n, partial or not, a new function F^* of dimension $n+1$ is defined as

(19) $\qquad \mathbf{v}(x_1, \ldots, x_{n+1}) \quad F^*(x_1, \ldots, x_n, x_i x_j) = F(x_1, \ldots, x_n)$ if $x_{n+1} = x_i x_j$
(20) $\qquad \mathbf{v}(x_1, \ldots, x_{n+1}) \quad F^*(x_1, \ldots, x_n, x_{n+1}) = nil$ if $x_{n+1} \neq x_i x_j$

with $i, j \leq n$, where *nil* means that the function is undefined for this input.

It is easy to see that this is always a partial function that gives the same weights $w_1, \ldots, w_n$ as F. The weight w_{n+1} is equivalent by definition to the non-linear term $a_{i,j}$. Indeed the output of the neuron is

(21) $\quad y = \Gamma[w_1 x_1 + \ldots + w_n x_n + w_{n+1} x_{n+1} - \theta] = \Gamma[w_1 x_1 + \ldots + w_n x_n + w_{n+1} x_i x_j - \theta]$

The *nil* correspond to cases which will not be considered for learning but for recognition after the learning procedure is performed.

Algorithms for direct learning and pattern recognition were developed from partial Boolean Tables and implemented in the software TurboBrain, see [7, 8] for details.

2.3 Polynomial Extension of the Non-linear Threshold Logic

In the original Dubois' Fractal activation function

$$(22) \quad f(x_1, x_2, ..., x_n) = g(\sum_{i=1}^{N} s_i x_i)$$

$g(y)$ is a polynomial form. When $s_i = 2^i$ and $x_i \in \{0, +1\}^N$ the function $\sum_{i=1}^{N} s_i x_i = M$ is the natural number X associated to the vector x_i of bits. With these conditions

$$(23) \quad f(x_1, x_2, ..., x_n) = g(M) = a_0 + a_1 M + a_2 M^2 + ... + a_n M^n$$

by the interpolation theorem, at any set of discrete values for f, it exists one and only one set of coefficients a_i that give the polynomial model of the assigned function.

For the XOR Boolean function $f(0,0) = -1$, $f(1,0) = 1$, $f(0,1) = 1$, $f(1,1) = -1$, we obtain

$$(24) \quad M(x_1, x_2) = s_1 x_1 + s_2 x_2 = 2^0 x_1 + 2^1 x_2$$

and $M(0,0) = 0$, $M(1,0) = 1$, $M(0,1) = 2$ and $M(1,1) = 3$

With the interpolation method, we can write the polynomial form for XOR

$$(25) \quad f(x1, x2) = -1 + 3M - M^2 = -1 + 2x_1 + 2x_2 - 4x_1 x_2$$

For the function $f(x_1, x_2, ..., x_n) = g(M) = a_0 + a_1 M + a_2 M^2 + ... + a_n M^n$ the Newton representation of the interpolation polynomial function with the data (M_k, f_k), $k = 1, 2, ..., n$, is

$$(26) \quad f(x_1, x_2, ..., x_n) =$$

$$b_0 + b_1 (M - M_1) + b_2 (M - M_1)(M - M_2) + ... + b_{n-1}(M - M_1)(M - M_2)....(M - M_{n-1})$$

where

$$(27) \quad b_0 = f_1$$

$$b_1 = \frac{f_2 - f_1}{x_2 - x_1} = \frac{\Delta f}{x_2 - x_1}$$

$$b_2 = \frac{f_3 - 2f_2 + f_1}{(x_3 - x_1)(x_2 - x_1)} = \frac{\Delta^2 f}{(x_3 - x_1)(x_2 - x_1)} \quad,$$

....

$$b_{n-1} = \frac{\Delta^n f}{\prod\limits_{k=1}^{n} (x_{k+1} - x_1)}$$

With the Newton representation the coefficients of the polynomial form is given directly by the output of the function $f(x_1, x_2, ..., x_n)$.

For the XOR, the data are

x_1	x_2	F_k	M_k
0	0	0	0
1	0	1	1
0	1	1	2
1	1	0	3

and we can compute

$$(28) \quad b_0 = f_1 = 0, \ b_1 = \frac{\Delta f}{M_2 - M_1} = 1, \ b_2 = \frac{\Delta^2 f}{(M_3 - M_1)(M_2 - M_1)} = \frac{1}{2},$$

$$b_3 = \frac{\Delta^4 f}{(M_4 - M_1)(M_3 - M_1)(M_2 - M_1)} = 0$$

The Newton representation of XOR is

(29) $f(x_1, x_2) = M - M(M-1)/2$

and with $M = 2^0 x_1 + 2^1 x_2$, we obtain the polynomial form

(30) $f(x_1, x_2) = x_1 + x_2 - 2x_1 x_2$

which is the Heaviside Fixed Function for XOR.
A minimal polynomial form can be computed with the Newton representation. In fact a polynomial form $P_{min}(x_1, x_2, ..., x_n)$ exists with a minimum degree for which

(31) $\Gamma(f(x_1, x_2, ..., x_n)) = \Gamma(P_{min}(x_1, x_2, ..., x_n))$

This theory of the minimal polynomial form is given in Dubois and Resconi [7].

Now, let us show some applications in neuroCAST with the software TurboBrain.

3 NeuroCAST with the Software TurboBrain

Let us consider a neural network with n inputs: x_1, x_2, x_3, ..., x_n, where $x_i \in \{0,1\}$ i = 1,2,...,n are Boolean variables and m outputs are: y_1, y_2, y_3, ..., y_m, where $y_j \in \{0,1\}$ j = 1,2,...,m are Boolean variables. The supervised learning considers samples of inputs and the associated outputs presented to the network. So the problem to be resolved is presented as the following Partial Truth Table with multiple outputs:

	x_1	x_2	...	x_n	*	y_1	y_2	...	y_m
1st sample:	0	0		0	*	0	1		0
2nd sample:	1	0		0	*	1	0		1
3rd sample:	1	0		1	*	1	0		0
other cases:	nil	nil		nil	*	nil	nil		nil

The direct learning of the input patterns in different positions or scales in the vision field is supervised, i. e. the wanted different outputs are coded by the end-user after his criteria for the future recognition of similar patterns. In the non-linear direct learning, the number of neurons is exactly equal to the number of outputs and does not depend on the number of inputs. All the non-linear neurons learn in a parallel mode. TurboBrain is able to recognise both the learnt patterns and similar patterns. It is able of identification and classification of patterns. TurboBrain can finally memorise and rebuild any presented pattern after a direct learning.
With TurboBrain, the end-user can choose between 3 types of logic: the linear logic (see Resconi and Raymondi [12] for details) or non-linear threshold logic and the fixed non-linear threshold logic, this 3rd logic doesn't use the Heaviside function. All the parameters, that is to say the threshold and the weights, of both the linear and non-linear neurons are given with integer values.

Let us give a few examples of the power of TurboBrain to give solutions to problems.

3.1 Solution of the Parity Problem

In the parity problem the output neuron y is 1 when the number of input neurons equal to 1 is odd and 0 otherwise. After a direct supervised learning, with 3 inputs x_1, x_2 x_3 and 1 output y. TurboBrain gives the 3 following solutions:

For the Linear Threshold Logic, TurboBrain computes the 4 hidden neurons y_1, $y2$, $y3$, $y4$ and 1 output neuron (which realises an AND operation) y

$$
\begin{aligned}
y_1 &= \Gamma(2x_1 - 2x_2 - 2x_3 + 3) \\
y_2 &= \Gamma(2x_1 + 2x_2 + 2x_3 - 1) \\
y_3 &= \Gamma(-2x_1 - 2x_2 + 2x_3 + 3) \\
y_4 &= \Gamma(-2x_1 + 2x_2 - 2x_3 + 3) \\
y &= \Gamma(2y_1 + 2y_2 + 2y_3 + 2y_4 - 7)
\end{aligned}
\tag{32}
$$

For the Non-linear Threshold Logic, TurboBrain computes one neuron

$$
y = \Gamma(2x_1 + 2x_2 - 4x_1x_2 + 2x_3 - 4x_1x_3 - 4x_2x_3 + 8x_1x_2x_3 - 1)
\tag{33}
$$

For the Heaviside Fixed Function, TurboBrain computes one non-linear neuron

$$(34) \quad y = x_1 + x_2 + x_3 - 2x_1x_2 - 2x_1x_3 - 2x_2x_3 + 4x_1x_2x_3 = \Gamma(x_1 + x_2 + x_3 - 2x_1x_2 - 2x_1x_3 - 2x_2x_3 + 4x_1x_2x_3)$$

3.2 Learning and Recognition of Characters

In the learning phase, 4 patterns of the digits 0, 1, 2 and 3 are presented as input to TurboBrain by 4 matrices of 3 by 5 pixels, and their corresponding outputs given by the binary representation of the digits, i.e. 00, 10, 01 and 11. Figure 1 shows the 4 learnt patterns and the correct recognition of corrupted patterns.

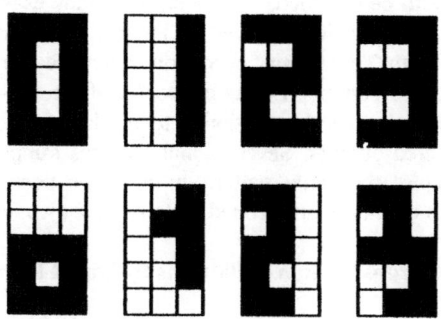

Figure 1: Learning of 4 input patterns and recognition of 4 corrupted patterns by TurboBrain.

The direct supervised learning of TurboBrain computes the two following output neurons

$$(35) \quad y_1 = \Gamma(-2x_{10}+1)$$
$$y_2 = \Gamma(-2x_{10}-2x_{12}+2x_{14}+1)$$

where x_i represents the value (1 = black and 0 = white) of the pixel at position i (from 1 to 15 in beginning by left up). The pixel x_{10} is 1 for 0 and 2, so $y_1=0$. Three pixels are necessary to classify the second output y_2. For example, for 0: $x_{10}=1$, $x_{12}=1$ and $x_{14}=1$, so $y_2=0$. Evidently if a pattern given by only one pixel $x_{10}=1$ will be recognised as 0 because $y_1=0$, $y_2=0$. This is due to the limited number of learnt patterns. For avoiding this case, it is evidently necessary to increase the number of samples or to increase the number of patterns as shown in Figure 2.

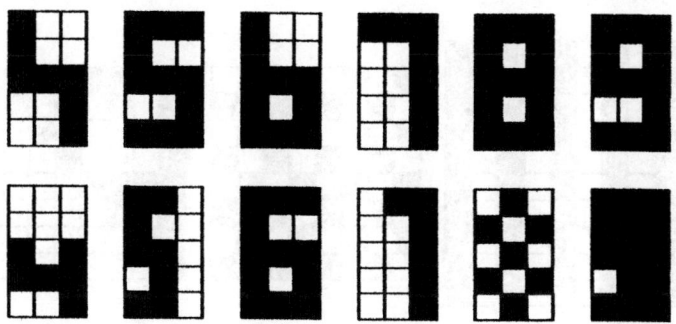

Figure 2: The patterns of digits 4 to 9 have been added to the patterns of digits 0 to 4 in Figure 1. The corresponding corrupted patterns have been correctly recognised by TurboBrain.

Indeed in adding the learning of the digits 4 to 9, as shown in the figure 2, TurboBrain computes 4 neurons corresponding to the coding of the outputs 0 to 9, i. e. 0000 to 1001,

$$(36) \quad y_1 = \Gamma\,[-2x_8 - 4x_{10} - 2x_{12} + 2x_{14} + 3]$$
$$y_2 = \Gamma\,[x_{15} + 2x_2x_{15} + (3/2)x_{14}x_{15} - 2x_2x_{14}x_{15} - 2x_4x_{14}x_{15} - x_6x_{14}x_{15} + x_8x_{14}x_{15} + x_{10}x_{14}x_{15} - 2]$$
$$y_3 = \Gamma\,[2x_2 - 4x_6 - 2x_{14} + 3]$$
$$y_4 = \Gamma\,[2x_4 + 2x_6 + 2x_8 - 5]$$

This is a very elementary example which is given for explaining the direct learning process. It permits to understand the meaning of the neural polynomial equations. Evidently, more sophisticated patterns are to be considered in pratice and many examples of different types of digits would enhanced the recognition power of TurboBrain, as classically made in neural network softwares. But with TurboBrain, even from only one sample of each pattern, it is possible to classify corrupted patterns. Let us notice that if the end-user does not agree with the recognition of patterns, he has the possibility to up-date the learning process in adding these cases with the wanted outputs. Several input patterns samples can be learnt with the same output: for example, the same pattern in different positions, orientations, scales. TurboBrain is also able to memorise and rebuild patterns as shown in the next section.

3.3 Memorisation and Reconstruction of Patterns by Direct Learning

Let us show explicitly the reconstruction of a pattern learnt by TurboBrain. The direct learning corresponds in this case to a direct memory, i.e. only one step of computation to obtain all the weights and thresholds of the neurons. These examples will show the power of the non-linear threshold logic.

In this first case, the first pattern "line" in Figure 2 is given as a 10 by 10 output matrix (white is 0 and black is 1) meanwhile the input matrix is given by the binary coding of the numbers, from 0 to 99, of the pixels of the output matrix line by line: the successive pixel are numbered as $x_6\,x_5\,x_4\,x_3\,x_2\,x_1$, the first pixel number, 0 in decimal, is then 000000 in binary, and the last pixel number, 99, is 1100011 in binary. Let us notice that this pattern corresponds to the parity problem: each black (white) pixel 1 (0) corresponds to an odd (even) decimal numbering, i.e. $x_1 = 1$ (0) in the binary coding. TurboBrain computes correctly the output neuron with only one binary digit x_1

$$(37) \quad y = \Gamma\,(2x_1 - 1)$$

Figure 3. These three patterns "line", "square" and "fractal" are learnt,
memorised and rebuilt with TurboBrain.

The second pattern "square" given in Figure 3 is an 8 by 8 output matrix which looks as simple as the preceding one. The same technics is used for the coding of the inputs.

With the linear threshold logic, TurboBrain finds 8 hidden neurons depending on 4 binary digits x_1, x_4, x_5 and x_6 and an output neuron y

$$
\begin{aligned}
y_1 &= \Gamma\ (2x_1 - 2x_4 - 2x_5 - 2x_6 + 5) \\
y_2 &= \Gamma\ (2x_1 - 2x_4 + 2x_5 - 2x_6 + 3) \\
y_3 &= \Gamma\ (2x_1 - 2x_4 - 2x_5 + 2x_6 + 3) \\
y_4 &= \Gamma\ (2x_1 - 2x_4 + 2x_5 + 2x_6 + 1) \\
(38) \quad y_5 &= \Gamma\ (-2x_1 + 2x_4 - 2x_5 - 2x_6 + 5) \\
y_6 &= \Gamma\ (-2x_1 + 2x_4 + 2x_5 - 2x_6 + 3) \\
y_7 &= \Gamma\ (-2x_1 + 2x_4 - 2x_5 + 2x_6 + 3) \\
y_8 &= \Gamma\ (-2x_1 + 2x_4 + 2x_5 + 2x_6 + 1) \\
y &= \Gamma\ (2y_1 + 2y_2 + 2y_3 + 2y_4 + 2y_5 + 2y_6 + 2y_7 + 2y_8 - 17)
\end{aligned}
$$

The number of neurons and parameters is great in comparison with the preceding case.

With the non-linear threshold logic, TurboBrain finds only one single neuron depending only on 2 binary digits, x_1 and x_4

$$(39) \quad y = \Gamma\ (-2x_1 - 2x_4 + 4x_1x_4 + 1)$$

Only 4 parameters are necessary in the activation function. The non-linear threshold logic realises a compression of data in comparison to the linear one. Moreover it gives some information about the symmetry of the pattern: in this case there are 2 periods in the pattern. For each line we have a succession of black and white pixels characterised by the binary digit x_1, and for the 8 lines of 8 pixels, we have also an alternation of 2 complementary patterns characterised by the binary digit x_4.

This last third case will consider the memorisation of the Sierpinski fractal pattern given as output in the figure 3, with the same binary coding of the inputs like in the two preceding cases. It is well-known that many natural images have a fractal structure.

With the linear threshold logic, TurboBrain finds the following 5 hidden neurons depending on all the binary digits and 1 output neuron y

$$
\begin{aligned}
y_1 &= \Gamma\ (-2x_1 - 4x_2 - 2x_4 - 6x_5 + 6x_6 + 9) \\
y_2 &= \Gamma\ (-2x_1 - 2x_4 + 2x_5 + 2x_6 + 3) \\
(40) \quad y_3 &= \Gamma\ (-2x_1 - 4x_2 - 4x_3 - 2x_4 - 10x_5 - 10x_6 + 23) \\
y_4 &= \Gamma\ (-2x_1 - 4x_3 - 2x_4 + 6x_5 - 6x_6 + 9) \\
y &= \Gamma\ (2y_1 + 2y_2 + 2y_3 + 2y_4 - 7)
\end{aligned}
$$

With the non-linear threshold logic, TurboBrain finds only one single non-linear neuron depending also on all the binary digits, but the number of parameters is dramatically low in comparison with the linear case

$$(41) \quad y = \Gamma\ (-2x_1 - 2x_4 - 4x_2x_5 + 4x_3x_6 + 3)$$

The fractal structure of the Sierpinski map is well reflected in the non-linear argument of the Heaviside function.

Remark: An other way to code a pattern is to divide it in a set of sub-patterns. For example, the 8 by 8 pixels fractal pattern, given in Figure 3, can be divided by 16 sub-patterns with 2 by 2 pixels. TurboBrain computes the 4 following neurons for the fractal pattern

$$
\begin{aligned}
y_1 &= \Gamma \left[\ -2*x_1 - 2*x_3 - 4*x_2x_4 + 3 \ \right] \\
(42) \quad y_2 &= \Gamma \left[\ -2*x_1 - 2*x_3 - 4*x_2x_4 + 3 \ \right] \\
y_3 &= \Gamma \left[\ -2*x_1 - 2*x_3 - 4*x_2x_4 + 3 \ \right] \\
y_4 &= \Gamma \left[\ -2 \ \right]
\end{aligned}
$$

where the sub-patterns are coded from 0 to 15 with the binary digits $x_4x_3x_2x_1$, from 0000 to 1111. The first 3 neurons are identical and the last one is always 0 corresponding well to the iterative construction of the fractal pattern. This procedure permits to memorise big patterns with a limited number of inputs but the number of neurons is greater, here there are 4 neurons with 4 inputs instead of 1 neuron with 6 inputs as in the precedinghigher.

3.4 Parallel Computation

One of the most important operation in computation is concerned with the arithmetic sum of two bits string. As Konrad Zuse ([17], p. 193) writes in his book : "Anyone who constructs a calculating machine starts in general with the adder unit. The difficulty is carrying digits." Classically, a certain number of steps are necessary before outputting the correct answer: for example, for a 2-bit sum S between X and Y, we write $X+Y=S$ as $x_3 x_1 + x_4 x_2 = y_c y_2 y_1$. To obtain the digits of S, we start adding from right to left the digits of X and Y, and using the carry in the next step. With the direct algebraic learning in TurboBrain, it is possible to design neural mathematical operations without carry. As an example (Dubois, Resconi & Raymondi, 1993), an adder for numbers given by two binary digits is given by the three non-linear Heaviside fixed functions:

$$
\begin{aligned}
y_1 &= x_1 + x_2 - 2x_1x_2 \\
(43) \quad y_2 &= x_3 + x_4 + x_1x_2 - 2x_3x_4 - 2x_1x_2x_3 - 2x_1x_2x_4 + 4x_1x_2x_3x_4 \\
y_c &= x_3x_4 + x_1x_2x_3 + x_1x_2x_4 - 2x_1x_2x_3x_4
\end{aligned}
$$

Let us consider the addition of 3+3, i.e. 11 and 11 in binary digits: $x_1=x_2=x_3=x_4=1$. From the above activation functions, we obtain: $y_1=0$, $y_2=1$, $y_c=1$, i.e. the binary number 110, which is 6 in decimal notation.

4 Conclusion

The parallemism of the neural brain is considered as an important property to explain the high speed of response of the brain, by example, in pattern recognition, although the neural dynamics is far slower than electronic processors. The McCulloch & Pitts model (1943), with a Heaviside function with a linear argument, related to the weights and thresholds of the neurons, have no solution for a parallel processing. This Threshold Logic, using linear argument of the Heaviside function, is not capable of resolving all Boolean functions with only one neuron. In the technology of neural networks, most methods are based on hidden neurons to resolve the Boolean functions and a big number of iterations are necessary for learning. There are some difficulties to obtain the global minimum as the number of neurons increases and the combinatorial explosion gives many problems for practical applications.

The Non-linear Threshold Logic, presented in this paper, shows the possibility to design a totally parallel neural architecture. Each neuron is connected to all the inputs and the number of neurons is equal to the number of outputs. The weights and the thresholds are computed from a direct supervised learning. The example of the Parity Problem with Exclusive OR shows the power of the direct learning with only one single non-linear neuron. The direct supervised learning of patterns generates neural polynomial arguments of the Heaviside function, which give informations about these patterns as in the example of the fractal map.

The theory and the applications presented in this paper have the purpose to show that the Non-linear Threshold Logic with Direct Supervised Learning is a new tool in the framework of neuroCAST, Computer Aided Systems Theory and Techonolgy.

Acknowledgement: D. Dubois would like to thank the Fonds National de la Recherche Scientifique for the grant related to CAST'94 at Ottawa. The authors thank the referees for the hints given for enhancing the understanding of the paper.

References

1. J. A. Anderson, Ed. Rosenfeld (eds.): NEUROCOMPUTING: Foundations of Research, The MIT Press Cambridge, Massachusetts, London, 1988, 729 p.
2. M. L. Dertouzos: Threshold Logic: A Synthesis Approach. Res. Monogr. no. 32, The MIT Press, Massachusetts 1965
3. D. M. Dubois: Self-organisation of Fractal Objects in XOR rule-based Multilayer Networks. In: EC2 (ed.): Neural Networks & their Applications, Neuro-Nîmes, Proceedings of the 3rd International Workshop 1990, pp. 555-557
4. D. M. Dubois: Le Labyrinthe de l'Intelligence. InterEditions/Paris-Academia/Louvain-la-Neuve, 2nd edition, 1990, 331 p.
5. D. M. Dubois: Mathematical Fundamentals of the Fractal Theory of Artificial Intelligence. Communication & Cognition - Artificial Intelligence, 8, 1, 5-48 (1991)
6. D. M. Dubois, G. Resconi: Mathematical Foundation of a Non-linear Threshold Logic: a New Paradigm for the Technology of Neural Machines. ACADEMIE ROYALE DE BELGIQUE, Bulletin de la Classe des Sciences, 6ème série, Tome IV, 1-6, 91-122 (1993)
7. D. M. Dubois, G. Resconi: Advanced Research in Non-linear Threshold Logic Applied to Pattern Recognition. COMETT European Lecture Notes in Threshold Logic. Edited by AILg, Association des Ingénieurs sortis de l'Université de Liège, D/1995/3603/02, 1995, 182 p.
8. D. M. Dubois, G. Resconi, A. Raymondi: TurboBrain 1.0: User's manual, D/1993/Dubois, Resconi, Raymondi: Editeurs, registered the 29th October 1993, 75 p.
9. K. Fukushima, M. Sei, I. Takayuki. IEEE, Transactions on Systems, Man and Cybernetics SMC-13:826-834 (1983)
10. W. S. McCulloch, W. Pitts. Bulletin of Mathematical Biophysics 5:115-133 (1943)
11. M. L. Minsky, S. Papert. Perceptrons. MIT, Cambridge, 1969
12. W. Pitts, W. S. McCulloch. Bulletin of Mathematical Biophysics 9:127-147 (1947)
13. G. Resconi, A. Raymondi: A New Foundation for the Threshold Logic. Quaderno n°3/93 del Seminario Matematico di Brescia, 1993, 46 p.
14. F. Rosenblatt: Principles Neurodynamics. Spartan, Washington DC, 1961
15. D. E. Rumelhart, G. E. Hinton, R. J. Williams. Nature 323:533-536 (1986)
16. D. E. Rumelhart, G. E. Hinton, R. J. Williams: Learning internal representations by error propagation. In: D. E. Rumelhart, J. L. McClelland (eds.): Parallel Distributed Processing Explorations in the Microstructures of Cognition, Vol. 1, Cambridge MA: MIT Press 1986, pp. 318-362
17. K. Zuse: The Computer - My Life. Springer-Verlag 1993

GENIAL: An Evolutionary Recurrent Neural Network Designer and Trainer

R.J. Duro[1], J. Santos[1], A. Sarmiento[2]

[1] Departamento de Ingeniería Industrial
Universidade da Coruña. Campus de Esteiro. 15403 Ferrol. SPAIN.
e-mail: richard@udc.es
e-mail: santos@udc.es
[2] Departamento de Matemáticas
Universidade da Coruña. Campus de Elviña s/n. 15071. A Coruña. SPAIN.

Abstract. We report here on GENIAL (GEnetic Neural network Implementation AppLication), a genetic based recurrent neural network design tool being developed for the automatic generation of trained application specific neural network architectures that can be tailored to specific constraints in actual operating systems. Strong emphasis was placed on the implementation of a genetic based design tool that could work efficiently with very small populations and obtain adequate solutions in a very reduced number of iterations when handling complex problems. GENIAL is capable of designing and training optimal (or quasi optimal) recurrent neural network architectures that can handle sequential data, minimizing the number of connections between neurons, the number of neurons, the number of processing cycles in synchronous operating mode, deciding on the optimum number of inputs per cycle, etc. It can even combine several of these constraints. It would be very simple to include further constraints and/or minimization criteria in GENIAL. Additional features allow it to work with complex neurons and even irregular combinations of different types of neurons in the same network. In order to obtain a design, GENIAL only requires a training set containing the input output pairs desired.

1 Introduction

One open problem in Artificial Neural Network (ANN) research is how to automate the process of obtaining an adequate, or optimal if possible, neural network architecture and train it in order to perform a specific task, usually represented as a series of relationships between a set of inputs and a set of outputs in a supervised training mode, taking into account the constraints imposed by the systems in which the networks will be operating or the technological limitations for their implementation (number of input lines, silicon areas, etc...). Several methods have been proposed for the automatic training of certain types of networks once their architectures have been defined. For a review see [1] [2]. All of

---

* This work has been funded by the Universidade da Coruña.

these methods, however, are very architecture specific and there are still many questions open as the type of architecture that best suits each application (number of hidden layers, feedback loops, etc..) [3] [4] [5] and as to how to efficiently train architectures which do not present regular feed-forward processing topologies [6] [7]. This design and training task becomes even more complicated when the data the ANN must process is sequential, that is, not all the data is available in the same instant of time, making it necessary to use topologies that incorporate some type of memory, usually by means of feedback loops. Even when the data can be input in parallel, we can question whether this is the best option or whether it would be better to sequentiate it somehow and use the properties of recurrent topologies in order to reduce the number of neurons and/or connections necessary for the implementation of the task. In this line, probably the most difficult problem in ANN design right now is how to design and train networks whose neurons are more complicated (more biologically plausible) than the usual sigmoid of the weighed sum of the inputs type [8], and consequently present and increased wealth of behaviours. The problem of biological plausibility is made even more untractable when we consider networks that incorporate different types of specialized neurons working together in the same task, or neurons that instead of acting on other neurons act on the connections between other neurons. In order to address these problems some authors have turned to genetic algorithms [9] [10] [11]. In this line we have created GENIAL (GEnetic Neural network Implementation AppLication), an evolutionary based ANN design and training tool which is able to produce quasi optimal (optimal in most cases) ANNs for the training sets it is presented using very little outside information. Obviously, the concept of optimality is a very relative concept, it depends on what we desire in each case. Do we want a very fast ANN or do we need an ANN that minimizes the number of neurons. Do we desire an ANN which minimizes connections, facilitating VLSI implementation or do we require an ANN which minimizes input lines. GENIAL caters to all of these needs and many more. It includes features that permit selecting many types of optimization criteria and even compromises between them. It is designed in a modular fashion making it very easy and cheap to include new premises or different types of neurons and operation modes. It is based on the Genetic Algorithms (GAs) paradigm [12] as an optimization strategy and it incorporates many new combinations of GA features and some strategies that allow it to overcome the classical handicap of GA based systems, that is, the fact that they require extremely large populations of chromosomes and many generations in order to reach a solution, leading to very long execution times and very large memory requirements for the execution of quite simple optimization problems. GENIAL employs small populations and very few generations in order to achieve the same and sometimes even better results. It reaches solutions in very short execution times using relatively little memory.

2 Genetic Algorithms

Genetic Algorithms (GAs) have been proposed for solving complex optimization problems for which traditional methods are inadequate or computationally too expensive. This type of algorithms are based on a model of evolution in nature whereby a population of solutions to the problem in hand procreate, mutate, and produce offsprings which take the place of their parents. All of these processes using a natural selection criterium hopefully lead to better solutions to the problem arising in successive generations until an optimum solution is achieved. The general strategy can be summarized as follows:

1. An initial population of coded solutions to the problem (chromosomes) is randomly generated. These chromosomes are lists of the data that participate in the solution in a given format.
2. All of the chromosomes in the population are evaluated and their fitness for solving the problem is quantified.
3. The chromosomes procreate. Fathers and mothers are randomly chosen from the population assigning a higher probability of being chosen to the fittest. These parent chromosomes combine to generate an offspring.
4. There is a process of random mutation of genes in the offspring chromosomes.
5. The offspring substitute the parents and a new generation starts in step 2.
6. This process continues until an optimum solution is found.

3 Genial

GENIAL is a programming environment written in C-language. It is built around a modular kernel that implements the necessary functions in order to make programming of GA based applications straightforward and easy. This kernel contains many functions for procreation (one point crossover, two point crossover, random gen from father/mother, etc), mutation (set to zero, random mutation of one gen, random mutation of one digit, etc...), migration (global migration, local migration, selective, etc...), and many other genetic mechanisms. On top of this kernel there is a module in charge of the evaluation of the different networks GENIAL is presented with. This module can be easily adapted to fit many different types of networks or needs as well as to guide the optimization process towards the solutions we desire. In addition to these basic operational mechanisms, GENIAL includes an X-Window based user interface which shows at all times how the problem is evolving. The windows it displays include graphs showing the fitness of the different chromosomes in each niche, a graph which displays the evolution of the fitness of best solutions, and a representation of the best network each generation, showing its weights, architecture and how adequate it is for the task in hand. In figure 1 we present how the user interface looks in one of the examples.

Other features are devoted to following the execution of a network obtained by GENIAL cycle by cycle. There are features for storing intermediate results,

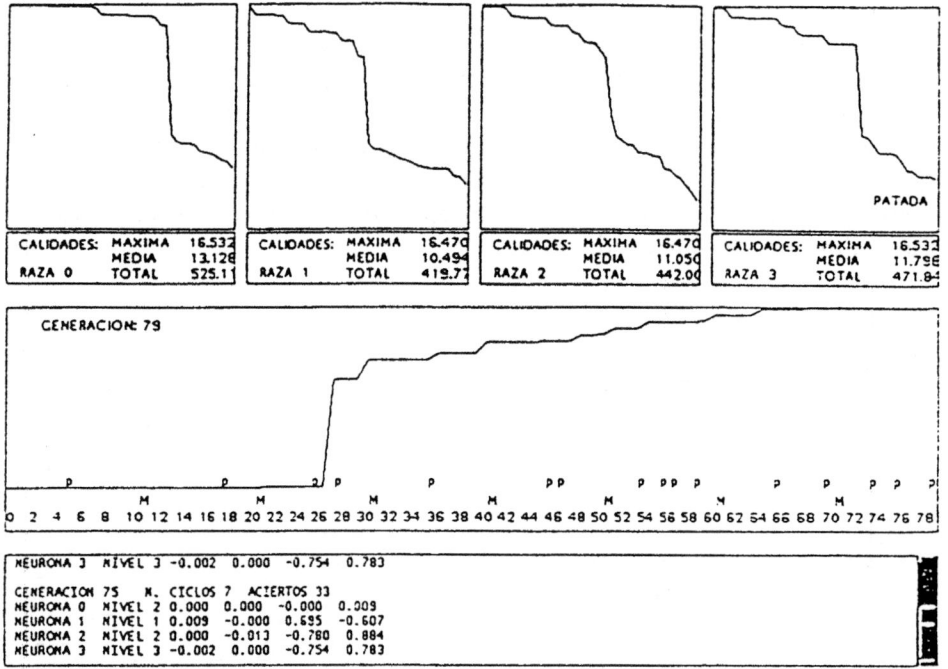

Fig. 1. GENIAL's user interface during the process of solving the five bit parity problem starting from a five neuron fully connected network. The top part of the interface shows the fitness distributions for each niche. The central window, displays the evolution of the best chromosome of each generation and the bottom one the best chromosome in the current generation.

for viewing previous networks, etc... GENIAL allows for any kind of distribution of the populations into niches that evolve independently except for migrations of fit individuals between them, which can be decided on by the user. This provides for a parallel search in different sections of the search space as well as for a selective increase in the variety of each niche. It also implements several solution acceleration mechanisms based on increasing the variety of the small chromosome populations it handles (mutation rate increases depending on the individual evolution of each niche, random chromosome generation mechanisms, etc.).

4 Representation of the Networks

The networks in GENIAL are represented as two lists, one for the weights (represented as floating point numbers) and one for the architecture and number of execution cycles (represented as integer numbers). GENIAL reaches an optimum (or quasi optimum) solution by carrying out the whole GA process taking both lists simultaneously into account and working with them as floating point and integer numbers instead of using the traditional binary representation. GENIAL can also allow for any kind of initial topology or any kind of neurons or combination of neurons, although in the implementation we present here we have narrowed down the problem to the design and training of a synchronous recurrent neural network. In this case GENIAL starts from a fully connected version of oversized networks and depending on the optimization criteria selected produces the optimum trained network.

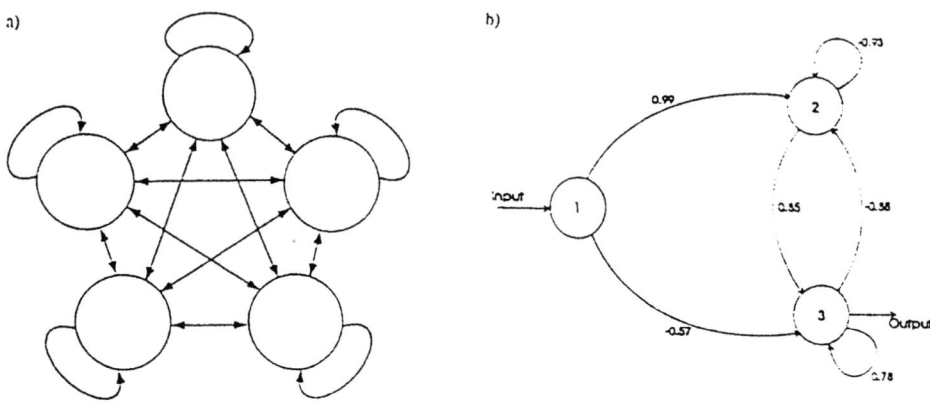

Fig. 2. a) Starting network provided to GENIAL for the five bit parity problem and b) network obtained by GENIAL after 64 generations. Note that the inputs are taken sequentially.

5 A Simple Example

We present a simple example of the operation and results produced by GENIAL for a typical problem found in the literature. This is, in the case of the five bit parity problem we ask it to generate a network that can input the data sequentially, minimizing the number of connections between neurons and the number of neurons. As a first example, we started from a fully connected three neuron network and GENIAL obtained a solution with 100% hit rate for the problem in hand in 20 generations using a population of 30 chromosomes per niche and four niches. The solution had three neurons and six connections, and it processed the five inputs in seven cycles. In order to demonstrate the pruning capabilities of GENIAL we have run another example starting from an overdimensioned fully connected synchronous network consisting of 5 neurons and 25 interneuronal connections which used 15 processing cycles and tried to obtain an optimum solution minimizing neurons and connections while achieving 100% hit rate for the same problem as before. As shown in figure 2, after 26 generations GENIAL found an architecture that obtained the appropriate outputs in every case, after 64 it had pruned the network to three neurons, 6 connections and 7 processing cycles maintaining the 100% hit rate. The resulting optimum network is shown in figure 2 and it is almost the same as the one for the previous case.

6 Conclusions

In this paper we present GENIAL, an automatic designer and trainer of recurrent neural networks based on an evolutionary strategy which leads to a very simple, robust and general process for obtaining the designs required from it. GENIAL uses floating point and integer number representations of the networks, it also includes many features for reducing the sizes of the populations and number of generations required, thus accelerating execution. GENIAL is an almost network independent designer-trainer, it can handle problems which require different types of neurons or different optimizations, even problems where combinations of neurons or optimizations are required. This is one of its most outstanding features, as in real system design many external constraints such as limitations in input lines, simultaneous processing capabilities, silicon area, etc, have to be taken into account. The next steps in the development of the GENIAL environment will be the inclusion of topologically irregular networks, networks whose weights are determined by other networks, higher order neurons, neuron clusters, networks which handle sequential inputs and produce sequential outputs, etc...

References

1. G.E. Hinton, "Connectionist Learning Procedures", Artificial Intelligence 40 (1989), pp. 185-234.
2. R.P. Lippmann, "An Introduction to Computing with Neural Nets", IEEE ASSP (1987), pp. 4-22.
3. S.E. Fahlman, C. Lebiere. "The Cascade-Correlation Learning Architecture" Advances in Neural Information Processing Systems, Vol. 2 (1990), pp. 524-532, Morgan-Kauffmann.
4. D.E. Rumelhart, J.L. Mclelland. Parallel Distributed Processing 1, 2. MIT Press (1986).
5. F. J. Pineda. "Generalization of Backpropagation to Recurrent Neural Networks", Phys. Rev. Letters, Vol. 59, n19 (1992), pp. 2229-2232.
6. J.J. Hopfield. "Neural Networks and Physical Systems with Emergent Computational Abilities". Proc. Natl. Acd. Sci. USA. Vol. 79, (1982), pp. 2554-2558.
7. S. Kirkpatrick, C.D. Gelatt Jr. and M.P. Vecchi. "Optimization by Simmulated Annealing". Science, Vol. 220 (1983), pp. 671-680.
8. L. Giles, T. Maxwell. "Learning Invariance and Generalization in High Order Neural Networks". Applied Optics, Vol. 26 n23 (1987), pp. 4972-4978.

9. S. Harp, T. Samad, and A. Guha, "Towards a Genetic Synthesis of Neural Networks". Proceedings of the Third Conference on Genetic Algorithms. San Mateo, CA: Morgan Kaufmann, (1989).

10. S. Harp, T. Samad, "Genetic Synthesis of Neural Network Architecture". In Handbook of Genetic Algorithms. L. Davis, ed., New York: Van Nostrand Reinhold, (1991).

11. D. Whitley, T. Starkweather, and C. Bogart, "Genetic Algorithms and Neural Networks: Optimizing Connections and Connectivity". Parallel Computing, 14 (1990), pp. 347-361.

12. J. Holland, Adaptation in Natural and Artificial Systems. Univ. of Michigan Press. Ann Arbor, 1975.

DASE: An Environment for System Level Telecommunication Design Exploration and Modelling

Oryal Tanir, V.K.Agarwal and P.C.P. Bhatt

McGill University - Bell Canada
2265 Roland Therrien Blvd., Longueuil, QC, Canada J4N 1C5

Abstract. Design automation (DA) has steadily contributed to improvements witnessed in the system design process. Initial applications were to address low level design concerns such as transistor layout; however the focus of tools has slowly progressed up the design abstraction scale. The current state-of-the-art provides solutions for synthesis issues at the register-transfer (high level) and lower levels. While DA has helped in reducing design timelines and re-design, a major source of design difficulties are just recently being addressed and promise to be the next wave in DA applicability. The problems arise within the architectural (or system) level of abstraction very early in the design cycle. They may be classified as those related to design specification and design capture. The former is a complex problem area and significant work has been accomplished in attempting to formalize specifications for modelling [1]. The latter, a relatively new research area that attempts to bridge the design process gap between specification and design, is the focus of this paper.

1. Introduction

Digital system design can be described at different levels of abstraction [2]. Although the terminology used may vary slightly, these levels are typically referred as (in increasing order of abstraction): *Circuit level, Logic level, Register transfer level, High level* and *Architectural (System) level* design. Architectural level design automation dwells in a very abstract domain. Transparent modelling of both software and hardware must be supported since design partitioning and co-design (the concurrent design of software and hardware) are common activities undertaken at this level. Due to these factors, ambiguities in initial specifications or poor choice of design architectures contribute to significant and costly design rework in the latter stages. Hence if better control and structure can be established early in the design process the potential savings are significant [4]. At the architectural level of representation, design input is directly taken from initial specifications. Even assuming that specifications are relatively stable, the possible design architectures and solution domain can be large. This makes it difficult to utilize rigidly defined component libraries (that are typically deployed in lower level design aids) specified in languages such as VHDL.

The objective of this paper is to present an environment to help designers represent, model and explore design trade-offs at the architectural level of abstraction and synthesize designs to lower level representations so that they may be usable by other DA tools. To do this, an environment must address three specific issues [12]: *(a) design capture, (b) design space exploration, and (c) design synthesis.*

These points are addressed by a specification driven Design Analysis and Synthesis Environment (DASE) [10]. Model libraries within the environment are directed at the telecommunications area, however the concepts are equally applicable to other domains. DASE allows a telecommunication designer to experiment with different design options and trade-offs at a significantly abstract but intuitive architectural representation level. The remainder of this paper will cover these issues as addressed by DASE.

2. DASE

The major components of DASE are shown in figure 1. Specifications are entered through a user interface and captured by the environment through the use of an internal Design Specification Language (DSL) and supported by an object oriented library system. DSL primarily captures specifications by use of abstractions and is free from rigid disciplines of simulator oriented hardware languages. However, since the language must bridge the abstraction gap between high level system design notions to low level hardware descriptors, it must possess the flexibility of re-defining and altering its model interfaces during design exploration.

The basic functional issues of the language motivated the use of Prolog as the language for implementation of DSL. It may be noted that DSL is a meta-language in the sense that it is based upon Logic programming semantics, utilizing built-in predicates to define its own constructs. The language is interpreted through a DSL processor which also interacts with a DSL simulator and user interface. When the design is found suitable, it is then synthesized to a lower level description. A formal (Predicate/Transition) Petri-Net based analysis of the DSL constructs is also available.

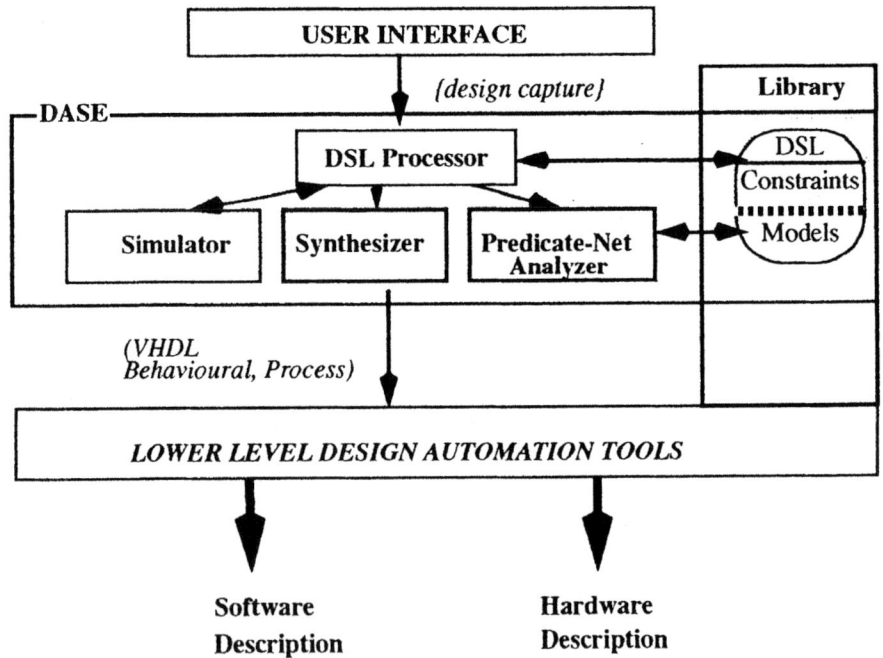

Fig. 1. DASE organization

2.1 Specification Capture

Specification capture has been an evolving discipline in software engineering. Today's engineer will find many different mechanisms for design capture. Of these, an underlying formal basis is desirable to ensure consistent properties of a model. For example, LOTOS and SDL [5, 8] utilize a CCS based formalism and have been developed to specify nuances related to telecommunication systems. More general formalisms such as Petri-Nets [9] and Temporal Logic [7] can also help capture and analyze different aspects of a design.

The requirements for architectural level automation dictate different needs for the specification language. The language must be able to satisfy the three basic criteria: *(a) capture a design intent in abstract terms, (b) possess a formal basis to permit analysis and verification of the system being designed, (c) initial design interfaces should be mutable or undefined until a final design architecture is selected for synthesis.* In other words, an interface specification will bind to a model as late as possible in this design stage. These criterion were used in the definition of DSL. The language addresses the three with module abstraction, predicate transition net analysis and port descriptors respectively.

DSL, the internal specification language within DASE is interpreted through a DSL processor. Modular object oriented design entities within the language called *modules* are the primitive building blocks of the language. These are constructs that possess a "name", a set of possible behaviors, and resources. The name of a module is a unique and meaningful identifier of the module. The behaviors of the module represent the actions undertaken by the module when an event occurs. The occurrence of an event implies the arrival of a message from another module to the module - which then attempts to execute the behavior associated with the message. Resources are local data entities associated with a module that could be used to store local variables, registers or state information. An important aspect of the environment is the ability of hierarchical storage of modules in a module library as "generic" entities which facilitate model re-use.

Modules can possess their own unique behavior or can inherit the behavior of other modules. This is facilitated with the use of the DSL "isa" predicate. For example, *isa(card(X), mux(4))* could imply that module *card(X)* inherits the behavior associated with a multiplexer *mux* instantiated with a parameter *4* (identifying the number of input lines).

Module Behavior description. Module behavior is a procedural description of actions which, within satisfaction of a set of constraints, may consist of:
 (a) communication initiation with other modules.
 (b) modification of resources or data structures associated with the module, and
 (c) a suspension of further operations for a specified time period.
A brief description of the major actions are described below.

Conditional constructs: Adhering to Prolog notation, behaviors can be viewed as a set of clauses. Each behavior can also be multiple clauses - each applying under different conditions. The head of a clause consists of the behavior name - a unary or N position predicate. The body is a set of conditions and actions - which are compound sub-goals of the clause. It should also be noted that DSL conditions are Prolog sub-goals to be satisfied, and the actions are sub-goals that are always satisfied, but may result in some desired side effects (such as message generation and resource manipulation).

For a given behavior, a set of conditions can be evaluated before further execution of the behavior is attempted. This facilitates the description of if-then-else type constructs as well as supporting state oriented behavioral description.The behavior selection may be through (relational Prolog) predicates based on conditional operators in DSL as well as some reserved statements for querying the state of resources associated with the module.

The *check_res(name(U1,...,Um),V1,V2...,Vn))* predicate will query any resource *name* with current parameters U1,...Um and values V1,..Vn. If there is no match, the predicate will fail and another satisfaction of the behavior will be attempted. DSL uses the Prolog programming style for resolving ambiguities and testing conditions.

Delay statements: A delay action suspends the module for a given period of simulated time *Delta*. After the elapsed time, the next action for the behavior is executed in the module.The syntactic form of this action is: *delay(Delta)*. The delay *Delta* can assume a discrete or probabilistic time value.

Data manipulation statements: Data manipulation actions can change the internal state of resources local to a module. The action *set_res(res_name(U1,...,Um), V1,...Vn)* will modify data fields of resource *res_name(U1,...,Um)* with values *V1,..Vn*. For example to reset a resource called *counter_value(1)* one could simply write: *set_res(counter_value(1),0)*.

Communication statements: Communication is established using a set of *send* statements.The form of the statement is: *send(Destination, Port, Message(P1,...,P2))* where *Destination* can be a name of a destination module (i.e. *module1*), a set of destinations (i.e. *[module1, module4]*), all destinations connected to module's port *Port*, or "Undefined". In the latter form, a message will be sent to the first module connected to *Port* capable of interpreting the message. *Port* is an optional port name of the module and *Message* is a message to be sent to the destination module.

DSL permits composition of modules in to higher order (HO) modules. Connections between modules are established through the use of ports. A port is a virtual communication channel between the module and its environment. The language deduces direction from information flow across ports. Port specification is not typed so that different levels of abstract information may flow through the same port.

To support these actions, modules contain data structures which facilitate the internal queuing and scheduling of messages as well as manipulation of internal variables. As in object oriented constructs, a module can inherit the behavior of another module or a standard template module.

2.2 Design Space Exploration

After design capture, an experimentation and model management mechanism must be available for re-using "generic" model components to support simulation of different design options. This is a major shortcoming of many simulation or modelling languages. Many provide mechanisms for specific model storage and retrieval, but lack the ability to easily re-use and adapt architectural level models.

DSL incorporates a versatile library support system managed by the DSL processor which enable the creation, storage and retrieval of module libraries in an organized manner. Libraries maintain all module or Higher order module information, as well as added rules regarding any constraints to be imposed on the modules, any configuration rules to be applied to the components of the library, and the interface specification of the library module. The interface specification is created by the library system to define exactly what ports are available for communication with the library module. The basic structure of a library module is depicted graphically in figure 2.

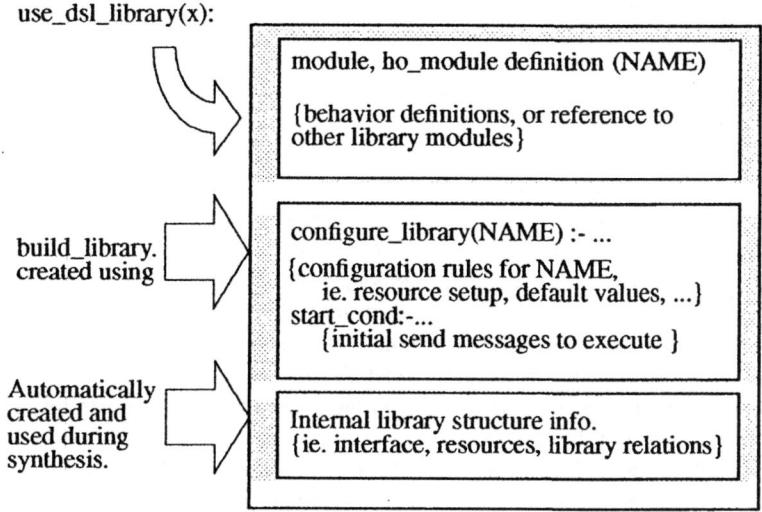

Fig. 2. Library module structure.

Library modules are created using the *"build_dsl_library(NAME)"* predicate which interactively constructs the necessary data structures for the library module "NAME". For example, a library module containing a *ho-module* named *processor(CLOCK, BUS, MODEL)*, can then be built using *build_dsl_library(processor)* and accessed with *use_dsl_library(processor)* statements. Instantiation of modules (or ho-modules) using this library module could then be invoked by: *isa(risc_processor, processor(25,64, typical_processor)*. The predicate proceeds through three phases of construction to create the library module structure as shown in figure 2. The first stage is to create the module and ho-module definitions of the library. The system will search for the specified module and if successful, it will recreate the behavior of the module. In the event that the module is undefined, it will ask the user to provide the behavioral description.

The second stage is the construction of the library configuration rules and starting conditions. These are created interactively throughout the DSL processor and stored as Prolog rules as the predicate *configure_library(NAME)*.

Configuration rules allow for the creation of module resources, interconnection definition (for ho-modules), definition of constraints and instantiation of other modules. During library invocation the rules are executed by the DSL processor to create the required resources for the library module. The rules may be parameterized, allowing for different possible types of path definitions. Conflicting or missing interconnections during configuration are identified by the DSL processor and reported back to the user.

The third stage is to capture the interface and resource information for the module.

This is achieved by grouping the required module resources, required ports and relations. In the latter case, a class relationship (if applicable) is associated with the module library and is invoked with the *isa* predicate as with modules. Hence our previous example could belong to the class *microprocessor* with *isa(processor, microprocessor)*. This library module would then be organized by the DSL processor with other modules belonging to the same class. The class relation allows library components to be identified and utilized by the system during simulation, design exploration and synthesis.

Upon creation of a library module, DSL also verifies that constraints are not in conflict and there is no inconsistency with the library interface and its associated modules. Library configuration rules allow a library to configure itself depending upon different conditions during invocation. These rules allow a generic library module to be reused in different ways as the design requirements demand.

Constraints are related to a module and are used to define limits upon parameters of a given module, resources or messages. These can be defined to be system wide constraints (affecting more than one module) or local constraints (restricted to individual modules). For example, the maximum size of a memory module, can be regarded as a local constraint. However, if the memory size of a library module depends upon its use in a DSL model, then this is a system level constraint. System level constraints are defined within the main DSL model, whereas local ones can also be established in library modules. After invocation of the library, DSL performs routine checks upon the constraints so that they are not violated during simulation.

The simulator uses DSL library modules to configure and set relevant constraints and resources to support DSL simulation. An abstraction hierarchy is automatically created by the simulator which represents the level of detail different modules define. For example, at the topmost level (level 0), the most abstract module would be defined. The next level down would consist of modules that are referenced and inherited by the topmost module - and similarly for subsequent levels. Configuration rules for a given module affect the structural definition (or configuration) of modules directly referenced by the module. Hence, each library module may be used in many different configurations by its parent module facilitating model re-use.

Abstracting hierarchy has often been employed to facilitate simulation and modelling at various levels of detail. For example, a system level analyst may not be interested in the lower level details of a given model, whereas a hardware designer would be more concerned with the lower timing details of the simulation [11]. The DSL simulator provides for dynamic alteration of the level of abstraction viewed by defining an observation level. The level can be defined to be limited to a given level or at a particular level and all levels below it. The observation level can be set at any time in the simulator with the "set_level" command.

Apart from the traditional simulator functions, the simulator manages the view and constraints. After each simulation step constraint breaches are checked and messages are scheduled for DSL modules. The simulator interacts with the user to relax a constraint (in which case the simulator will commence at the same time point) or request the simulator to suggest alternate design modules (belonging to the same module class) from the existing module library and rerun the simulation. Hence design exploration is guided by the environment.

2.3 Petri-Net Analysis

Module behavior can be translated to a Petri-net based netlist. Petri-nets [9] have been used in many areas to analyze hardware or software systems. The formalism provides for a relatively simple way for defining asynchronous concurrent communications. The main complaints voiced about Petri-nets have been the problem of state-space explosion and complexity of the generated graphs.

The above mentioned problems are avoided in DASE by observing two restrictions to the general Petri-net formalism. The first restriction is that modules exhibit a safeness property akin to that of Petri-nets. This is observed from the fact that a module processes one behavior at a time - which is equivalent to a known bounded maximum number of tokens in a Petri-net. The second restriction is that the possible types of message a module can accept or generate is finite and can be parsed. This implies that a finite number of input and output places for each module can be determined.

A module's behavior can be translated into a Predicate-Transition net as depicted in figure 3. A parse of a module generates the types of messages it accepts, resources it uses and messages it generates. Hence a static Petri-net model can be generated per module based on the structure defined in the figure. A busy place ensures a safe net operation for the Petri-net. Incoming tokens are given names of relevant messages and a set of module specific messages ensures that the module accepts only tokens it can process. Predicates are used in the net to establish constraints and access resources (which are tokenized places per resource). Finally output transitions enable interconnection to other modules' nets.

Each module's Petri-net representation can be analyzed using reachability or invariance techniques. Complexity is minimized since analysis is performed at a module level. This analysis can detect inconsistent behavior structures, unused DSL code and undefined message ports. Analysis of the entire DSL model can be achieved subsequently by abstracting the behavior of each module to a Predicate-Net place and transition and restricting the output message sets to those acceptable to a module. Although the resulting net is not necessarily safe since messages can queue at modules, the complexity is reduced considerably and in most cases can be analyzed. A statistical analysis can also be possible at this level to determine input buffering requirements during synthesis to VHDL. A general purpose simulator can be used to gather statistics regarding buffer sizes.

2.4 CO-Design Within DASE

The DASE environment is amiable to co-design of telecommunication systems. The representation mechanism provided by DSL facilitates the description of generic modules. These modules are defined entirely in terms of their potential behavior with respect to the environment and are free to bind to software or hardware constructs. There is no default notion in DSL of hardware and software, hence the user may impose restrictions to generic modules as library constraints which are used by the environment during synthesis to control the target synthesizeable model.

Co-design concepts within DASE are illustrated in figure 4. Modules can represent behavior and structure (HO-modules) so that many different abstraction levels of hardware can be described. Software behavior can similarly be captured by the behavior of a

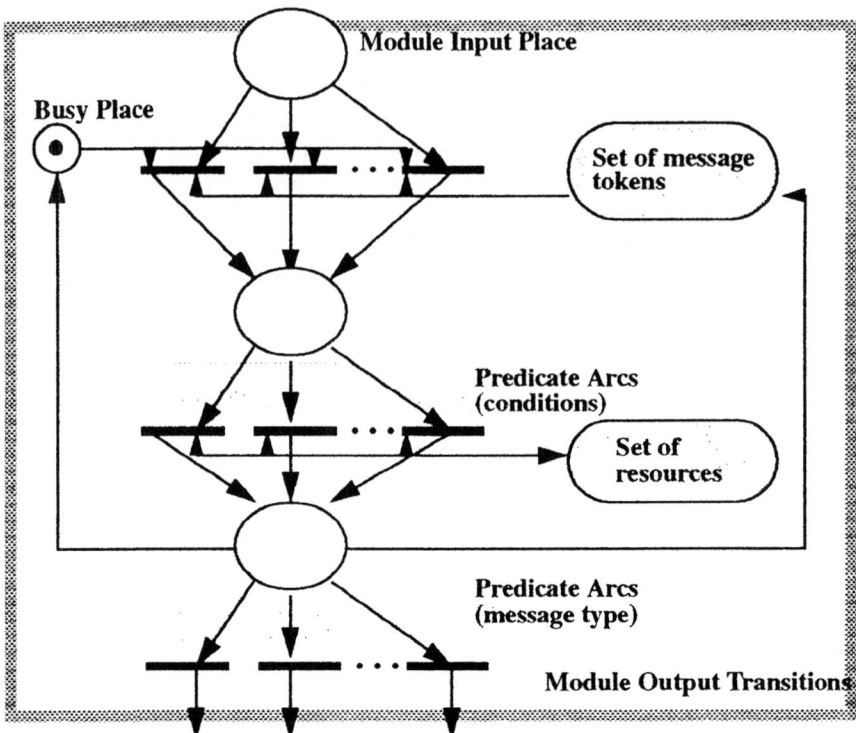

Fig. 3. Petri-net representation of a module.

module. Since each module may execute one particular behavior at a time, it is straightforward to describe serial program execution in terms of delays and program control. This is depicted within figure 4(a) by the shaded generic module S (for software). The ellipses graphically represent a particular behavior, while the arrows indicate the program flow control. The names within "< >" indicate a name of a behavior defined within the module. In this simple example, module S will accept a message with parameter X and either increment or decrement the value, depending on the condition X<Y (where Y is assumed to represent a local variable). Concurrent software/hardware can also be described with a set of modules, confined within a HO-module. An example of this is given in figure 4(b). Higher level program control is described by the module interconnects, while local module control flow is described by the behavior of each module.

In this example, a module *PROC(X)* spawns three identical messages indicated by the *fork* behavior. This behavior would typically be described in DSL in the form:

  fork(X):- send([proc(1), proc(2), proc(3)],_, X).

The three concurrent processors (hardware or software) inherit a part of their behavior from the generic *module S*. Consequently each processor will process the message X concurrently. It should be noted that the processors *PROC(1-3)* can represent both hardware processors or software processes. Inheritance is the typical way in which a hardware module may absorb the behavior of a software module.

If a library is maintained with generic software modules representing specific func-

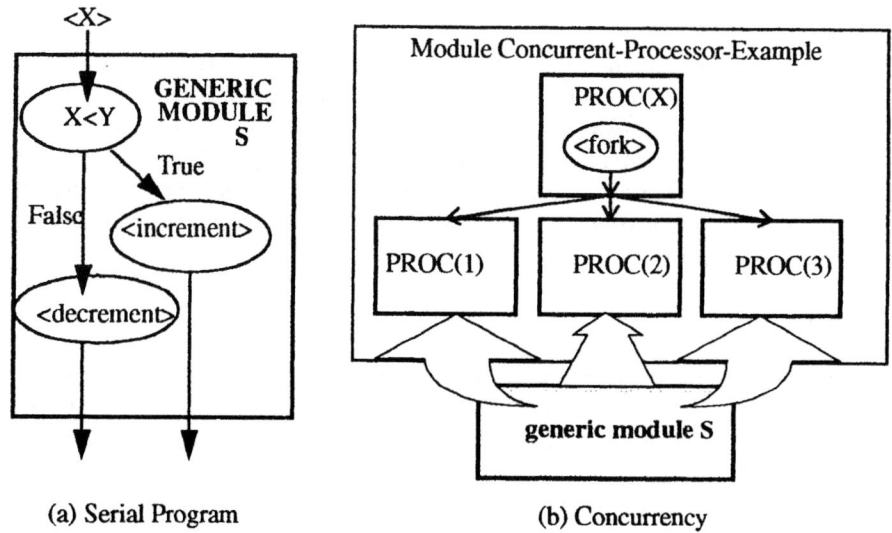

(a) Serial Program   (b) Concurrency

Fig. 4. CO-design construct examples.

tions such as branching, looping and sorting, then these modules can be inherited by other modules as required. Parameterized generic modules allows for the dynamic definition of module resources and delays upon invocation. This facilitates the description of re-usable software modules such as "sorter" which may accept messages such as *quick-sort(SIZE)* and *binary-sort(SIZE)*. The parameter *SIZE* may be used by the generic sorter module to determine the delay of the particular sort algorithm.

2.5 Synthesis

When the designer is satisfied with the simulation results, the synthesis [6] stage can occur. A translator within the DASE environment translates the DSL constructs into concurrent entities in VHDL under user guidance. By default, the environment will assume that casualty of messages arriving to modules are satisfied by the DSL design. Most synchronous DSL designs will satisfy this assumption. In such a case, translation from DSL to VHDL follows direct rules where DSL ports and messages translate to VHDL signals and values respectively. Each DSL module corresponds to a VHDL entity and DSL behaviors map to VHDL processes. State or variable values and types are also defined through the DSL environment.

    For asynchronous design components, the user may direct the environment by setting a DSL predicate. This has the effect of indicating that the implied casualty and ordering of DSL messages should be maintained in the VHDL model. In such a case, the corresponding VHDL model will include an input buffer and control for each entity so that the order of arriving signal values (which correspond to DSL messages) are maintained. Each signal value within the buffer in turn are processed by the VHDL entity. Such designs may be costly in terms of hardware, but VHDL post synthesizers can optimize it further. Note that initial estimates of buffer sizes can be defined with a petri-net analysis of the DSL models.

3. Design Examples

The use of concepts introduced will be demonstrated with two examples of architectural system modelling. The first example is directed at modelling a distributed computing protocol based upon the ABCAST protocol used in the distributed fault-tolerant *ISIS* environment [3]. The example will present the software modelling capabilities of the environment and DSL. The second example is that of the modelling of a digital telephone switch. This example will present the complex design exploration issues that can be addressed with the level of library support.

3.1 The ABCAST Protocol

The ABCAST protocol is part of a set of broadcast process primitives that support distributed computing in a reliable manner.Two other process primitives exist (CBCAST and GBCAST), providing added functionality and support, however due to space limitations only an ABCAST based model will be presented here.

The ABCAST primitive ensures that the order of a broadcast message received at multiple destinations from a source process is the same even though the order is not pre-determined. The algorithm presented in [3] for ABCAST is summarized as follows:

1. The sender transmits a message (Msg) with a unique label to its destinations.

2. Recipients add Msg to a priority queue associated with label, marking it as undeliverable. A priority is assigned to the message (NPri) larger than the largest in the queue, with a process identification (PROC_ID) added as a suffix. This suggested priority is then transmitted to the sender.

3. The sender waits for all the suggested priorities from the destinations, computes the maximum of all the values and sends this value to all the destinations.

4. The destinations update the priority for Msg to the new value and mark the message as deliverable and re-sort the priority queues. The destinations then move messages in order of increasing priority from the priority queues to a delivery queue. This continues as long as the priority queue remains non-empty and there is a deliverable message at the top cf the queue.

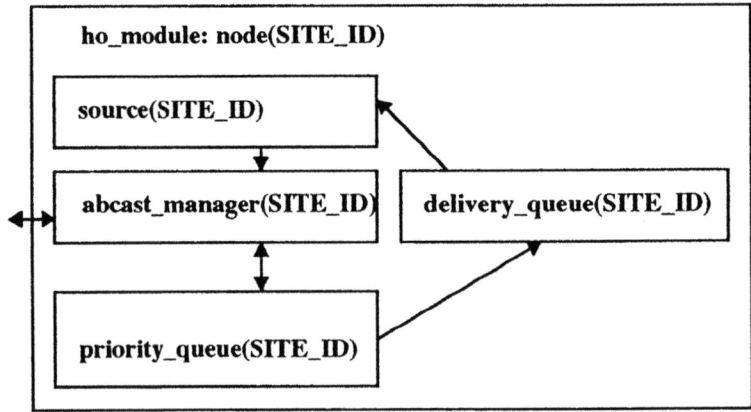

Fig. 5(a). DSL representation of ABCAST nodes

The corresponding DSL model will assume model entities called nodes represent distributed sites. A high level program (source) will generate processes that request ABCAST communication with other nodes. To support the ABCAST primitive, each node will have an abcast_manager, a priority_queue_manager, and a delivery_queue. The structural representation of this model is given in figure 5(a).The abcast_manager provides the protocol interface between the application (source). The priority_queue_manager maintains the priority queue functions while the delivery queue processes each entry to the local destination.

The behavior of the abcast_manager module is described in terms of its equivalent Petri-net given in figure 5(b). The protocol functionality is separated into two sets for clarity; the sender and receiver behavior portion. Each handles the respective functionality of the protocol. Input places (identified by the darkened circles) will allow their respective token types (such as 'form_abcast(D,M,P)). These tokens arrive from other sub-nets (representing the other modules). A controlling net (omitted here for sake of clarity) ensures that the specified incoming tokens are directed to the appropriate input places as well as ensuring safeness within each sub-net. Operations on transitions are indicated between a "◇" while predicates are labelled on arcs. By convention, labels beginning with a capitalized letter is an unassigned variable.

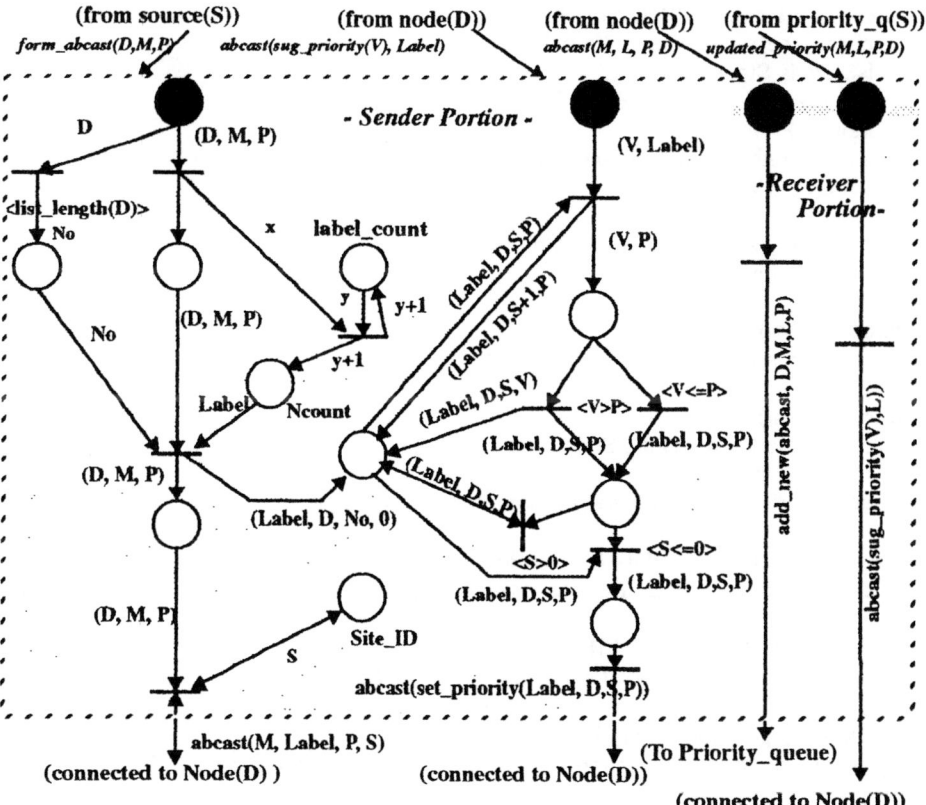

Fig. 5(b). Petri-net representation of DSL ABCAST Manager Module.

Even at this level of representation, some design partitioning of the software complexity is being assumed. For example, the protocol and priority queue management aspects are separated into different modules which implies that the eventual design would also favor such a distinction. The corresponding internal DSL representation of the abcast_manager is provided in figure 5(c). The close relation between the Petri-net representation can be observed by looking at some of the important details of the two. For example, in both representations the input messages to the module are in bold while the output messages are italicized. The resources defined within DSL are defined as tokenized places in Petri-nets, and message parameters are broken into predicate arcs.

```
module(abcast_manager(SITE_ID),[
(form_abcast(PROC_ID, Dest, Msg):-
    check_res(label_count, COUNT),
    NCount is COUNT + 1,
    list_length(Dest, No),
    set_res(label_count, NCount),
    create(NCount, (Dest, No, 0)),
    send(node(Dest), _, abcast(Msg, NCount, PROC_ID, SITE_ID))),

(abcast(sug_priority(Value), Label):-
    check_res(Label, (Dest, State, Prior)),
    NState is State -1,
    set_res(Label, (Dest, NState, Prior)),
    Value> Prior,
    set_res(Label, (Dest, NState , Value )),
    send(check(Label))),

(abcast(sug_priority(Value), Label):-
    send(check(Label))),

(check(Label):-
    check_res(Label, (Dest, State, Prior)),
    State>0),
(check(Label):-
    check_res(Label, (Dest, State, Prior)),
    remove(Label, X),
    send(node(Dest), _, abcast(set_priority(Prior), Label))),

(abcast(Msg, Label, PROC_ID, Sender_ID):-
    send(_, pq_port(SITE_ID), add_new(abcast, Sender_ID, Msg, Label, PROC_ID))),

(updated_priority(abcast, SITE, Label, NPri):-
    send(node(SITE), _, abcast(sug_priority(NPri), Label))),

(abcast(set_priority(Value), Label, Proc_ID):-
    send(_, pq_port(ID), change_priority(abcast, Value, Label, Proc_ID)))]).
```

Fig. 5(c). DSL code for abcast_manager

3.2 Digital Switching Example

This example depicts a telephone switch where telephone calls are handled through interface units and the switching and call scheduling is performed by a generic switch element (shown in figure 6(a)). The latter is automatically configured according to the number of interface units it must support. The function of the generic switch element is to accept a stream of serial input time multiplexed channels from the interface units, packaged as frames, and direct them into any output channel stream. The size of the switch is dependant upon the number of input/output serial lines, as well as the number of channels per frame.

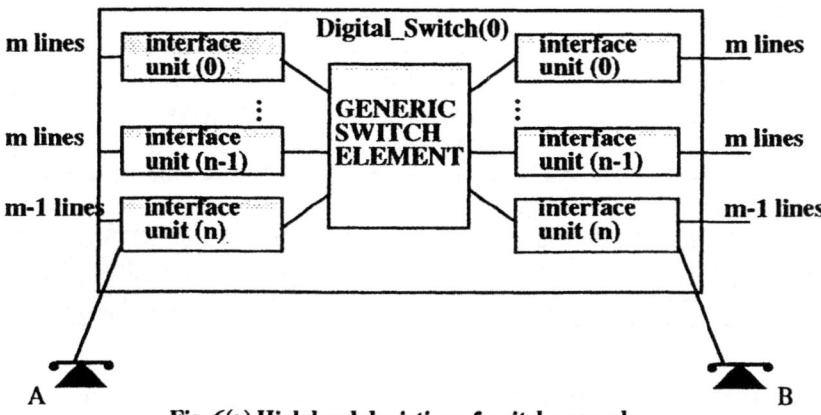

Fig. 6(a).High level depiction of switch example.

The figure illustrates the perspective a system level designer (early in the design cycle) may be interested in modelling. In this particular case two telephone lines are isolated (A and B) which are of interest to the modeler. For example, the particular response time for a call from A to B can then be observed. The other telephone lines can be provided with random stimuli mimicking "background" telephone traffic to load the digital switch.

For a designer later in the design cycle, a more detailed model of the generic switch element would be of interest. Such a DSL representation is given in figure 6(b). While its details are not significant here, the figure is intended to illustrate the levels of hierarchy within this design. All the sub-modules within the generic switch element are configured through the upper level module (which is the digital switch). For example the number of incoming lines (2n) will generate appropriate "line-in" and "line-out" modules, allocate message buffers, and determine the internal time delays for the modules. In addition, the shaded modules represent software components that the designer is concerned about (such as telephone call set-up). The behavior of the required software is also represented in terms of DSL modules, however during synthesis, these are not necessarily translated into hardware descriptions in VHDL.

In the example we wish to describe potential software defined operations such as call processing, error processing, billing and call routing routines. The software components are defined to be related to the control processor. Potential software functionality is presented as the shaded modules in figure 6(b) and the arrows depict a relationship with a parent module - in this case the control processor and memory modules. Additional software

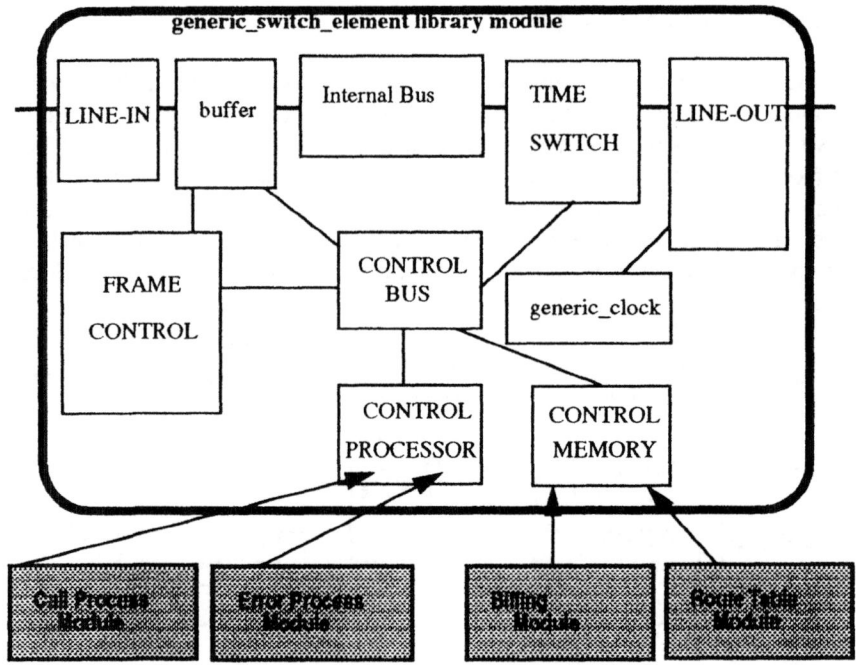

Fig. 6(b). DSL structural representation of library modules.

can be described using modules which are inherited by the control processor.

The switching element can be invoked with a number of input/output line combinations adhering to built-in library constraints. Hence a designer can experiment with different size of switching elements to obtain pertinent information such as performance measurements, switching times, and buffer requirements.

A major portion of the model consists of modular library modules instantiated by a generic library module named *generic_switch_element*. The module can be instantiated under any name such as: *isa(switch(0), generic_switch_element(5, 10)*. This would imply that *switch(0)* is instantiated with the respective parameters applied to the library module. In this particular example, the parameters define the size of the switch (the number if input and output lines) and the number of voice channels carried by each line in the switch.

When the library module is instantiated, configuration rules attached to the module will interconnect necessary modules such as *line-in, time switch* and *frame control* as well as instantiate the modules and resources. The modules accessed by the library module may also be generic library modules - thus allowing hierarchical definition of the modelled system. The design abstraction hierarchy generated by the simulator is given in figure7 (there are further sub-modules associated with the interface units which are omitted for sake of clarity). Levels 1 and 3 would be of interest to the system and hardware modelers respectively. The shaded boxes represent re-usable library modules. The levels of the hierarchy indicated in the figure are automatically defined by the DSL simulator. As mentioned before, during simulation observances and simulation clock increments can be limited to a particular set of levels - thus avoiding unnecessary levels of detail at the output. Finally, at level4 there is also a reference to software modules such as *call process* and *error process*.

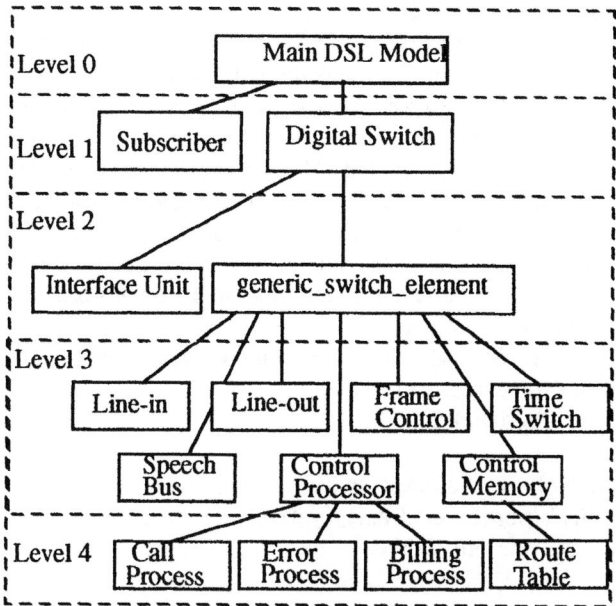

Figure 7: Hierarchy example

To illustrate the mapping of specifications to specific modules, a sample configuration of a size 5 switch with 10 channels per line will require 5 *line-in* and *line-out* modules (to interface to the input and output lines of the switch). The "frame control" module will be configured to generate a timing sequence for the internal modules such that a 10 channel per line sequence can be exercised. Also, any constraints attached to the modules will also be identified and instantiated by the system. The designer may also impose additional constraints if so desired.

The DSL simulator can exercise the model to determine if any of the constraints are violated with this design. When a constraint is not met, the system will request user guidance to relax the constraint. Alternatively, an equivalent module that is linked to the constraint may be requested from the library system by the DSL processor. In this case. the system can substitute an alternative and reconfigure the system and restart the simulation. In doing so, the system can explore different design possibilities.

In the course of a simulation, a user can step through a given scenario one simulation "clock step" at a time, until a given end time, or progress through until a certain constraint is satisfied. In this context, a "simulation clock step" does not correspond to the traditional notion of a simulation clock. In this context, a simulation time increment corresponds to the closest time an event is scheduled at the set observational level. The observational level can be changed at any given time during a simulation (in fact it can be changed automatically by the simulator upon a condition invoked by a given constraint). In this manner, the user may "zoom" in and out into different levels of detail depending upon the interest.

Constraints are check points for the simulator to test and validate. They correspond to hard system specifications inherent in the design. Hence, any violation detected by the simulator is considered to require some action where the user will be prompted for some decisions. A constraint may have been too restrictive, in which case the designer may opt to relax it. This is performed through the DSL simulator with predefined commands such as *set(X)* - where *X* is a constraint. The simulator will request clarification as to which constraint is to be changed (since there could be more than one constraint violation in a given clock step). The environment will then test the new constraints, and if there is no violation, it will continue with the simulation. Alternatively, the user may indicate no further changes to the constraints. At this point the simulator would request the DSL processor to search for an alternative model that belongs to the same class as the module which generated a constraint violation. If a suitable module exists, the system will reconfigure with the new module and restart the simulation. If no further replacements are available, the user is flagged and the simulation continues.

As a sample of the DSL code, the behavioral code for a generic clock to control the framing of the channels is presented below:

```
module(generic_clock(Index, No),
  [
  (clock_count(No):- send(clock_count(0)),
      send(_,clock_port(Index),frame_cycle)),
  (clock_count(NEW):- NEW<No,
      Clock_rate is (125/No),
      delay(Clock_rate),
      COUNT is NEW+1,
      send(_,clock_port(Index),clock(NEW)),
      send(clock_count(COUNT)))
  ]).
```

In this module, there is one behavior *clock_count(X)* where *X* can be bound to *No* (the number of channels/frame) or to a value determined by *Clock_rate* (a variable that calculates the clocking rate based upon the number of channels). At each clock period a *clock(Count)* message is broadcast on its port *clock_port(X)*. This message signals the beginning of a period as well as the channel number (*Count*). There is also a *frame_cycle* message generated when the clock *Count* is reset. This is used to synchronize the destination modules with the beginning of a frame.

Both hardware and software modules are treated in the same manner during simulation - the distinction is made only during synthesis is imposed by the designer through a constraint on the module. The constraints can be stored as part of a library module or defined interactively by the user. Hence delays defined by a module representing software functionality represents processing times for given code segments, procedures etc. whereas the hardware module delays define timing relations between different messages or modules (such as bus delays, memory access times, etc.).

4. Conclusions

This paper has introduced an environment to assist in the design exploration of architectural level modelling of (mainly telecommunication) systems. An internal design capture language was introduced providing a bridge between initial models and more complete ones suitable for lower level synthesis. Two examples were provided to illustrate the capabilities of the environment.

With the use of the above mentioned representation mechanisms, a co-design model of a system can be described. Design exploration can proceed as usual with the aid of the environment, using library components as required. The current work in progress is to synthesize DSL models to a behavioral VHDL description which can be used by design automation tools for structural synthesis. Although not currently implemented, it is expected that software synthesis would also be feasible from the behavioral model (which is can be viewed as concurrent communicating processes) to target CASE methodologies

References

1.   IEEE Software Magazine, Special issue on formal methods, September 1990.
2.   C.G. Bell and A.Newell, Computer Structures:Readings and examples. New York, NY, McGraw Hill, 1971.
3.   K.P. Birman and T.A. Joseph, "Reliable Communication in the Presence of Failures", ACM Transactions on Computer Systems, Vol. 5, No. 1, February 1987, Pages 47-76.
4.   B.W. Boehm, Software Engineering Economics, Prentice Hall, N.Y., New York, 1981.
5.   C.J. Koomen, The Design of Communicating Systems: A System Engineering Approach, Kluwer Academic Publishers, Boston, 1991.
6.   M.C. McFarland, A.C. Parker and R. Camposano, "The High-Level Synthesis of Digital Systems", Proceedings of the IEEE, Vol. 78, No. 2, February 1990, Pages 301-318.
7.   B. Moszowski, "A Temporal Logic for Multilevel Reasoning about Hardware", IEEE Computer magazine, February 1985, Pages 10-19.
8.   F. J. Ramming, "System Level Design", Fundamentals and Standards in Hardware Description Languages - NATO ASI series, Kluwer Academic Publishers, 1993, Netherlands, Pages 109-151.
9.   W. Reisig, "Petri-Nets - An Introduction", EATCS Monographs on Theoretical Computer Science, Springer-Verlag, NY, 1982.
10.   O. Tanir, V.K. Agarwal and P.C.P. Bhatt, "Specification driven architectural design environment", *IEEE Computer*, Vol. 28, No. 6, June 1995, Pages 26-35.
11.   O. Tanir and S. Sevinc, "Defining the Requirements for a Standard Simulation Environment", *IEEE Computer*, Vol. 27 No. 2, February 1994, Pages 28-34.
12.   O. Tanir, V.K. Agarwal and P.C.P. Bhatt, "The Design of a Library Support System for a Telecommunication System Synthesis Environment", Proc. of the fourth International Workshop on Rapid System Proto-typing, Research Triangle Park, N.Carolina, June 28-30, 1993, pp. 54-67.

VOMDraw - A Tool for Visual Object Modeling

Gabriele Oberlik *
Stefan Preishuber *
A Min Tjoa * ◊
Roland Wagner *
Klaus Wolfmayr *

* FAW
Research Institute for Applied Knowledge Processing
Johannes Kepler University of Linz, AUSTRIA
email: [klaus, rrw, spr]@faw.uni-linz.ac.at

◊ Department of Information Systems
Technical University of Vienna, AUSTRIA
email: tjoa@ifs.tuwien.ac.at

Abstract. The weakness of an application often has its roots in the analysis phase. This is why current software engineering concentrates on appropriate systems analysis methods and tools and tries to get away from code programming by code generation out of an abstract model. This paper describes a graphically oriented object-oriented analysis tool with the possibility to describe the structure part and the behavior part of an application. A class schema in the object-oriented database system GemStone can be generated from the structure part.

1. Introduction

Systems analysis is the basic requirement for building an application. There are two kinds of systems which are essential: the environment in which the application should be embedded and the application itself. An application system can be described as a number of interacting objects. Systems analysis serves for understanding and documenting the main characteristics of these interacting objects.

There are a number of approaches to systems analysis, whereby the object-oriented analysis is a relatively young method among them. To describe a system as well as to understand interaction between objects, it is necessary to simplify the system. As systems have grown very complex, and they contain many objects and many interactions between them, success of more traditional systems analysis methods like natural-language has become uncertain. Object-oriented programming languages have the purpose to structure an application in order to correspond better to natural structures.

The problem of managing complexity of systems and the employment of obejct-oriented programming languages are the main reasons why analysts prefer the object-oriented approach.

Since the mid-1980s many object-oriented analysis (and design) methods have been developed. The most common methods are *Object-oriented Modeling and Design (OMT)* by Rumbaugh [11], *Object-oriented System Analysis (ooSA)* by Embley/Kurtz/Woodfield [9], *Object-oriented Analysis (ooA)* by Coad/Yourdon [4] and *Object-oriented Design with Applications (ooD)* by Booch [2]. All these methods have in common that a structural part as well as a dynamic part can be modeled.

The structural part contains at least classification, inheritance and association. *Visual Object Modeling (VOM)* [12] [13], on which VOMDraw*) is based, provides the definition of object groups like ooSA and ooD do. In addition, they can also be specified as derived or constrained (cf. section 2.1). Even the role concept is realized in VOM, OMT and ooSA.

Summing up the VOM method can be viewed as graphical description language for an object-oriented model designed out of a part of the real world. A VOM model should be the basis for an application embedded in the universe of discourse, i. e. the VOM model describes the static and dynamic structures of existing (concrete or abstract) objects as well as a control flow which controls these objects and their interaction.

The process of object-oriented analysis is a creative process, i. e. the user designs parts of the model, adds classes, changes them later etc. Thus, for efficient use of an object-oriented analysis method a tool is necessary. VOMDraw is such a tool for the VOM method.

Since the analysis process is a very creative process, an analysis tool should not be a handicap for the user, but a useful guide not to lose the general view of the model. Thus, the main purpose of VOMDraw is to support the analyst to make use of the VOM method and to concentrate on the model and not on the method.

Section 2 of this paper describes the user interface of VOMDraw, referring to the three model parts of VOM: structure diagram, state transition diagram and interaction diagram. Section 3 deals with implementation details of VOMDraw, namely the realization of the model-view-controller concept and the realization of consistency checks. Comparisons with related work to VOMDraw are made in section 4, and in section 5 some additional functions of VOMDraw, which are in the planning stage, will be mentioned.

*) The support by FWF under grant No. P09369 is gratefully acknowledged.

2. User Interface

According to the VOM method, the tool consists of three kinds of editors: The *Structure Editor* is used for modeling the static structure of the model. Dynamic aspects of each "class" in the Structure Editor are modeled in several *State Transition Editors*. A control flow is designed with the help of the *Interaction Editor* where also the user interface parts of the model are designed.

The main window of VOMDraw (see Fig. 1) reflects the division into the three editors and the connection to the GemStone data base.

Each editor is made up of a specific tool palette and a drawing area. The tool palette contains symbols for "class", "inheritance relation", "base transition" etc.

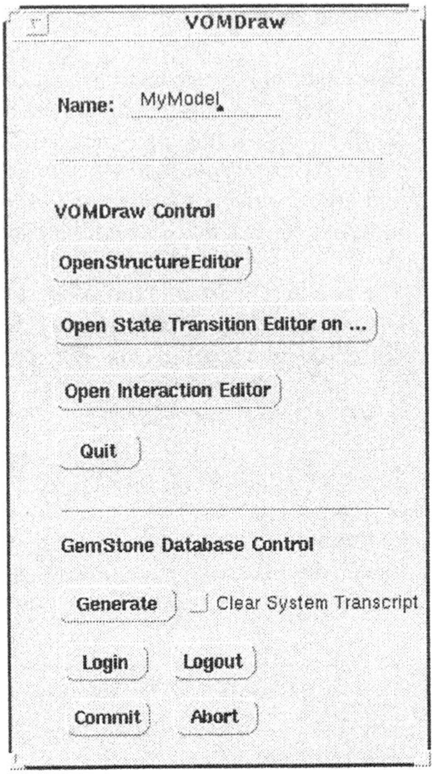

2.1. Structure Editor

Fig. 1: Main window of VOMDraw

The Structure Editor of VOMDraw serves for modeling the static part of a VOM-model, i. e. classes and atomic instance variables as well as relationships to each other (is-a-relationships or class hierarchy and association relations or class composition hierarchy, for a detailed description of the expressions being used cf3] and [8]).

In order to create classes and relationships, a tool palette is provided which contains the necessary symbols for managing the structural part (see Fig. 6). Depending on the elements being created, an additional dialog opens where the user has the possibility to specify more information (e. g. when creating a class, the class name and its atomic instance variables have to be named, see Fig. 2).

The main construct in the Structure Editor is the class construct, represented by a rectangle. Before creating a class, atomic instance variables can be added. Each atomic instance variable is characterized by a name and a datatype. Additionally, it can be constrained (e. g. the age of a STUDENT has to be more than 18 years) or defined as derived (e. g. the age of an ASSISTANT is calculated by subtracting his birthdate from the actual date).

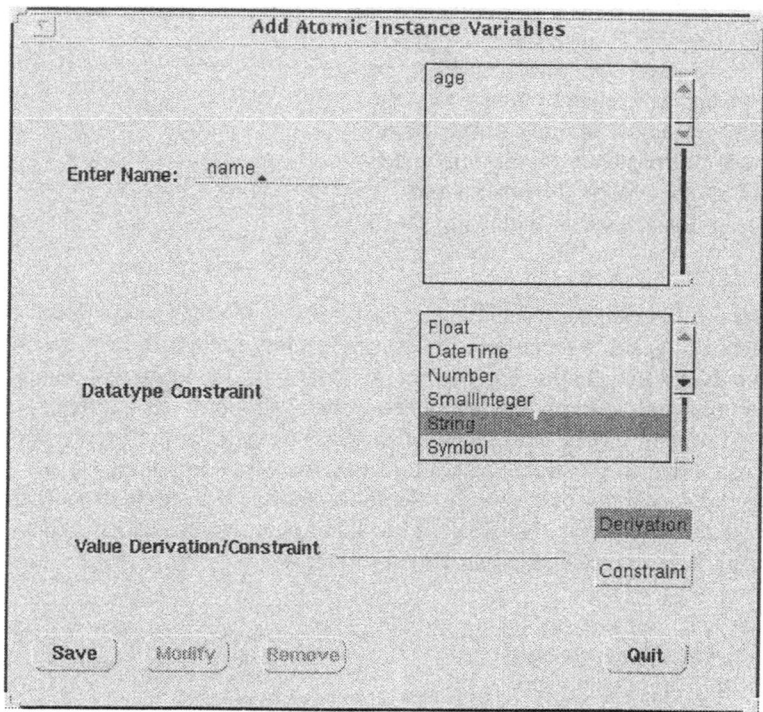

Fig. 2: Dialog for Atomic Instance Variables

When the class has been created, a new rectangle appears within the drawing, containing only the class name. Thus, it is avoided to overcharge the drawing with minor relevant information, and the user can concentrate on the essentials.

Once classes are created, they can be related to each other by creating inheritance relationships and associations.

The inheritance relationship serves for modeling the class hierarchies. In order to make it possible to specify a semantically correct model and to avoid multiple inheritance, besides the ordinary is-kind-of-relationship VOM also allows the is-role-of-relationship [14] (e. g. STUDENT is role of PERSON).

When modeling associations between several classes the user can choose between strong and weak dependencies [2]. By using a strong dependency between two objects BOOK and CHAPTER (see Fig. 3), the referenced object (object of class CHAPTER) is existiantially dependent of the referencing object (object of class BOOK), i. e. if BOOK is deleted, CHAPTER is deleted, too. In other words, a CHAPTER cannot exist without a BOOK. Strong dependency is represented by a black arrow, weak dependency is represented by a grey arrow in the Structure Drawing.

In contrast, a weak dependency as modeled between class BOOK and class COVER in Fig. 3 expresses the independency of objects of the two related classes, i. e. the deletion of an object of class BOOK has no impact on the existence of the referenced object of class COVER. In other words, a COVER can exist without belonging to a BOOK.

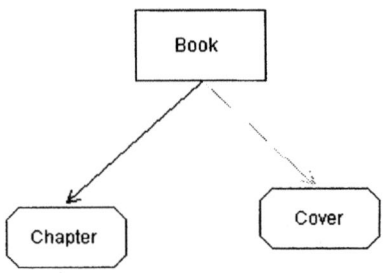

Fig. 3: Strong and weak dependency

Inheritance-relationships (building the class hierarchy) and associations (building the class composition hierarchy) can be orthogonally applied. In this context, the modeler has to comply with one of the most important concepts, namely the strong hierarchical structure. As a direct consequence, in a VOM model only several independent class trees (e. g. the tree below Person in Fig. 6) can exist. A construct like a DAG (a Direct Acyclic Graph) is not supported. Because this concept would prohibit the multiple use of a single class (because only one reference to the class is allowed), an additional symbol called "alias" has been introduced. It is represented by a rounded rectangle.

The alias can be seen as a container which may contain one or a collection of objects of a certain class. Like an atomic instance variable, the elements of a container may be derived or constrained (see Fig. 4). STUDENT is a constrained alias which means, that only students which are older than 18 years are allowed on this university. The derived alias TECHNICSSTUDENTS is a container of all objects of class STUDENT who study informatics.

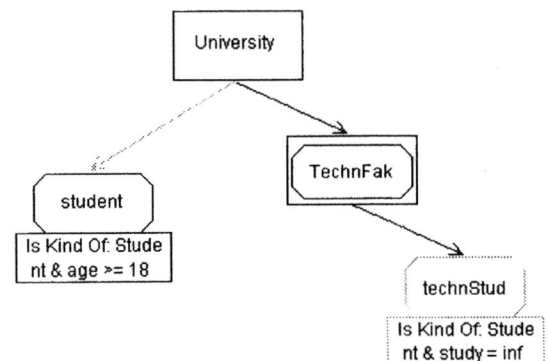

Fig. 4: Constrained and derived aliases

As many aliases as necessary can be created out of a class either by selecting the "Define Alias" button in the "Create Class"-Dialog (see Fig. 5) or by selecting the appropriate symbol for an alias (rounded rectangle) in the tool palette of the Structure Editor (see Fig. 6). Aliases can then be

Fig. 5: Create Class Definition Dialog

included in the class trees for the purpose of building the class composition hierarchy.

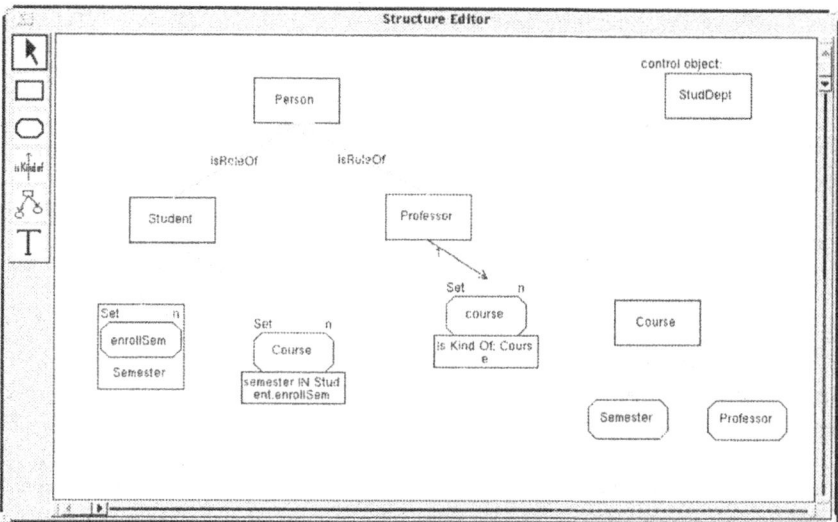

Fig. 6: Structure Editor

Concluding, the structural part of a VOM model consists of several independent, hierarchically structured class trees. Each class tree consists of a root-class which may recursively reference either subclasses (the class hierarchy) or aliases of other classes (class composition hierarchy).

To support several levels of abstraction, the user has the possibility to combine several object classes to one subject. A VOM subject always includes all object classes which are related to the selected object classes via the class tree. When the user wants to extract the subject, some unoccupied space on the Structure Drawing must be found. The needed space can be indicated by a dotted rectangle.

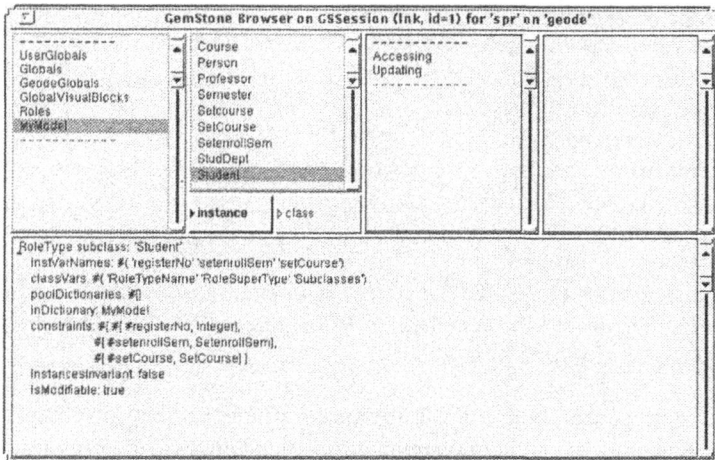

Fig. 7: Generated classes in GemStone

After finishing the modeling process of the structural part of the VOM model, its classes and relationships can be generated in the object-oriented database system GemStone (see Fig. 7). In this way, the data model, including role hierarchy, access methods for the atomic instance variables, and methods for managing the specified constraints and derivations are automatically generated. For control purposes, the generation process is documented in the system transscript. Thus, the basis for implementing the application in OPAL - the database programming language of GemStone (cf. [5]) - is created. For a detailed description of the schema generation see section 3.4.

2.2. State Transition Editor

In addition to the definition of a class as part of the static structure of the model, the user can define the dynamic behaviour of each class. The behaviour model of a class is designed with the help of the State Transition Editor. A State Transition Diagram consists of states representing the life cycle of an object and transitions which transfer an object from one state into another. The life cycle of an object starts with a transition called constructor and ends with the destructor. These transitions can never be deleted.

There are two possibilities to open a State Transition Editor for a certain class. The modeler can use either the menu of the class figure in the Structure Diagram or the appropriate button in the main window of VOMDraw.

In the State Transition Diagram, transitions and states must alternate. They are connnected by simple arrows. States are either set by a transition or they are derived from other states or from the value of some attributes respectively. For instance, the state DRUNKEN in Figure 8 is derived from the instance variable BLOODALC, i. e. the object is in state DRUNKEN when the instance variable

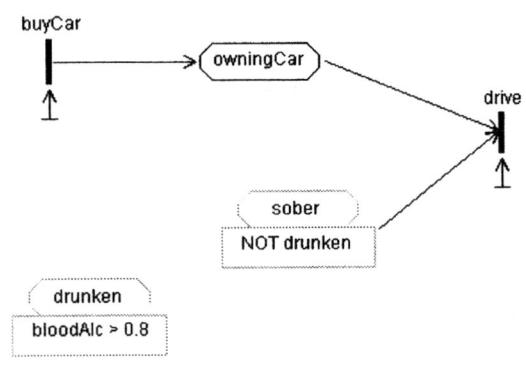

Fig. 8: Derived states

BLOODALC is greater than 0.8. The object is in state SOBER when it is not drunken. This is called "the state SOBER is derived from state DRUNKEN". The object can drive when it is OWNINGCAR and SOBER.

An object state can be bound to an instance variable (atomic or complex), i. e. the defined state is only valuable in combination with another object which - in most cases - is part of a container. The example in Figure 9 such a relationship state: A STUDENT is only enrolled for one certain SEMESTER.

Every base transition is triggered by a control transition which represents an action of a control object. The responsible control object is shown in a rectangle below the verical bar which is the symbol for transition. If the user does not know the control object yet, a simple horizontal line is shown instead of the rectangle. The control object can be added later.

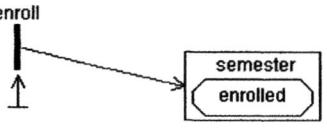

Fig. 9: relationship state

When a control object triggers a base transition, a number of parameters can be submitted. These parameters are specified in the "Add Parameters" window shown in Figure 10. A parameter is stored in an instance variable of the corresponding class. This may be an (inherited) atomic or complex instance variable of the class.

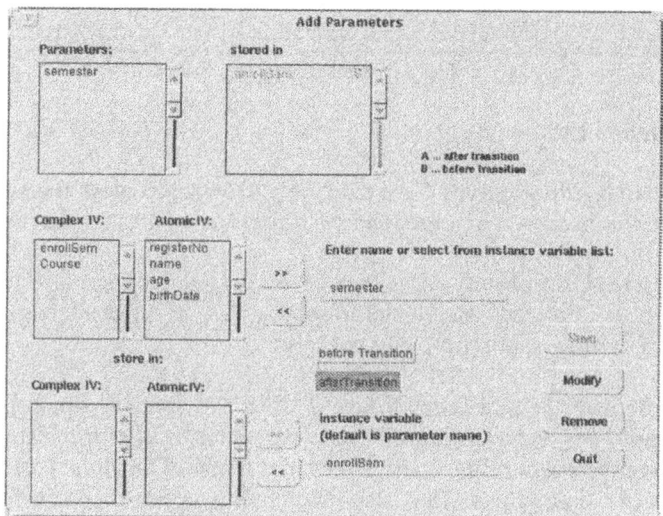

Fig. 10: Parameter Definition Window

As a default for the parameter name the name of the instance variable is used, but the modeler has also the possibility to specify a different name for the parameter. He can choose wether the parameter value is stored into the instance variable before of after execution of the

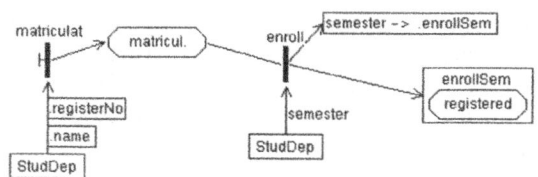

Fig. 11: Assignment before/after transition

transition. If the assignment is executed before the transition, both, the parameter and the instance variable are shown in a rectangle between control object and base transition (cf. Fig. 11, transition MATRICULATE). Otherwise the assignment is shown in a rectangle above the transition figure, and the parameter name is placed between transition and control object (cf. Fig. 11, transition ENROLL).

Figure 12 shows an example of a State Transition Diagram for the class STUDENT (cf. Fig. 6).

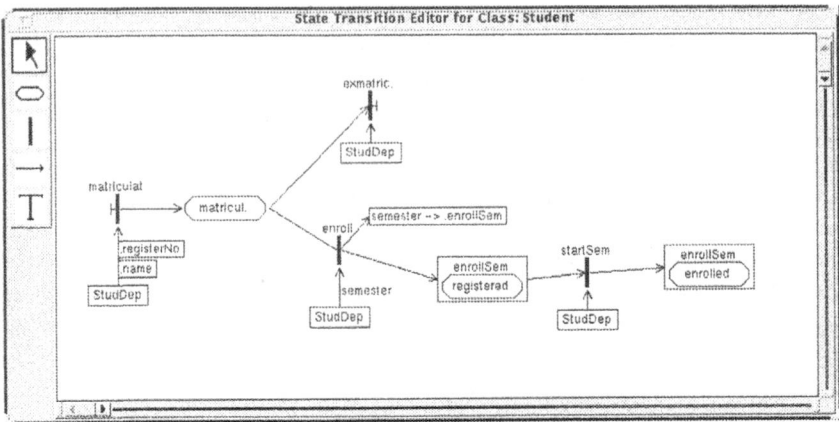

Fig. 12: State Transition Editor

2.3. Interaction Editor

The Interaction Editor serves for modeling a flow of control transitions which are responsible for the interaction between the objects as well as for the user interface.

Interactions between objects can either be output interactions, i. e. the trigger on a base transition, as already mentioned in section 2.2, or input interactions constituting the retrieval of information from a base object.

The tool palette of the Interaction Editor (cf. Fig. 13) contains, among others, symbols for modeling states and transitions. A state can eihter be a control state, which is equal to a name for the status of the control flow, or a copy of an already defined base state, i. e. a state of a base object. This base state copy is necessary to model an interaction input. A control transition selects some objects or values of object attributes for further processing of the objects which are in the chosen state. Thus, for an interaction input the user can select instance variables of an object class. This is equivalent to a projection in a data base. Also, a selection constraint can be specified which means that only object that meet the constraint are selected. In addition, the number of objects to select can be specified. The user must choose between manual and automatic selection. Interaction with the user of the application is needed when the selection should be performed manually.

A transition in the Interaction Diagram can be a control transition or a copy of an already defined base transition. A control transition is equicalent to an action of a control object which reads input from a base object (input interaction) and/or triggers a base transition (output interaction). To insert an output interaction, a copy of a base transition has to be inserted into the Interaction Diagram. The submission of parameters with an output interaction is equivalent to the one defined with the trigger of a base transition in the State Transition Diagram (cf. Figures 12 and 13, base transition ENROLL). Thus, the specified parameters of the base transition are adopted for the output interaction. When these parameters are modified in the Interaction Editor, the corresponding base transition in the State Transition Diagram is updated.

A control transition must belong to a control object. Since the VOM method allows objects being both, base objects and control objects, the user can choose among all defined classes defining a control object for a control transition. In order to ensure clearness of the Interaction Diagram only one input interaction for a state copy and one output interaction for a base transition copy is allowed. Finally, a base transition can only be triggered by one control transition, i. e. only one copy of a base transition in the Interaction Diagram is allowed as well.

Figure 13 shows the Interaction Editor for our example in the Structure Editor (Fig. 6) and State Transition Editor (Fig. 12).

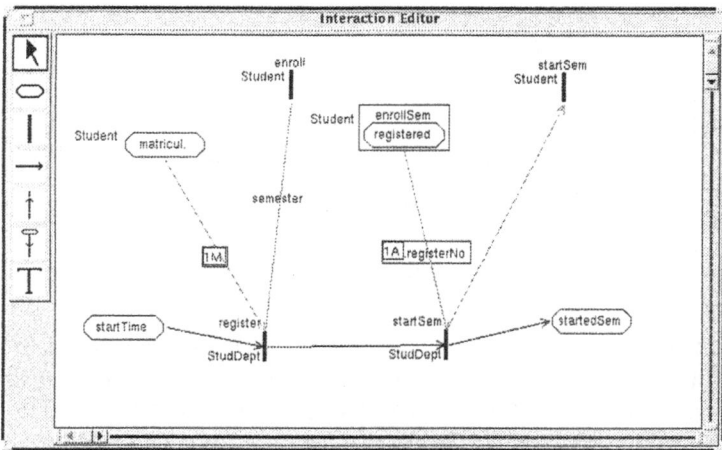

Fig. 13: Interaction Editor

3. Implementation Details

VOMDraw was implemented in Smalltalk-80 [6], using the application frameworks Visual Works 1.0 [15] and HotDraw 4.1 [7] and the object-oriented database system GemStone 3.2.5.

3.1. Model View Controller (MVC) Concept

HotDraw is an Application Framework having the functionality of a graphic editor. An editor is made up of a tool palette and a drawing consisting figures like rectangles, lines, arrows or text. The tool palette can be adapted to own purposes by inserting icon symbols. The corresponding figures for the drawing are created by selecting the appropriate symbol in the toos palette and a mouse click in the drawing. HotDraw only provides "flat" figures, i. e. only the view part of an MVC concept can be realized (cf. [10]); the figures only have attributes concerning the view. Additional information is stored in base classes which represent the model part of the MVC concept.

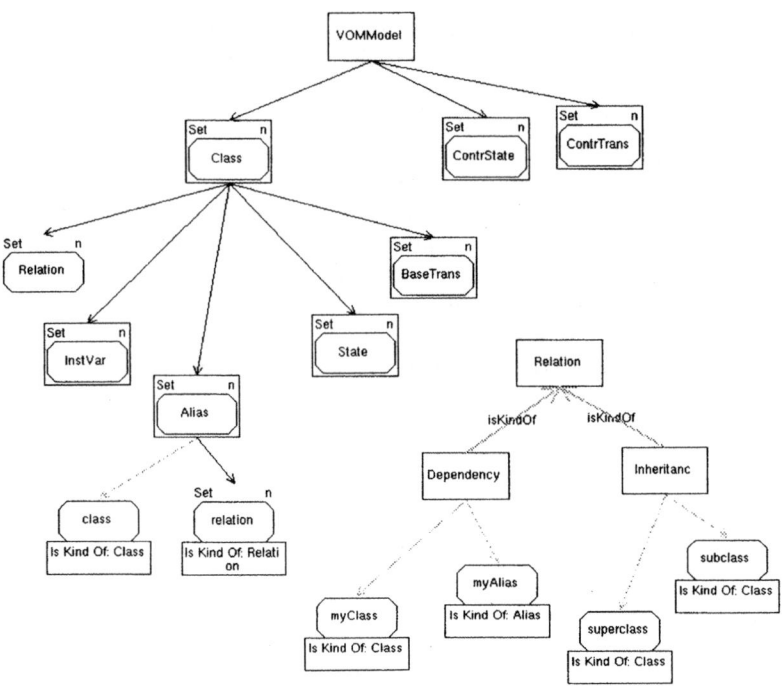

Fig. 14: Model part of VOM model

Figures 14 and 15 - Structure Drawings in VOM notation - show the class hierarchy of Smalltalk base classes for the model part and HotDraw classes for the view part of a VOM model. Each base class like MYCLASS, INHERITANCE or BASETRANSITION has its corresponding class in the HotDraw hierarchy, like CLASSVIEW, INHERITV or BASETRANSV.

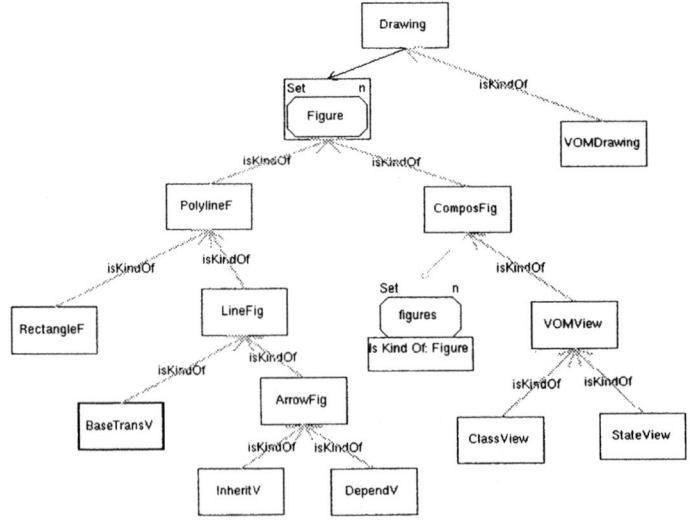

Fig. 15: View part of VOM model

3.2. Update Propagation

An important aid of VOMDraw is the update propagation of names in any context. For example, a class name is used as a contol object in a base transition. When the class name is changed in the Structure Drawing, the modeler would have examine the whole model - including State Transition Drawings and Interaction Drawing - and update the class name wherever used.

In VOMDraw, name changes - concerning clases, instance variables, parameters, base transitions and states etc. - are only allowed at one specific point in the editors, namely where the corresponding figure is created. In case of a change, every figure that uses this name text is updated in its drawing.

3.3. Consistency Checks

A very similar problem is the prohibition of deletion of a figure if some information concerning this figure is used in another context. This includes that an atomic instance variable or a class may not be deleted, if a subclass needs this instance variable; the corresponding inheritance relationship may not be removed either. A tool for object-oriented systems analysis should also perform some consistency checks such as uniqueness of class names etc.

The update of figures and the prohibited deletion of figures that are still in use are realized with an instance variable of the base class (like MYCLASS in Figure 14). The variable is called "usedIn" and represents a container for all base classes which use instance variables - e. g. the name - of this class. In case of an update, the "usedIn" set is run through and a message is sent to every corresponding to update itself. The class may not be deleted as long as the "usedIn" set is not empty, which means that the class is still needed in another context. To delete the class, all other references to it must be deleted first.

Additional consistency checks like the identification of cyclic inheritance have already been explained in section 2.1.

3.4. Code Generator

The code generator of VOMDraw creates classes in the object-oriented database system GemStone according to the structure part of a VOM model. Since VOM contains definitions which cannot be realized in GemStone, a protocol is written into the system transscript containing eventual error messages and warning messages.

Roles are mapped to GemStone, too, provided that the role concept has been installed before [14].

To analyze the structure model and to generate a class scheme in GemStone, VOMDraw follows seven steps:
1. A dictionary with the name of the VOM model is created in the database.

2. The classes are created according the specified class hierarchy.

3. For each class the atomic instance variables are added with a constraint on the selected data type.

4. Aliases are realized in GemStone as collection classes whose elements are constrained to the class, the alias is defined for.

5. According to the class composition hierarchy, complex instance variables are added to each class. That means, a referencing class gets instance vareables for each alias which is referenced by this class:

6. If the alias represents a single object, the name is created as "ref" plus the name of the alias. For example, the alias SEMESTER referenced by class COURSE in Figure 6 becomes an instance variable named "refSemester" for class COURSE.

7. If the alias represents a collection of objects, the name for the instance variable is created as the collection type plus the alias name. For example, the alias ENROLLSEM referenced by class STUDENT in Figure 6 becomes an instance variable named "setenrollSem" for class STUDENT.

8. For each instance variable (atomic and complex) access and update methods are created.

9. Constraints or derivations on the instance variables are added: First, the syntax of the string containing the constraint or derivation is examined. If it is correct, the update and/or access method is overwritten. If the syntax of the constraint or derivation is not correct, the code is inserted in the update and/or access method as comment, and a corresponding message is wirtten to the system transscript.

The so created class schema is syntactically correct and can be used in an application. It contains as much semantics as are supported by the database system GemStone. Additional semantics must be realized within the application.

4. Related Work

Since object-oriented analysis methods are relatively young, they are still being developed and many of them are only described by giving examples. Only very few tools to support these methods have been implemented yet.

For instance, Booch's ooD method is supported by the tool *Rational Rose*™*) , which is available since mid 1992. Rational Rose is a graphically oriented tool allowing to specify and examine syntax and semantics of a Booch model. A code generator is in the planning stage.

Prograph is a visual programming tool for the MacIntosh including object-oriented structures like classes and inheritance relations. Behavior is represented by the application flow.

In order to support the OMT method, *OMTool* has been implemented. The only semantical consistency check OMTool provides is to prohibit cyclic inheritance. A structure part as well as a behavior part can be modeled. OMTool also contains a code

---

*) Rational Rose ™ is a trade mark of Rational®.

generator which creates text files out of the structure part of the model. These text files represent header-files containing C++ class definitions for a C++ application.

The *GemStone Visual Schema Designer (GS Designer)* is a graphical schema designer for the GemStone object-oriented database management system [1]. It only supports the design of a static structure, behavior cannot be modeled.

VOMDraw is also a graphically oriented tool based on the VOM method. VOM already performs syntax checks in the analysis phase, i. e. when classes, states, relations etc. are created. VOMDraw is able to generate a class schema for the database system GemStone out of a complete VOM model.

5. Future Work

During the modeling process the modeler should have the possibility to rum some simulations by means of a *VOM Interpreter*. The VOM model should be saved in an object-oriented database already during the analysis process. Model updates should be immediately propagated to the database. It should be able to simulate the control flow of an object: The life cycle of created objects should be shown graphically in the drawings. Even the simulation flow should be saved in the database.

Finally, an *Application Generator* should be implemented creating a full application of a VOM model. The creation of the structure part in GemStone with constraints is already realized. But the behavior part will have to be adjusted for an application. Some additional information might be needed like the method part of a transition (e. g. specified with programming code).

Otherwise, there is probably information contained in the VOM model which cannot be usen in an application. For example, the life cycle of an object only indicates the sequence in which transitions (methods) may be performed.

References

[1] J. Almarode et al., "Issues in the Design and Implementation of a Schema Designer for an OODBMS", Proceedings of ECOOP '91, European Conference on Object-Oriented Programming, Geneva, Switzerland, 1991.

[2] G. Booch, "Object Oriented Design with Applications", Publishing Company Inc.

[3] R. G. G. Cattell, "Object Data Management, Object Oriented and Extended Relational Databases", Addison Wesley, Reading, Mass., 1991.

[4] P. Coad, E. Yourdon, "Object-Oriented Analysis", Yourdon Press, Prentice Hall, Englewood Cliffs, New Jersey, 1990.

[5] GemStone 3.2.5, Programming in OPAL, Servis Corporation Inc., 1993.

[6] A. Goldberg, D. Robson, "Smalltalk-80 - The Language", Addison-Wesley, ParcPlaceSystems 1989.

[7] R. E. Johnson, "Documenting Frameworks using Patterns", a documentation included with the HotDraw Application Framework.

[8] W. Kim, "Introduction to Object-Oriented Databases", MIT Press, Cambridge, Mass., London, England, 1990.

[9] B. Kurtz, S. N. Woodfield, D. W. Embley, "Object-Oriented Systems Analysis and Specification", Hewlett Packard & CS Dept., Brigham Young University, 1991.

[10] W. R. Lalonde, J. R. Pugh, "Inside Smalltalk", Volume II, Prentice Hall, Englewood Cliffs, New Jersey, 1991.

[11] J. Rumbaugh et al., "Object-Oriented Modeling and Design", Prentice Halle, Englewood Cliffs, New Jersey, 1991.

[12] R. Schauer, "Entwicklung einer visuellen Objektmodellierungsmethode und deren Anwendung im Bereich der rechnerintegrierten Fertigung", Diploma Thesis, FAW Research Institute for Applied Knowledge Processing, University of Linz, Austria, 1992.

[13] R. Schauer, S. Schönberger, "VOM - Visual Object Modelling" in: A M. Tjoa, I. Ramos (eds.), "DEXA 92: Database and Expert Systems Applications", Proceedings of the International Conference in Valencia, Spain, 1992.

[14] M. Schrefl et al., "The Role Concept in Smalltalk", Technical Report, University of Linz, Austria, 1993.

[15] "Visual Works - User's Guide, Release 1.0", ParcPlaceSystems 1992.

A Computer Aided System for Developing Graphical Telematic Applications *

Franco Arcieri[1], Michelangelo Fossa[2], and Enrico Nardelli[3,4] **

[1] Algotech Sistemi s.r.l., Via Biella 10, 00185 Roma, Italia.
[2] Saritel S.p.A., S.S. Pontina km.29,100, Pomezia, Roma, Italia.
[3] Dipartimento di Matematica Pura ed Applicata, Università di L'Aquila, Via Vetoio, Loc. Coppito, I-67010 L'Aquila, Italia, e-mail: nardelli@univaq.it.
[4] Istituto di Analisi dei Sistemi ed Informatica, Consiglio Nazionale delle Ricerche, Viale Manzoni 30, I-00185 Roma, Italia, e-mail: nardelli@iasi.rm.cnr.it.

Abstract. In this paper a computer aided system for developing graphical interactive telematic applications is described. The system is novel in the sense that it integrates ideas from software engineering in a harmonious and goal-oriented way to obtain a powerful and high productivity development tool for the production of telematic applications. It features a visual programming language which allows the parametric instantiation and customization of generic objects and functions, explicitly designed for the interaction with remote database services. It also provides the possibility of cross-compiling to many different platforms the prototyped application.

1 Introduction

The European market of on-line services offers a large amount of information, available from different services provided by several suppliers. This information, however, is little used due to the absence of standards regarding query languages, the structure of databases, data formats and many other aspects. Moreover, interaction is based on query languages, different from service to service, very often difficult to learn and tedious to use. These aspects make the use of telematic services difficult for the end-users and the production of telematic applications complex for developers. Major problems for the end-user are: to find which databases are available, to identify the bases containing the needed information, and to learn the particular query language of the chosen service(s) [7].

The ease of use is very important since the end-user of a telematic application is generally a person with no or very little knowledge about Computers or Telematics. Hence an application with a very high degree of usability is required [1, 9].

---

* This work has been partially supported by the "TOOTSI" project (EP2109) of the Commission of the European Communities 'ESPRIT' Research Program.

** Partially supported by the "MULTIDATA" project of the 'Sistemi Informatici e Calcolo Parallelo' Research Program of the Consiglio Nazionale delle Ricerche.

The human computer interface (HCI) must be based upon *typical interaction objects*, whose usability is well known and tested. Obviously the telematic context puts additional constraints and make it necessary to design ad-hoc telematic objects for interacting with on-line services [6]. These objects, in the context of a software development environment, can be identified as GTP (Graphic Telematic Primitives) for what concerns the conceptual telematic aspect and as GTO (Graphic Telematic Objects) for what regards their actual istantiation.

It is important to remind that the interaction with on-line services is currently obtained, in most of the cases, by means of *command languages* [7]: it is so possible to assert that the use of the GTOs allows a mapping from a graphical interaction to an interaction based on command languages. Another requirement that has to be taken in account is the consistency of these objects both mutually and referring to the platform on which the final application is supposed to run.

Better applications can be obtained if a graphical interactive framework is used for the application's user interface. But, even with current high-level software development environments, the time required to produce and tune to end-user needs an application may be too high in this rapidly changing market. A rapid prototyping approach can be used, but only if it provides the means to quickly and easily produce the (more efficient and commercially safer) machine language version of the application.

The aim of **TOOTSI** project (acronym for Telematic Object Oriented Tools for Services Interfaces)[1] has been the overcoming of these barriers, that is the development of a set of software tools able to support the implementation of telematic applications independently from the specific service which is accessed. Within the context of TOOTSI it was acknowledged the need of a visual programming language [3, 12] to allow the rapid development and prototyping [1] of telematic applications and their user interfaces: the Visual Tootsi Programming Language (VTPL).

VTPL and the related VTPL Development Environment represent the highest level tool in the set of ones foreseen in the TOOTSI project [14, 15]. The application to be developed by means of VTPL are multi-platform, that is portable to different host environments (e.g., Ms-Windows, X-Windows, OSF-Motif, TINTO[2]) and highly user-friendly. Using VTPL it is possible to developers to implement telematic applications, by means of a prototyping approach, with a high quality graphical interface, indipendently from the specific accessed service and from the specific application platform. Applications can be delivered to the end-user as executable modules and their visual interface makes it very easy to use them focusing on the task to be done instead of on the peculiarities of the command language of the specific service. Moreover, applications provide a hypertextual interaction style which allows the browsing of functionalities and

---

[1] TOOTSI project (EP2109) has been carried out by a european consortium composed by: Algotech Sistemi (I), Desarrollo De Software (E), Ifatec (F), Infotap (L), Saritel (I), Telésystéme (F) and Université de Nice (F).

[2] TINTO is window-manager able to run on a *cheap* PC (i.e. a 80286 machine with 256KB RAM), developed by Desarrollo De Software.

information nodes (corresponding to different on-line services) [5, 10].

Developers of telematic applications are provided in the VTPL development environment with a complete support, taking into account that they may vary, on one hand, from a designer with a high degree of knowledge in telematics but with no experience about human-computer interaction to, on the other hand, an HCI developer without any competence about on-line services. As a consequence of this wide range of skills VTPL must warrant a high usability to its users based on a direct manipulation approach [11].

The structure of the paper is the following: we first describe the reference framework for design activity (Sect. 2) and provide a high level description of the VTPL development environment (Sect. 3). Next, in Sect. 4, we discuss the Graphic Telematic Primitives which are the building blocks for the development of telematic applications. In Sect. 5 we then present the structure of the object oriented classes which are the framework for cross-compilation activities and finally discuss in Sect. 6, through an example of a work session, the functions of the prototyping subsystem.

2   The Design Reference Framework

Roughly speaking, VTPL is made up by a prototyping environment and a cross-compilation module which allows to translate the developed prototypes into applications portable to different host environments. In Fig. 1 a context diagram [17] for VTPL is shown. The end-user defines the requirements for a telematic application and gives feedback to the designer on intermediate prototypes.

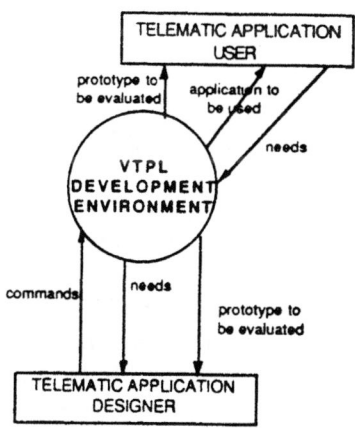

Fig. 1. Context diagram

The designer (which is a VTPL user) analyses the requests in order to derive a specification for the telematic application. Such a specification is used to produce an interpreted prototype, whose behaviour is tested by the end-user.

When eventually the prototype satisfies the requirements of the end-user, an intermediate (*linearized*) representation of it is automatically produced.

The linearized representation is then cross-compiled to produce C++ code based on an object oriented framework explicitly designed. The C++ code, that can be further manipulated and refined, if needed, is used to delivery the final application to the end- user. A more refined Data Flow Diagram (DFD) presenting how telematic applications developement is carried out within VTPL is shown in Fig. 2 (flows are not labeled because they are self-explanatory).

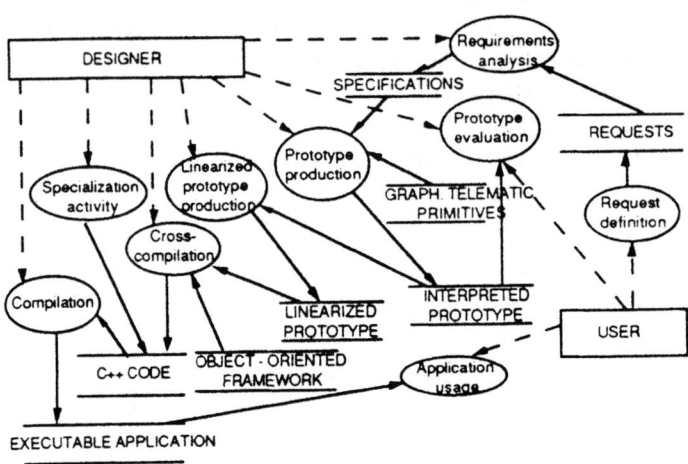

Fig. 2. The development process

3 A High Level Description of VTPL Environment

The VTPL environment provides the possibility of building a prototype of a visual telematic application and the possibility of cross-compiling the prototype producing C++ code.

The basic idea for the development of VTPL has been to avoid the ex-novo definition of a graphical application development system and to customise instead an existing hypertextual environment to the telematic context. In practice, the VTPL architecture foresees subsystems to support all the designers activities, that is prototyping, evaluating and cross-compiling (see Fig. 3) [6]. The hypertextual environment chosen as basis for the prototyping and development environment is ToolBook[3]. The customization basically consists of a library of GTPs and of an application for managing the dialogue with the designer.

The prototyping subsystem allows the istantiation of GTOs, the call of some utilities and the generation of the intermediate (*linearized*) representation that

---

[3] ToolBook is a trademark of Asymetrix Corporation.

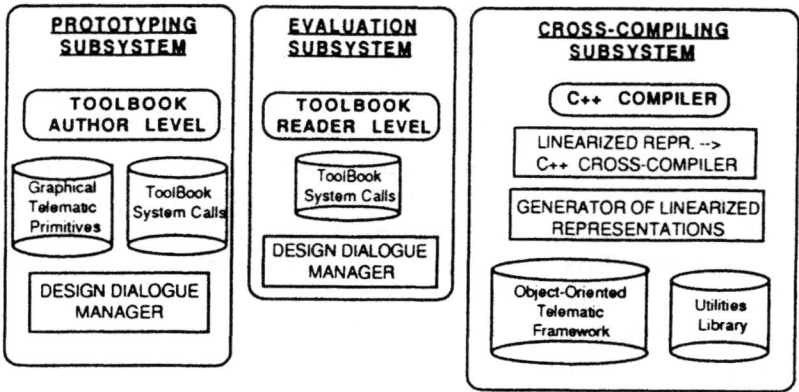

Fig. 3. VTPL software architecture

constitutes the basis for the integration with the cross-compiling module. The evaluation subsystem is in practice the same as the prototyping one. The most relevant difference is that here ToolBook is used at reader level, while in the prototyping subsystem it was used at author level.

Prototyping terminates with the production of an intermediate representation of the prototype, which is then processed by the 'Generator of Linearized Representations'. This intermediate representation includes:

1. the hierarchical structure of the hypertext, that is the relations among the objects;
2. the interface lay-out;
3. the functional links among the nodes of the hypertext;
4. the current state of the application.

The generator of linearized representations and the cross-compiler are the modules of the cross-compiling subsystem.

4 Graphic Telematic Primitives (GTP)

In this paragraph a description of the graphical telematic primitives is given. For each of them it is described:

- the visual appearance;
- the properties;
- the services provided.

Services are implemented in OpenScript[4] code. The low-level functionalities for interacting with the on-line services and for managing communication protocols

---

[4] OpenScript is the interpreted programming language of ToolBook.

are both realized in a purposely designed Dynamic Link Library of Ms-Windows. The implemented GTPs are six:

1. **Query Panel** that allows to issue a query in a visual and non procedural way (that is, by means of a query-by-example approach);
2. **History Box** that allows navigation among the most recently issued queries;
3. **Command Box** to send commands different from the querying ones (monitoring of the application status, processing of the output of an issued query);
4. **Connection Box** to connect and to disconnect from the services and to hang-up the communication;
5. **Output Box** to show the results of the issued queries and any other message incoming from the remote database;
6. **Contextual Help Box** that shows information about the behaviour of the object currently pointed by the mouse.

In the following a more detailed description of each of the GTP is given. References to figures in this section are to figures in the appendix, where screen dumps of a working session are presented.

4.1 Query Panel

It is constituted by a matrix, which is filled by the end-user to issue a query (query-by-example approach), and by a set of buttons to execute commands (top left in Fig. VI). Each column corresponds to an attribute of the universal relation[5] representing the view of the accessed service. Each element of the matrix is divided into two cells: in the smallest one the end-user chooses from a pop-up menu an operator acting on the attribute, in the largest one the end-user edits (or chooses from a pop-up menu) the search value for the attribute. The predicates of each element in a row are AND-composed, the resulting row-predicates are OR-composed. The buttons allow to show the History Box, to send the composed query to the service, to clear all the fields and to shift the columns.

Properties of Query Panel include the ones of a ToolBook group (i.e. the position, the bounds, the visibility), the name of the telematic service, the list of operators acting on attributes, the set of the default values for attributes, the list of already issued queries. The available services makes it possible to choose (or to edit) an operator and/or a value for an attribute, to send the query to the service, to show the History Box, to shift the columns of the panel.

4.2 History Box

It is constituted by buttons to recall the already issued queries from a catalogue. The total number of issued queries and the order number of the currently dis-

---

[5] The term 'universal relation' has to be interpreted in the sense of the database theory as the collection of all attributes defined in the relations present in the scheme of the accessed database.

played query are shown (bottom left in Fig. VI). It is possible to browse among the last twenty issued queries and to show and issue a previous query.

4.3 Command Box

A list-box makes it possible to choose the command to be sent, whose parameters can be specified in a specific field (right in Fig. X). The Command Box owns all the properties of a ToolBook group. Other noteworthy properties are the name of the telematic service, the list of commands and the relative parameters available for the accessed service.

4.4 Connection Box

It is constituted by just three buttons, namely for connecting to the service, for disconnecting and for hanging-up (top left in Fig. IX). The only relevant property is the name of the telematic service.

4.5 Output Box and Help Box

The Output Box is a field used to show the output of an issued query or to monitor the messages incoming from the service. The Help Box is instead used to give information about the object currently pointed by the mouse (right in Fig. XI). Properties and services of these objects are trivial.

5 The Object Oriented Framework for Cross-compilation

The aim of the cross-compiling subsystem is to analyse the linearized representation of the application and to generate the corresponding C++ code. Roughly speaking, this requires to generate a set of classes and a set of methods. The set of classes contains the implementation of the abstract data types representing the objects of the prototyped application, the set of methods includes the member functions translating the OpenScript methods of objects themselves. The definition of the classes is made by specialising the leaf classes [13] of an object-oriented framework designed on purpose. The methodology followed for the definition of the objects is based on an Abstract Data Type approach [8, 4]. The cross-compiler generates:

- a set of classes and a set of objects corresponding to the ToolBook application objects;
- member functions for each class corresponding to the ToolBook methods;
- the **Start** function that reads the representation of the application and implements it by means of a structure based on multiple lists;
- the **Close** function that saves the preordered representation of the application when a work session ends.

The C++ code to be compiled is constituted by the predefined framework added to a set of four files generated at cross-compilation time, namely: vtpl.cxx containing the member functions, the Start and the Close function, vtpl.hxx containing the classes corresponding to the objects in the prototype, mater.cxx and mater.hxx containing some utilities to realize the structural links among the classes of the framework. The inheritance structure of the object-oriented framework is shown in Fig. 4, while the software architecture of the cross compilation subsystem and other design issues have been discussed in [6, 2].

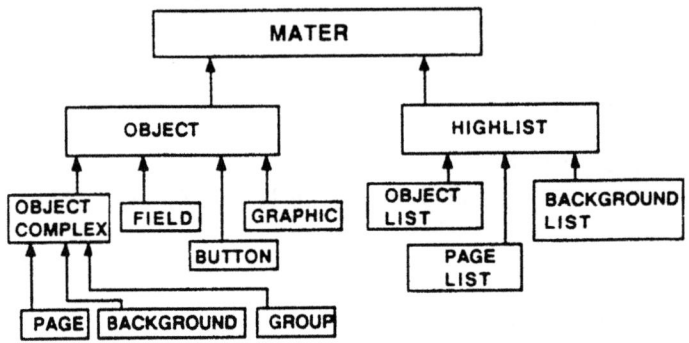

Fig. 4. Inheritance relations of the object-oriented framework

6   The Prototyping Subsystem of VTPL

The designer builds the telematic application in a interactive and completely guided way with **SEVEN T**[6], the prototyping subsystem of the VTPL Development environment. SEVEN T contains also some utilities, implemented in OpenScript code, shared by all the possible telematic applications [6, 16]. In the appendix pictures of an example of a work session with the prototyping subsystem are shown. The work session is discussed here below.

Step 1. Let us suppose the designer wants to develop an application interacting with the PGE (Italian Yellow Pages) service. First of all he/she launches the 'SEVEN T' application that shows a mask with two buttons to create a telematic application and to open an already existing one (Fig. I). If the user clicks the 'Create Telematic Application' button a new ToolBook application is launched. In this application the 'TOOTSI' menu is available (Fig. II). The user chooses the Connection Box command and the system asks, by means of a dialog box,

---

[6] SEVEN T = The Toolbook Tool To Taylor and Test Telematic applications. A prototype of this application has been shown at the 1991 'ESPRIT' Technical Week (Bruxelles 26/30 November 1991).

which is the service desired (Fig. III). After the answer of the user the system first builds the Connection Box and then builds also an Help Box and an Output Box, if they do not already exist in the current page (Fig. IV).

Step 2. The designer creates a new page in the ToolBook application and chooses from the 'TOOTSI' menu the 'Query Panel' item: the system shows a mask for the parametric instantiation of the object (Fig. V) and, after the 'OK', builds the Query Panel, the History Box, the Help Box and the Output Box (Fig. VI).

The mask for the parametric instantiation (Fig. V) proposed to the designer, gets all the information about the available services and the related properties from a global dictionary offered by the services' suppliers. The 'Choose Service' field is a list-box containing the available services: the designer makes the choice and immediately all the attributes of the service are shown in the 'Available Fields' list-box (it is important to remind that a service is viewed as a single universal relation). The *status* of an attribute can be modified by double-clicking the attribute itself: there are three possible values: 'YV', 'YI', and 'NO'. YV means *Yes Visible*: the attribute will be shown to the end-user who can choose a value for it. YI means *Yes Invisible*: the attribute will participate to the query with a default value chosen by the designer. NO means *NOt involved* by the query.

Furthermore when an attribute is double-clicked a list of available operators acting on it and a list of default values for it are shown respectively in the 'Operators' field and in the 'Feasible Values' field: the designer chooses which of these operators and default values will be make available to the end-user. Finally the designer can also choose the name of the Query Panel, the number of its rows and the number of its visible columns.

Step 3. The designer chooses the item 'Command Box' from the 'TOOTSI' menu and builds this object, but not the Output Box and the Help Box which are already present in the page (Fig. VII and Fig. VIII). Like for the Query Panel, a special template (Fig. VII) is presented for the parametric istantiation of the Command Box. All the information about the available services and their characteristics are got from a global dictionary.

The user interacts with the template exactly as with the Query Panels instantiation template. Once a service has been chosen the list of the possible commands acting on it is shown in the 'Available Commands' list-box. It is possible to choose if each single command will be shown or not to the end user, furthermore when a command is clicked it is shown the number of its parameters and, possibly, a list of default values for the parameters.

Step 4. At this point the designer makes some refinements on the application using the standard ToolBook tools. In our case buttons to hide and to show the Help Box and the Output Box have been cut, the Help Box of the second page has also been cut and two buttons to navigate among the pages have been created.

Moreover the istantiated GTOs can be further manipulated in their visual appearance by means of a direct manipulation approach. The final application consists of the two screens shown in Fig. IX and Fig. X, which is a real working application for the interaction with the PGE on-line service.

Step 5. Finally the designer cross-compiles the application: chooses the item 'Intermediate Code' from the menu and generates the C++ code, which is then compiled with a commercial C++ compiler[7] to obtain a stand-alone Windows application. The obtained application is shown in Fig. XI and Fig XII: note that is similar to the ToolBook prototype even if the telematic object are implemented as independent windows in order to force the consistency with other Windows applications.

The linearization is obtained by means of ad-hoc algorithms and it is made according to a special syntax defined in TOOTSI project's framework. The syntax is a superset of the OpenScript's one (all the rules recognized by the OpenScript interpreter can be recognized also by the VTPL Linearization Module). Both the structural description of the application and the description of its behaviour are provided in the linearized representation. Further details on this point can be found in [6].

7   Conclusions

The VTPL development environment is a hypertextual environment for developing graphical interactive application, whose most relevant and remarkable aspects are the innovative software architecture for engineering user-interface prototypes and the possibility of developing high quality interfaces for on-line services.

Regarding the development of high quality user-interface interfaces for telematic services we already noticed that the greater part of the on-line services currently available on the European market allows only a human-computer interaction based on *command languages* that are difficult to learn and tedious to use. Thus, the possibility of a rapid development of graphical interfaces for these services represents a first attempt to shift the attention from the *needs of service suppliers* to the *needs of users*. From this point of view we can include in the category 'user' both the application developers and the end users: the designers will find it useful the possibility of developing telematic application independently from the accessed service, the end users will be helped in using the services by the powerful graphical interfaces. Moreover the use of this environment allows the use of the same user interface metaphore (based on the same GTPs) for different services, so that standardization and consistency are encouraged.

---

[7] In our case Glockenspiel C++ pre-processor plus Microsoft C compiler v6.0.

Regarding the software architecture it is important to remark that VTPL is a generator of complete multi-platform graphical applications. This means that an application developed in this environment can be compiled to run in different host environments (e.g., Ms-Windows, OSF-Motif, TINTO). This approach is quite innovative because usually the applications developed in an hypertextual environment need the environment itself to run. The proposed architecture overcomes this problem and shows the way to develop an environment complete with a cross-compiling module. From this point of view we can define VTPL as a complete Graphical Application Development System.

A last remarkable aspect is the use of commercial products such ToolBook, Commonview or the Glockenspiel C++ compiler that are integrated by means of ad-hoc designed modules. This choice helps the usability, the standardization and the consistency both of the VTPL and of the developed application.

Acknowledgements. We thank the following people for useful discussions during the development of the TOOTSI project: Gregorio Lella, Giorgio Quaglieri, and Benedetto Serra.

References

1. F. Arcieri, G. Ausiello, E. Nardelli, M. Talamo, Strumenti per la specifica di interfacce utente, Proceedings of the workshop "Conoscenza per immagini", Il Rostro, 1989, Roma.
2. F. Arcieri, M. Fossa, E. Nardelli, A prototyping and development environment for graphical interactive telematic applications, Int. Conf. on Human Computer Interaction (HCI'93), Orlando, Fl., USA, August 1993.
3. S. Chang, T. Ichikawa, P.Ligomenides, Visual languages, Plenum Press, New York, 1986.
4. P. Coad, E. Yourdon, Object Oriented Analysis, Prentice Hall, Englewood Cliffs, NJ, 1991.
5. J. Conklin, Hypertext: an introduction and survey, IEEE Computer, September 1987.
6. M. Fossa, Interazione visuale con sistemi informativi: integrazione di un ambiente ipertestuale con un framework object oriented per lo sviluppo rapido di applicazioni grafiche interattive, Electronic Engineering Master Degree Thesis (in italian), University of Rome 'La Sapienza', February 1992.
7. G. Lella, F. Bartolomucci, Stop C.A.O.S.: a common access to on-line services, Proceedings of the national conference of the Associazione Italiana per il Calcolo Automatico (A.I.C.A.), Siena, September 1991.
8. B. Meyer, Object Oriented software construction, Prentice Hall, 1988.
9. D. Murray, Embedded user models, Proceedings of the Human Computer Interaction conference (Interact'87), Stuttgart, September 1987.
10. J. Nielsen, Hypertext and hypermedia, Academic Press, London, 1990.
11. D. Norman, S. Draper, User Centered System Design, Lawrence Erlbaum Associates, London, 1986.
12. N. Shu, Visual programming, Van Nostrand, New York, 1988.
13. B. Stroustrup, What is Object Oriented Programming?, IEEE Software, May 1988.

14. Technical Annex of TOOTSI Esprit Project 2109, Tootsi Consortium, December 1988.
15. Scenario and Trends: analysis of on-line services, Deliverable 6.1 of TOOTSI Esprit Project 2109, Tootsi Consortium, September 1989.
16. Semestral Report of TOOTSI Esprit Project 2109, Tootsi Consortium, February 1992.
17. E. Yourdon, Modern structured analysis, Prentice Hall, Englewood Cliffs, NJ, 1989.

Appendix

In the following pages, figures referred to in Section 6 are reported. They present dumps of the screens relative to the work session discussed there.

Figures I to XII are presented four per page, with a landscape orientation, in a left to right, top to bottom order.

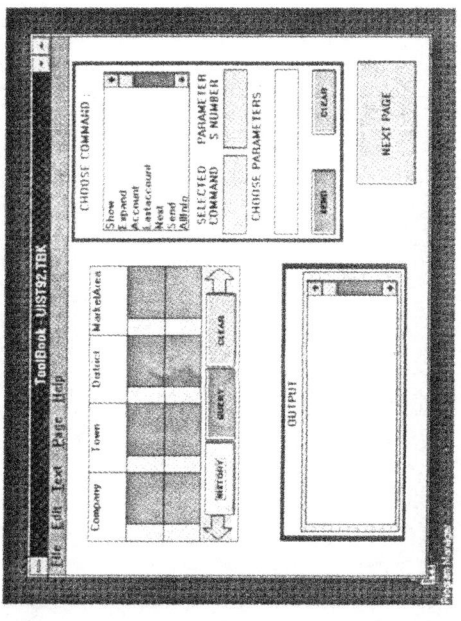

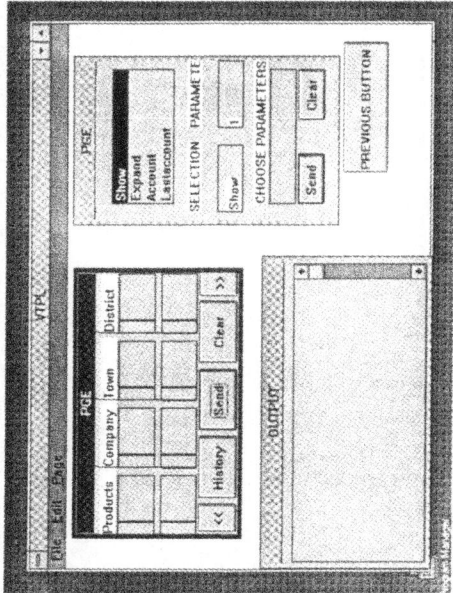

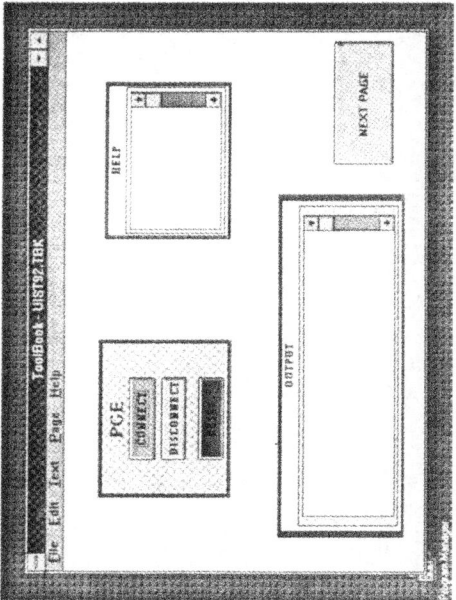

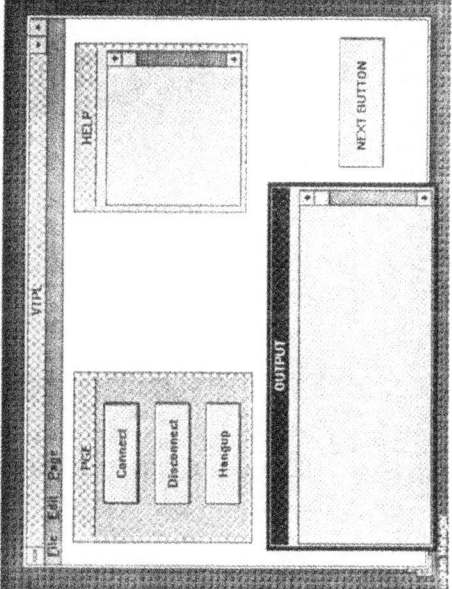

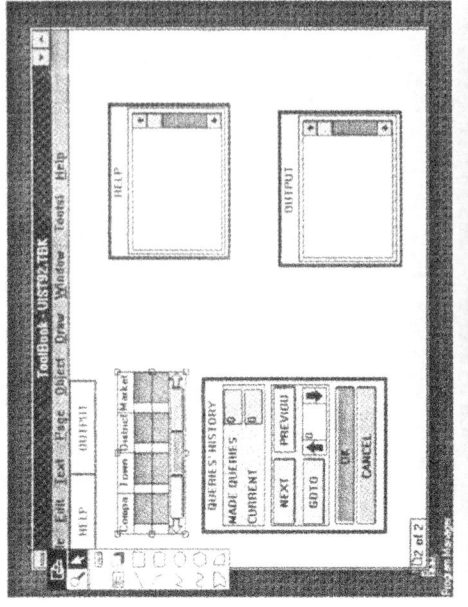

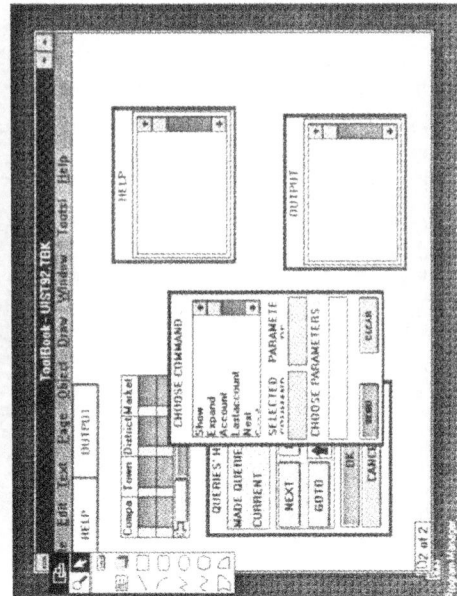

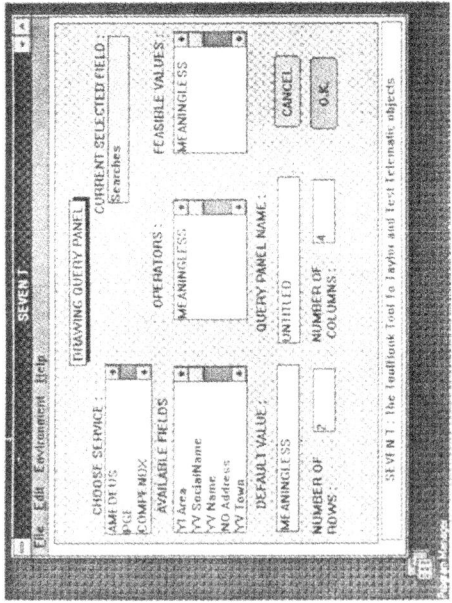

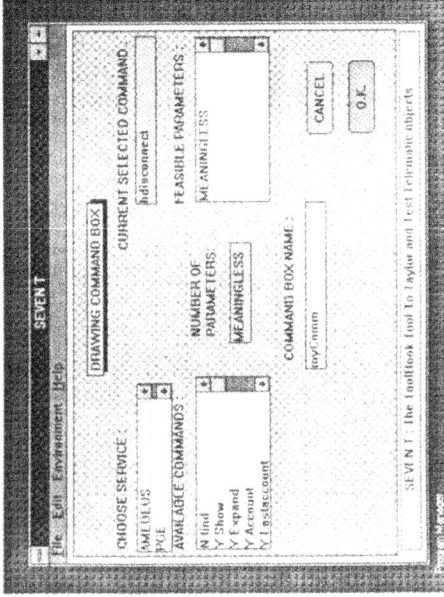

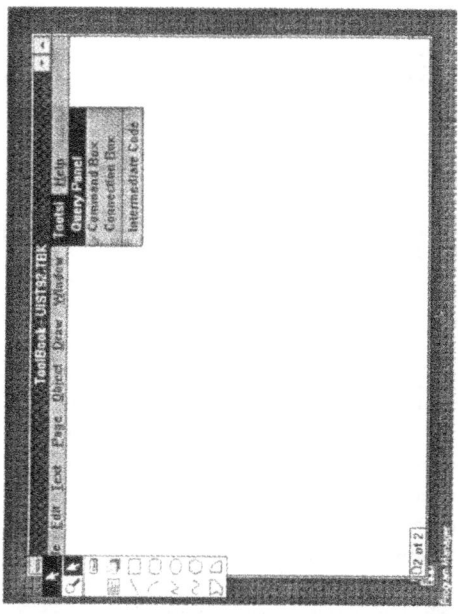

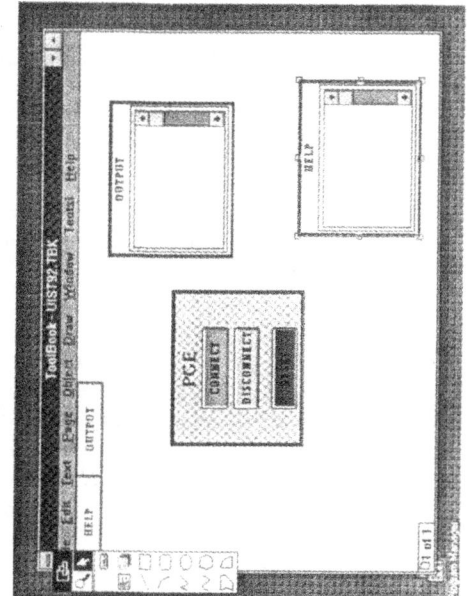

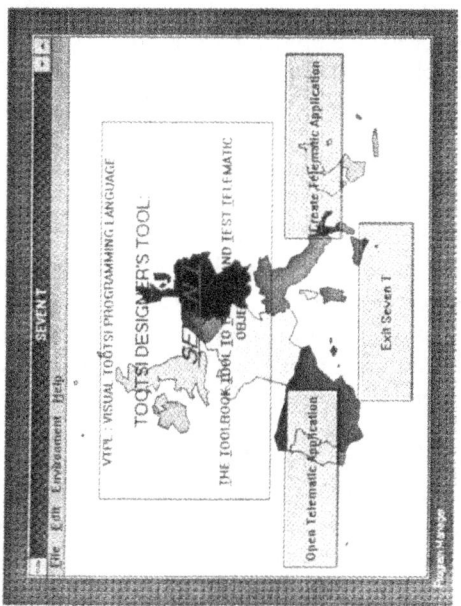

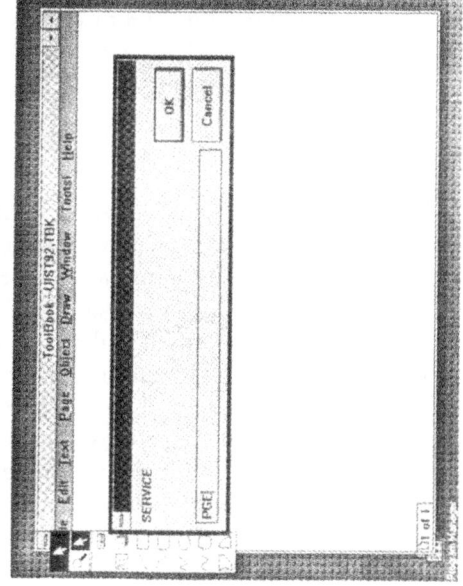

A Simple Approach to Improve the Abstraction Level of Object Representation

Clenio F. Salviano*, Mario A. Nascimento** and Murat M. Tanik***

Department of Computer Science and Engineering, Southern Methodist University, Dallas, Texas 75275-0122 USA. {clenio,mario}@seas.smu.edu

Abstract. The management of complexity is an essential problem in software development and abstraction is one of the best mechanisms dealing with complexity. In order to increase productivity in software development, we should provide formal notation to support high-level abstractions. In this way, the semantic gap between the designers' conceptual model and the execution model is diminished. This article describes an approach to improve the abstraction level of encapsulation. This approach supports the expression of the class interface as data and functions. Although each one of these constructs (encapsulation, and interface as data and functions) is provided by most programming languages, the combination of both for the same attribute is not provided. In addition, the level of abstraction of data representation is increased allowing partial automation of its implementation.

1 Introduction

The management of complexity is an essential problem in software system development. Abstraction is one of the best mechanisms dealing with complexity [3]. Abstraction allows humans to focus to some specific properties of a problem. In an interactive problem solving process, the complexity of each interaction should be in a level of difficulty that a human can manage. Hierarchy of abstractions allows the organization of these interactive processes, making possible the management of the total complexity.

Computer-Aided Systems Technology (CAST) should support high-level abstractions in order to improve the development process of complex systems. Notations supporting high-level abstractions are necessary to provide an effective usage of abstractions in computerized environments [11]. High-level abstraction means that the abstraction is close to the designer's conceptual model. The difference between the designer's conceptual model and the execution model is called

---

* On leave from CTI, Campinas, Brazil (clenio@ic.cti.br) and supported by CNPq (Process No. 200801/87-2,) Brasilia, Brazil.

** On leave from CNPTIA–EMBRAPA, Campinas, Brazil (mario@cnptia.embrapa.br) and supported by CNPq (Process No. 260088/92.7,) Brasilia, Brazil.

*** Current address: Department of Electrical and Computer Engineering, University of Texas at Austin, Texas.

the *semantic gap*. This article describes an approach to diminish this semantic gap.

The level of abstraction supported by the constructions of programming languages has increased continuously since the creation of the first assembly language. Since then, some constructions have been consolidated as good representation for useful abstractions. Among these constructions are encapsulation and the expression of an object in terms of its structure and its behavior. We often use data type declaration to represent the structure, and function declaration to represent the behavior.

The Object-Oriented approach has been used to design systems from the perspective of system theory. In [7], for example, Mittelmann describes the object-oriented design of a CAST prototype named CAST:FSM (Computer Aided Systems Theory, Finite State Machine). In that design, system types were mapped to classes, systems to objects (instances) and transformations to methods.

There are many definitions of encapsulation, but all of them agree in the essential point, which is the enforcement of abstraction by the separation of an object into two parts: interface and implementation [2, 10]. Booch defines encapsulation as "the process of compartmentalizing the elements of an abstraction that constitute its structure and behavior; encapsulation serves to separate the contractual interface of an abstraction and its implementation" [2, p.50]. A good interface should contain all information needed to use the object and should not contain any information about how this object is implemented. The advantages of encapsulation are well argued in the literature [2, 6, 10]. Most programming languages support some form of encapsulation. Among them we can consider Ada, Ada9X, C++, Smalltalk, Modula-2 and Modula-3 as a representative subset.

Encapsulation also plays an important role in the design of complex systems, where heterogeneous software representation formalisms are used. We should not expect a single formalism to be appropriate for all kinds of systems or even for all subsystems of a specific complex system. Rather different formalisms tend to be appropriate for different subsystems or different levels of abstractions of a complex systems [13]. For example, Daniel Bobrow stated that "no single [programming] paradigm is appropriate to all problems, and powerful systems must allow multiple styles" [1, p.1401]. Therefore, in order to allow the usage of the "right formalism" for the "right subsystem," we need to realize the interoperabilty of subsystems expressed in different formalisms. One approach for this interoperability is the encapsulation of each subsystem using a uniform interface formalism (e.g., [14] uses this approach). Object-Oriented approach has been used for this purpose. For example, Nicol, Wilkes and Manola [8] describe its usage in heterogeneous distributed computing systems.

The rest of this paper is organized as follows: Section 2 describes background and related work. Section 3 describes our approach. First we describe an example, the concept of a date. Then this example is used for the description of the traditional approach, our approach and an extension of our approach. Finally, Section 4 describes some conclusions.

2 Background and Related Work

Although the way programming languages support encapsulation varies among them, they have at least one point in common: In order to use encapsulation, the interface must be defined in term of function declarations, and the implementation must be composed of data declaration and the implementation of these functions. Some of these languages (e.g., C++) allow the definition of the interface with function declarations and data definition. In this case, however, the data also means the implementation, which implies that there is no encapsulation. Therefore, there is a conflict between these two styles of constructions: a software designer either uses encapsulation with the expression of an object only in terms of functions, or expresses an object as data and functions, without encapsulation.

Based on our experience and observations [9, 13], we hypothesize that the conceptualization of an object in terms of data and functions, rather than only functions, is close to the user's conceptual model. When we conceptualize a date, for example, we usually do it in terms of data abstractions (e.g., day, month and year) and functions (add a number of days in a date, etc.). The modeling of a date as functions, without data abstraction, forces the usage of artificial functions (e.g., get and set the value of the date) that actually only simulate the data abstraction, instead of perform a "real" function.

Our approach supports the expression of an object interface as data and functions, with encapsulation technique. In addition to that, the level of abstraction of the data representation is increased in order to allow a partial automation of its implementation. Although our approach is described from the object-oriented point of view, it is suitable for abstract data types and modular programming as well.

Our approach is an evolution and a generalization of the one used in the Object Management System (OMS), named Signo [9]. The main difference is that constraints are introduced into the data specification and the syntax is simpler, yet still powerful. As an OMS, Signo can be defined as a system that provides a Computerized Software Environment to support the development, execution and evolution of Object-Oriented Software. Signo also supports creation, execution, evolution and persistent storage of objects and classes of objects. Three prototypes of Signo were developed in the Brazilian Software Plant Project (SPP), where the main goal was the creation of the necessary conditions to enable Brazil and other nations to pursue software factories [4].

In order to explain the differences between our approach and the traditional one, we show the model of a date as a C++ class and as a class in our approach. Although we compare our approach only with C++, similar comparison can be made with other programming languages (e.g., ADA, Smalltalk, and Modula-2) because they do not differ among them in terms of the conflicts we mention.

"The C++ programming language is designed to: be a better C, support data abstraction, and support object-oriented programming" [12]. Descriptions of Object-Oriented Approach and C++ Programming Languages are in many books and articles, e.g., [2, 5, 6, 12]. For those readers who are not familiar with C++, we describe in the next paragraph some characteristics of a class declaration, in order to allow them to understand our examples.

A class definition in C++ is composed of: *class* keyword, name of the class and definitions. Each definition can be *private, public* or *protect.* Private means that the declaration is part of the implementation and it can only be accessed in the implementation of the class. Therefore it can not be accessed from "outside" of the implementation. Private is the "default." Public means that the declaration is part of the implementation and part of the interface. Protect means that the declaration is private in terms of the use of the class and it is public for a subclass of the class. A definition could be either a function or a data. Data is defined with a name and a data type. A function is defined with a name, a data type and a list of parameters (zero or more). Each parameter is defined with a name, a data type, and, optionally, the symbol &. Without this symbol &, the parameter is a value parameter and with the symbol &, the parameter is by reference.

3 Description of our Approach

In our approach, the class interface can be expressed with data and functions, without commitment to a specific implementation. This means that the encapsulation technique can be used. In order to use encapsulation in C++, for example, the interface of a class must be expressed as functions, because the usage of data is always associated with the concrete representation of the object.

3.1 Example: the Concept of Date

A date can be defined as a particular point of time, where time is made discrete in periods of days and stated in terms of day, month and year. A day can be represented as a natural number, in a range [1..31], and this could be narrowed to range [1..28], [1..29] or [1..30], for some months or some combination of month and year. A month can be represented as a natural number, in range [1..12]. A name is associated with each month, which in English is either *January, February, ..., November,* or *December.* Finally, a year can be represented as a natural number. In this example, for the sake of simplicity, only years in range [1800..2099] are considered. In order to complete the definition of a date, it is necessary to describe the set of basic operations for date. Some of them can be the following:

- Calculate the number of days between two dates;
- Calculate the number of complete months between two dates;
- Add a given amount of days to a date; and
- Get the value of a date, represented as month name, day number and year.

3.2 Example in Traditional Approach

In this section, we show that there are two basic approaches to express data abstraction in C++ interface: (a) as data, without encapsulation; or (b) simulated as *get* and *set* functions, with encapsulation.

In Figure 1, there are three data elements (**day**, **month** and **year**) that represent the definition and also the implementation of the data abstraction. The level of abstraction is preserved, because data abstractions are expressed as data elements in the notation. However, there is no encapsulation, because these three data elements expressed in the notation already represent their implementation.

```
class date {
    public:
        int day;
        int month;
        int year;
        int numDays(date anotherDate);
        int numMonths(date anotherDate);
        void addDays(int nDays);
        void getSpell(char &month, int &day, int &year);
};
```

Fig. 1. Expressing attributes as data, without encapsulation, in C++.

An example of the usage of an object of this class is depicted in Figure 2.

```
date ADate; // declaration of ADate as an object of class date
// set values to attributes
ADate.day = 16;
ADate.month = 5;
ADate.year = 1994;
(...)
int day, month, year;
// get values of attributes
day = ADate.day;
month = ADate.month;
year = ADate.year;
```

Fig. 2. Using an object with attributes as data, without encapsulation, in C++.

In Figure 3 the three pairs of *set* and *get* functions represent the definition of the data abstraction. The *set* function is used to set a new value for the attribute and the *get* function is used to get the value of the attribute. Therefore there is encapsulation but the level of abstraction is not preserved because the attribute is simulated as *set* and *get* functions. In this case, the concrete implementation of these three attributes is represented by the variable *ymd*. The value of **day**, **month** and **year** are encoded as a single value in this variable *ymd*. The value of **ymd** is the result of the expression ((**year** * 13) + **month**) * 32) + **day**. The mapping between the value encoded and the three attributes is made by the implementation of the *set* and *get* functions.

```
class date {
    ulong ymd; // ymd is private, can not be accessed outside the object
    public:
        int getDay(); int setDay(int day);
        int getMonth(); int setMonth(int month);
        int getYear(); int setYear(int year);
        int numDays(date anotherDate);
        int numMonths(date anotherDate);
        void addDays(int nDays);
        void getSpell(char &month, int &day, int &year);
};
```

Fig. 3. Expressing attributes as get and set functions, with encapsulation, in C++.

An example of the usage of an object of this class is depicted in Figure 4.

```
date ADate; // declaration of ADate as an object of class date
// set values to attributes
ADate.setDay(16);
ADate.setMonth(5);
ADate.setYear(1994);
(...)
int day, month, year;
// get values of attributes
day = ADate.getDay();
month = ADate.getMonth();
year = ADate.getYear();
```

Fig. 4. Using an object with get and set functions, and encapsulation, in C++.

3.3   Example in our Approach

Our approach combines the advantages of these two traditional approaches. It supports the expression of attributes as data in the interface and supports en-

355

capsulation. The notation to express the interface of a class is composed of the keywords *def class*, followed by the name of the class and by 3 sections: *atts, pars* and *meths*. In these sections the data, parameters and functions are described. The parameters of the methods must be declared either as an attribute or as a parameter in theirs respective section. The reasons and advantages of this are described in section 3.4.

In Figure 5 and Figure 6 these three attributes (**day, month** and **year**) represent the definition of the attributes. The concrete representation is decided only in the implementation part. Therefore the level of abstraction is preserved and there is encapsulation.

```
def class date {
        atts {
                int day;
                int month;
                int year;
        }
        pars {
                date anotherDate;
                int nDays;
        }
        meths {
                int numDays(anotherDate);
                int numMonths(anotherDate);
                void addDays(nDays);
                void getSpell(char &month, &day, &year);
        }
}
```

Fig. 5. Expressing attributes as data, with encapsulation.

An example of the usage of an object of this class is depicted in Figure 7 and Figure 8, where it is shown that in order to access the attribute a program deal with it as a field of a structure.

In Figure 9 and Figure 10 a possible implementation of class date in our approach is depicted. The notation to express the implementation of a class is composed of the keywords *impl class*, followed by the name of the class, and by 3 sections: *vars, mapps* and *funcs*. In these sections the data, parameters and functions are described, respectively. In the *vars* section, the attributes are implemented using variables. In the *mapps* section, the functions that do the mapping of the attributes of the interface to the variables in the implementation are described. Finally in the *funcs* section the methods which were specified in the interface are implemented. In this example, the variable **ymd** encodes the

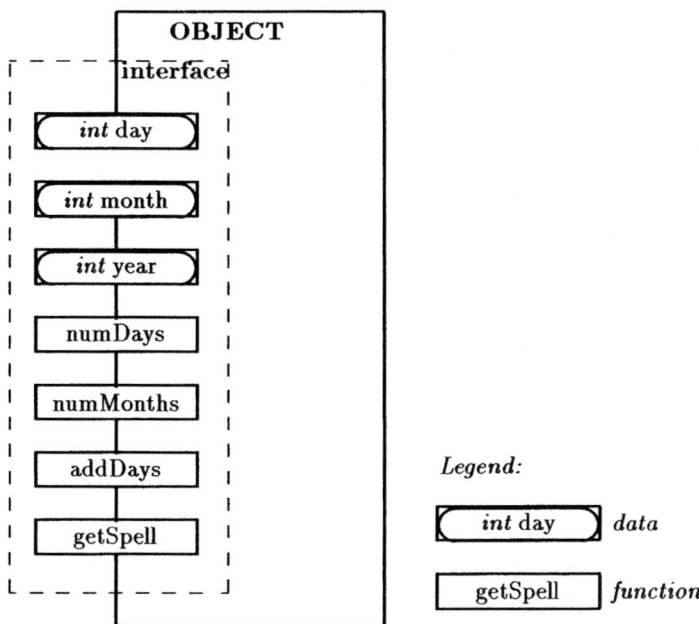

Fig. 6. Interface of an object.

attributes **day**, **month** and **year** as an integer number. This is an example of how the attributes implementation can be different from the attributes definition. Any other correct way to implement **day**, **month** and **year** could be chosen without changing the interface of this class. The mapping between the value encoded and each attribute is defined by the implementation of the *set* and *get* functions for each attribute. For example, the function _set_day maps the attribute day (defined in the definition part as an integer) to a part of the encoded value of the variable **ymd** (in this case the day is encoded as the modulo of **ymd** by 32). The function _get_day maps the encoded value to the attribute.

One way to implement this approach is through the generation of equivalent C++ function calls for each usage (*set* or *get*) of an attribute. Figure 11 shows this correspondence for our example.

3.4 Example in Our Extended Approach

Two extensions are described in this section. One to allow the automation of the implementation, and another to allow the expression of a concept just once.

Sometimes the best implementation of an attribute is its counterpart data structure. For example, the attribute day, that is an integer, can be implemented as an integer variable called day. This characteristic does not imply that we do not have encapsulation, because this implementation could be changed later on without impact on the interface. However it suggests a desirable feature. In this

```
date ADate; // declaration of ADate as an object of class date
// set values to attributes
ADate.day = 16;
ADate.month = 5;
ADate.year = 1994;
(...)
int day, month, year;
// get values of attributes
day = ADate.day;
month = ADate.month;
year = ADate.year;
```

Fig. 7. Using an object, with attributes as data, and encapsulation.

case the designer do not need to describe the implementation but only indicate it. The system already knows how to implement the attributes, thus it can generate this implementation automatically. Figure 12 depicted an example of the usage of this feature.

However, the notation used to express the attribute does not have enough semantics. Expressing day as an integer is not enough, because day is actually a subset of the integer numbers. A day must be between 1 and 31 (inclusive). This specialization should be captured by the notation in order to increase the level of abstraction. For this purpose, we allow a simple constraint specification. A month must be in the range [1..12]. There are also constraints related with the combination of attributes. In month 11, for example, a day must be in the range [1..30] instead of range [1..31]. When the complete date is provided at once, the system can check all constraints. If we want to allow the date to be provided by parts, the system can not check the constraints related with the whole date. In this case we decided to create a method to check these constraints. This method can be automatically generated from the specification of the constraints.

In Figure 13, the declaration *int* **day** {#[1..31]} means that the value of a day must be between the values 1 and 31. We can use the following expression as constraints: range, logical, comparison and arithmetic. The basic elements can be constants or attributes. Special constructions are available for comparison with multiple values. For example, the sentence **month == 4,6,7,9,11** means that the variable *month* is tested if it is equal to *4, 6, 7, 9* or *11*.

The other extension is based in the criteria that each concept, or design decision, should be expressed just once. If the notation requires the expression of a decision more than once, the notation does not support the abstraction used in this decision. In our C++ example, the concept of day is modeled as an integer parameter called **day**. The functions **setDay** and **getSpell** use **day** as a parameter. For this example, in the designers' conceptual model, these two parameters and the attribute called **day** model the same concept. However, in

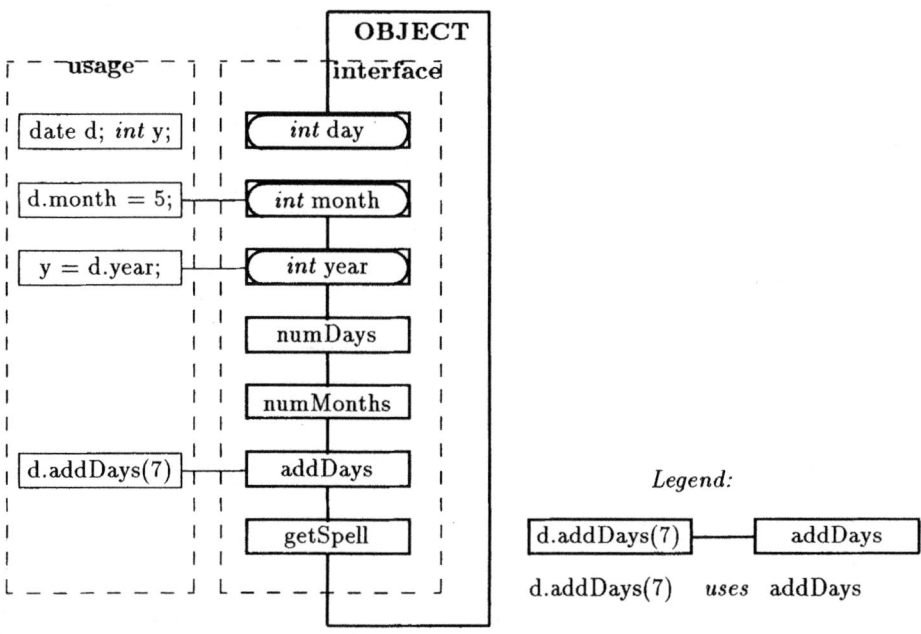

Fig. 8. Usage and interface of an object.

```
impl class date {
        vars { ulong ymd; }
        mapps {
            set day {
                if (day < 1 || day > 31) return error day_out_of_range;
                this->ymd -= this->ymd % 32; // this is the object
                this->ymd += day;
                return;
            }
            get day { return this->ymd % 32; }
            (...) // similar get and set for month and for year
        }
        funcs {
            (...) // implementation of the methods
        }
}
```

Fig. 9. Implementation of a class.

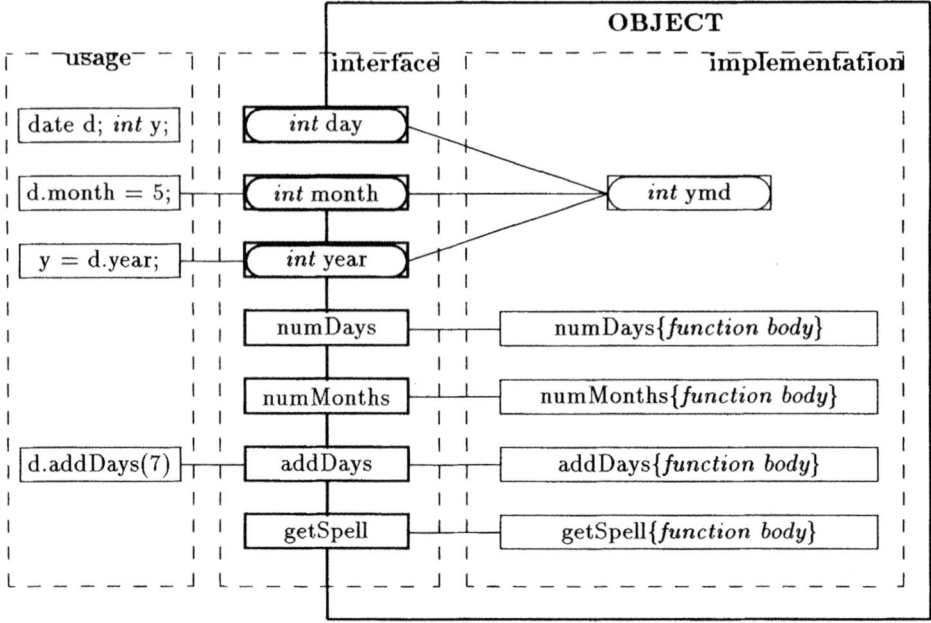

Fig. 10. Usage, interface and implementation of an object.

the notation there is no connection between them. The parameter `month` is the attribute `month` with its own data type, that means another representation for the same concept. In this case the attribute `month` is represented as an integer number (e.g., 5) and the parameter `month` is represented as a string (e.g., *"May"*). The parameter `day` is the attribute `day`, with the same data type, which means the same representation.

We solve this problem by assuming that the usage of the same name always implies the same concept. For example, we have to define day only once and because every usage of the word day in the interface refers to the same concept. We do not have to define it again. Another aspect is that when anyone reads this name, she or he can be sure about its meaning, because it is captured by the representation. If the name is the same, it always means that it is the same concept.

4 Conclusions

Some advantages of the approach described in this article are the following:

– The notations to set the value of an attribute (i.e., *attribute = expression*) and to get the value of the attribute (i.e., *variable = attribute*), are the same one that deal with variables. As these two kinds of operations (set and get attribute, and set and get variable) are essentially the same, they should be provided through the same notation.

```
date ADate; // declaration of ADate as an object of class date
// commands to set values for attributes // equivalent code in C++:
ADate.day = 16;                          // ADate._set_day(16);
ADate.month = 5;                         // ADate._set_month(5);
ADate.year = 1994;                       // ADate._set_year(1994);
(...)
int day, month, year;
// commands to get values from attributes// equivalent code in C++:
day = ADate.day;                         // day = ADate._get_day();
month = ADate.month;                     // month = ADate._get_month();
year = ADate.year;                       // year = ADate._get_year();
```

Fig. 11. Using an object and equivalent code in C++.

```
impl class date {
        vars {
                #day;
                #month;
                #year;
        }
        mapps { }
        funcs {
                (...) // implementation of the methods
        }
}
```

Fig. 12. Alternative for expressing attributes as data, with encapsulation.

- It is not necessary to simulate attributes with artificial functions in order to achieve encapsulation. Therefore the notation can support encapsulation.
- All advantages of encapsulation are kept.
- The implementation of this approach can be done as a layer in the traditional approach that uses only functions. Because the traditional approach is already well understood and available for most of the programming languages, we envision that our approach will be easy to implement.

Some disadvantages of this approach are the following:

- The community is already familiar with the traditional schema (interface as functions).
- Sometimes it is not clear when a concept should be modeled as an attribute or as an function. In this case, functions should be used.
- There is a small overhead in the compilation time.

```
def class date {
        atts {
            int day {#[1..31]};
            int month {#[1..12]};
            int year {#[1800..2099]};
        } { if (month == 4,6,7,9,11) day[1..30];
            if (month == 2) {
                if ((year%100 == 0) && (year%400 != 0)) day[1..29]
                                                    else day[1..28];
            }
        }
        pars {
            date anotherDate;
            int nDays;
        }
        meths {
            int numDays(anotherDate);
            int numMonths(anotherDate);
            void addDays(nDays);
            void getSpell(char &month, &day, &year);
        }
}
```

Fig. 13. Expressing attributes as data, with encapsulation (final and complete version).

The expression of an interface with attributes facilitates the design of query languages and browser. The query and the browser should deal with an object using just its interface. Our approach is a simple way to use data representation as interface, without commitment with implementation and without simulation as a parameter of a function.

Acknowledgment

We would like to thank Ted Pedersen from SMU for his review, specially in the usage of English, and the anonymous referees for their comments, which helped to improve this paper.

References

1. Daniel G. Bobrow. If Prolog is the answer, what is the question? or What it takes to support AI programming paradigms. *IEEE Trans. on Software Engineering*, SE-11(11):1401–1408, November 1985.
2. Grady Booch. *Object-Oriented Analysis and Design with Applications*. Benjamin/Cummings Publishing Company, Inc., second edition, 1994.

3. Frederick P. Brooks, Jr. No Silver Bullet: Essence and Accidents of Software Engineering. *IEEE Computer*, 20(4):10–19, April 1987.

4. Fuad Gattaz Sobrinho. The Model for the Brazilian Software Plant. In R. Yeh, editor, *Modern Software Engineering*, pages 638–650. Van Nostrand Reinhold, 1990.

5. Jorgen L. Knudsen and Ole L. Madsen. Teaching Object-Oriented Programming is more than teaching Object-Oriented Programming Languages. In G. Goos and J. Hartmanis, editors, *Proceedings of ECOOP'88 - Object-Oriented Programming*, volume 322 of *Lecture Notes in Computer Science*, pages 21–40. Springer-Verlag, 1989.

6. Bertrand Meyer. *Object-Oriented Software Construction*. Prentice Hall International, 1988.

7. R. Mittermann. Object-Oriented Design of CAST Systems. In *Proceedings of Computer Aided Systems Theory - EUROCAST'89*, volume 410 of *Lecture Notes in Computer Science*, pages 69–75. Springer-Verlag, 1989.

8. John R. Nicol, C. Thomas Wilkes, and Frank A. Manola. Object orientation in heterogeneous distributed computing systems. *IEEE Computer*, 26(6):57–67, June 1993.

9. Clenio F. Salviano, Wagner R. de Martino, Pedro L. P. Correa, Adriano de Carvalho, and Fuad Gattaz Sobrinho. Signo: The SPP Object Management System. In *PD-Vol. 43, Computer Applications and Design Abstraction ASME 1992*, pages 141–149, 1992.

10. R. Sethi. *Programming Languages, Concepts and Constructs*. Addison-Wesley Publishing Company, 1988.

11. Mary Shaw. Larger scale systems require higher-level abstraction. In *Proceedings of Fifth International Workshop on Software Specification and Design*, pages 143–146. IEEE Computer Society, 1989.

12. Bjarne Stroustrup. *The C++ Programming Language*. Addison-Wesley Publishing Company, second edition, 1991.

13. Murat M. Tanik and Eric S. Chan. *Fundamentals of Computing for Engineers*. Van Nostrand Reinhold, 1991.

14. Jack C. Wileden, Alexander L. Wolf, Willian R. Rosenblatt, and Peri L. Tarr. Specification-level interoperability. *Communications of the ACM*, 34(5):72–87, May 1991.

On the Integration of CAST.FSM into the VLSI Design Process

T. Mueller-Wipperfuerth

Institute of Systems Sciences
Johannes Kepler University, Linz, Austria

Abstract. The integration of CAST.FSM within a VLSI design environment is presented. Currently very few algorithms for FSM manipulation and optimization are available in state-of-the art VLSI design tools. We propose a framework of tools where methods from state machine theory become applicable within a VLSI design flow and give results of a method application targeting FSM testability to illustrate the practical applicability of the approach.

1 Introduction

Industrial logic designs more and more rely on hardware description languages for the specification and implementation of functionality. "Schematics are now the output of the CAD system, rather than the input to it" [1]. So, a circuit design is conceptualized by means of a Hardware Description Language (HDL) on a behavioral level of abstraction rather than on the gate level.

As design data is now available at a high level of abstraction, methods of state machine theory for decomposition, state assignment, state minimization, shiftregister realization, etc. become applicable. These methods have been implemented in CAST.FSM [2, 3] and are currently being applied within a state-of-the-art VLSI design environment.

A closer look at most industrial synthesis and CAD tools reveals that very few methods of state machine theory are available for optimizing designs, although those tools define control units (finite state machines) and describe them separately from the data path operations being controlled.

Previous work showed the application of CAST.FSM to the MCNC benchmark set [3, 4]. However, results did neither consider wiring and floorplanning aspects nor realistic target libraries.

The remainder of this paper is organized as follows. In Section 2 we give an overview of the process of controller synthesis within the synthesis process of digital systems. The interoperating tools of the proposed framework are presented in Section 3 and a successful application, the FSM shiftregister realization, is presented in Section 4. Within our framework an initial version of the method has been developed, tested and evaluated, and is currently being refined to meet actual VLSI realization constraints.

2 Controller Synthesis for Digital Systems

To discuss the role of CAST.FSM within a state-of-the-art VLSI design environment and to obtain a basis for the classification of tool and method dependencies, the synthesis process for digital systems is illustrated in Figure 1 by means of a multi-strata [5] representation.

Synthesis is commonly defined as the translation from a behavioral form of description into a structural one. It is usually explained by means of the Y-chart [6], which depicts behavorial, structural, and physical domains of description on different levels of abstraction. A synthesis step is illustrated then by a shift of descriptions from the behavioral to the structural domain. The evolving structural description consists of components and component couplings, where each component is described independently using a behavioral form. A component's structure can be obtained by a further synthesis step on the next (lower) level of abstraction. The classification of synthesis tasks is based on the level of abstraction of underlying descriptions.

Two levels of synthesis are suited for CAST.FSM method application and are explained in more detail:

– *Register transfer level synthesis* transforms an initial description of a controller (defined by its states and state transitions) into a network of combinational logic blocks and storage elements. The controller's architecture is determined by structuring methods (e.g., decomposition). Additionally, binary codes are assigned to symbolic state names. RTL synthesis explicitly defines the timing model of the form of representation used.
– *Logic Synthesis* – Input to this level is a network of combinational blocks and storage elements. A technology independent optimization of combinational logic blocks is performed, whereby area only is optimized. It is measured in PLA size (two-level logic) or number of literals (multi-level logic). The synthesis task for multi-level logic on this level is denoted technology (library) mapping: The network of abstract gates is mapped to a network of physical library elements (standard cell, gate array, FPGA) considering timing and area constraints.

The last step of controller realization, the generation of the layout, is not a true synthesis step, but has to be considered for optimization and evaluation purposes only. It deals with the transformation from the structural to the physical domain. Tools for a final layout generation have to be included in the framework to have realistic evaluation criteria for design and optimization decisions on higher levels of the controller synthesis procedure. Currently a standard cell design after actual placement and routing is achieved by means of the proposed framework.

One topic of state machine theory is the problem of state assignment. Whereas those algorithms address, for example, the decoupling of next state logic, modern state assignment techniques take into account the characteristics of subsequent logic minimization procedures to decrease area and signal delays. Symbolic states are encoded with binary values which enlarge the potential of further

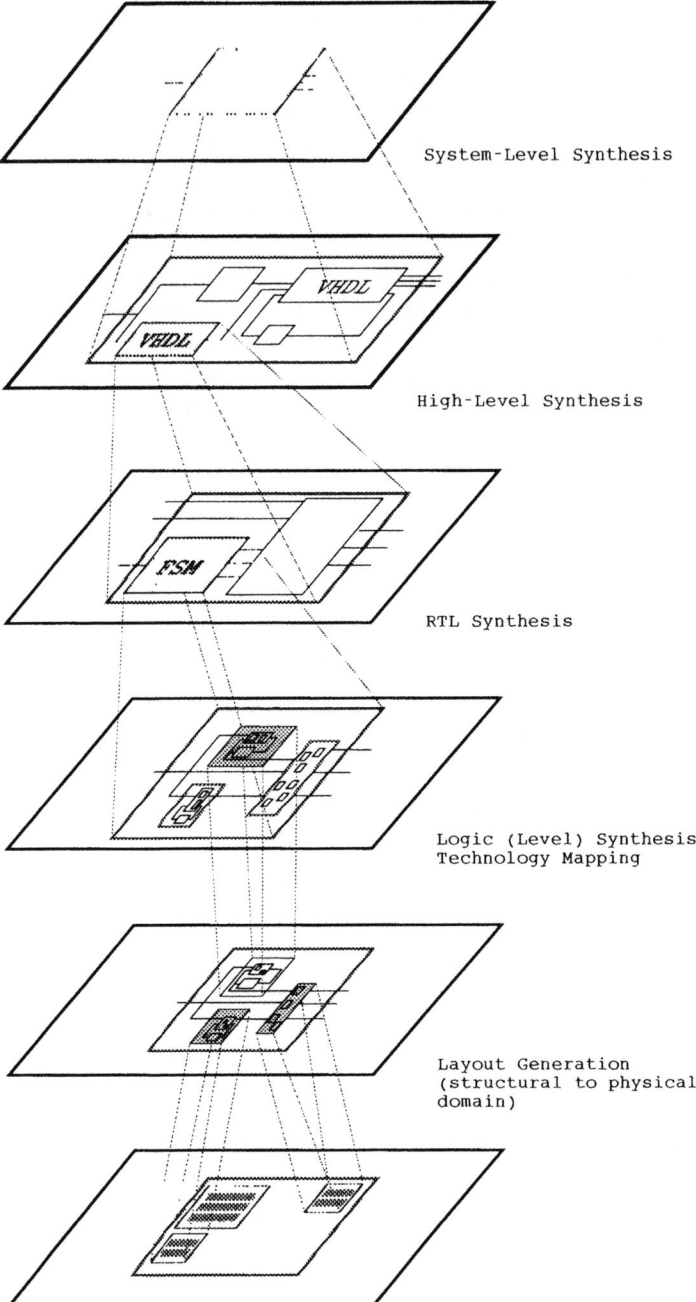

Fig. 1. Multistrata Representation of Controller Synthesis within the Synthesis Process for Digital Systems.

optimization steps. This implies that the cost-function and the optimization procedure of lower levels have to be considered already at a higher level of the design process.

Synthesis steps and design activities on a certain level must not be undertaken autonomously, but have to be adaptive according to transformations and cost-functions of methods on subsequent levels of abstraction. To make this interaction between different levels feasible, we propose a framework of interoperating tools.

3 Interoperating Tools

The synthesis process is covered within the proposed framework by:

1. *A high level synthesis tool*, which transforms behavioral descriptions into structural ones ([7]). Its main function is the transformation and management of different kinds of design data representation. It integrates VHDL representations, state tables, and boolean functions into a standard format for electronic design data (EDIF: Electronic Design Interchange Format), which is fed into the layout generation tool. Besides that it provides features for FSM optimization as follows:
   - *FSM extraction* creates a state table format from a netlist format
   - *State minimization*
   - *State assignment* (one-hot, binary, gray, manual, and auto)
   - *Read/write operations* for FSMs in state table format.
   - *Compare FSMs*

   But no more state assignment algorithms (e.g., aiming at the improvement of testability), state splitting techniques, structuring or decomposition methods are applicable in a straightforward way. These functionalities are covered by CAST.FSM.

2. *CAST.FSM* provides mathematical objects within a CLOS (Common Lisp Object System) class hierarchy. Mathematical operations on these objects correspond to CLOS generic functions:
   - *Basic mathematical objects and functions*
   - *Domain Specific System Classes* – FSMs/FSMDs, switching circuits, boolean functions, shiftregisters, and hierarchical networks are system types of the VLSI application domain domain. They are modeled according to the object oriented paradigm and are therefore easily extensible. Using that system classes CAST.FSM currently supports the Computation of Lattices of FSMs, Algebraic Lattice Operations, Finite Memory Test, Information Losslessness of FSMs, State Reduction of FSMs, Inversion of FSMs, Parallel/Serial Decomposition, Mealy to Moore Transformation, Shiftregister Realization, Input/Output Experiments, Computation of Homing, Diagnosis, and Synchronizing Sequences.
   - *User Interface* – Common Lisp is CAST.FSM's integrated command language for user specific extensions. No artificial macro language had to be

defined, the Common Lisp interpreter provides a homogeneous environment for users and programmers. User defined code is fully integrated, there are no differences to predefined system code.

– *Design Manager* – CAST.FSM is an interactive tool and supports users in decision making. Design histories, already explored design alternatives, and related data and files are handled by the system. Curently CAST.FSM is being extended with an experiment manager, which allows to define tasks and to let them be executed in batch mode. This includes the automatic generation of command files for foreign tools. CAST.FSM handles tool execution using the appropriate command files and cares for piping a tool's output data into the execution of a next tool.

3. The OCTTOOLS collection was developed by the UC at Berkeley and provides tools for most levels on abstraction for VLSI design activities. Of special interest for the proposed framework are the well known tools for FSM state assignment (JEDI and NOVA), logic minimization (ESPRESSO and MISII), and optimizations of sequential logic (SIS). Further information can be obtained in [8] and [9].

The role of CAST.FSM within the design environment is illustrated in Figure 2. Design Data is transferred into CAST.FSM where appropriate system classes are predefined. Within CAST.FSM the user is free to use standard systems algorithms [10] or to experiment with CAST.FSMs building blocks. For low level optimizations a subset of the OCTTOOLS collection is directly accessible.

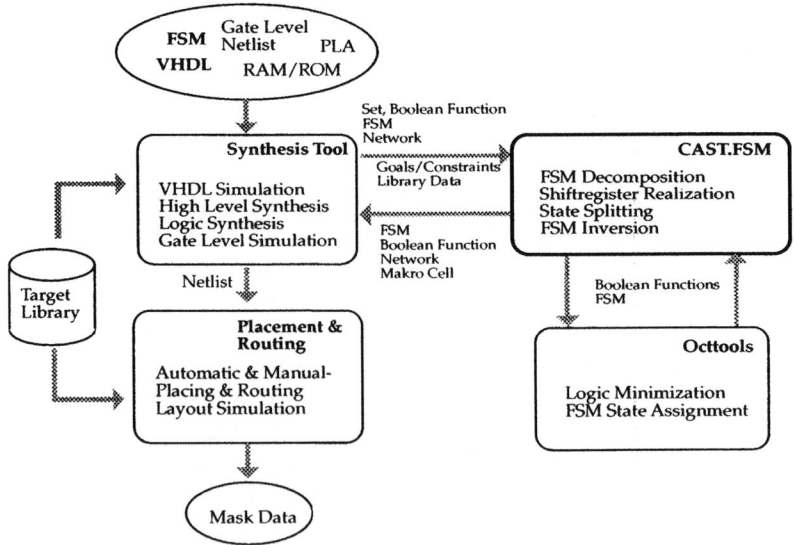

Fig. 2. Framework of Interoperating Tools

Various interfaces to CAST.FSM's application domain and client tools ensure real interoperability. We use a small subset of design representation formats for that purpose: *State-table-format* can express FSMs with symbolic input, output, and states as well as switching circuits. A *Pla-format* is used to represent functions (combinational logic), *BLIF* (Berkeley Logic Interface Format) can handle delay and clock constraints, and *structural VHDL* represents networks (components and couplings).

Currently CFI (CAD Framework Initiative) is developing different standards for tool communication and integration. As none of available tools comply with CFI 1.0 standards right now, these standards do not bring about a increase of functionality today (but are being considered for CAST.FSM's interfaces).

The tool communication within the proposed framework is file-based only. A tool based integration (given a specific foreign tool appropriate files are generated automatically), and a task based integration are currently being specified. In a task based integration CAST.FSM allows an intelligent tool and method application. It serves as a design assistant by selecting the most suited tool, by adding necessary parameters for a given task, and by considering additional constraints (eg., problem size and available resources).

4 Application to FSM Shiftregister Realization

The proposed framework is being evaluated by a FSM optimization method targeting testability: The shiftregister realization of FSMs is an approach for the FSM state assignment problem, where FSMs are transformed into a shiftregister structure, i.e., where memory elements are put together to shiftregisters targeting a scan path architecture for testability purposes. Figure 3 illustrates the general structure of a shift register realization of a FSM addressed in this section. Using a shiftregister realization the hardware overhead of scan path architectures can be reduced by

- a decreased number multiplexers
- a decreased number of output lines of combinational logic implementing state transition functionality and
- a smaller wiring area as not all memory cells have to be loaded from the combinational logic.

The framework allows to extract state transitions which prevent good shiftregister realizations (transisitions to a reset state, and transitions from a state into itsself) and do the shift register realization for the remaining (essential) FSM. If a good realization is obtained the extracted functionality is reinserted by additional hardware. Here we make use of special properties of elements of the target library, especially of reset- and shift-enable functionality.

Figure 4 shows the design flow and involved design formats for the realization of a FSM within the framework. For reasons of simplicity the special treatment of reset and shift-enable is not considered here. The results of the shiftregister

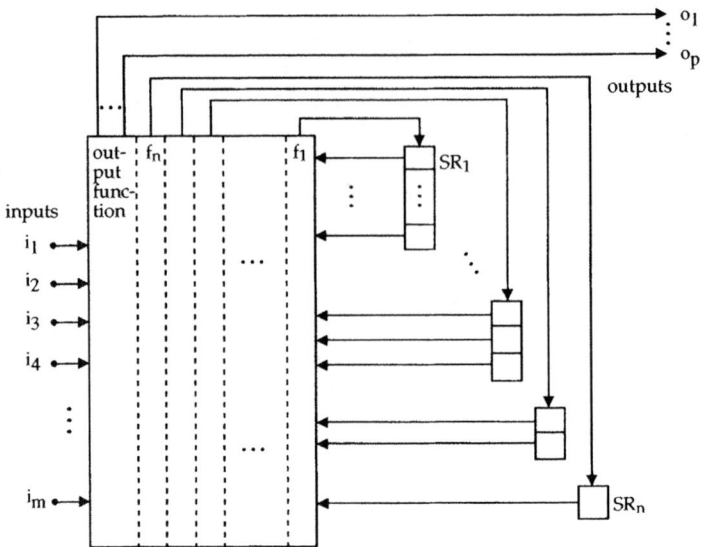

Fig. 3. General Shift Register Structure

realization are given in terms of PLA-area and literals for multi-level implementations, respectively. Table 1 shows area changes between -14% and +22%.

Memory elements, additional multiplexers, and wiring aspects are not considered here. The combinational logic only of achieved shift register structures are compared to the combinational logic of state assignments calculated with JEDI.

The minimal number of bits for state encoding are shown in column *bits*. *SR* illustrates the achieved shift register structure and *Time* shows the algorithm's runtime in seconds on a Sun ELC, running Lucid CommonLisp. *Lits* are given in *factored* form prior and after library mapping using *stdcell2_2* [9]. In [4] results of two-level implementations are given.

The machines *scf**, *kirkman**, and *s1488** are marked because their reset-functionality is separated and not considered. For *s208* a reset state r is defined, but there exists another state r', which also shows reset-functionality. Hence, in *s208*** not the transitions into the specified reset state r but the state transitions into r' have been excluded. The realization algorithm has been applied to the modified machines and all area results refer to the essential state transition functions.

Using the flow of Figure 4 not only the size of combinational logic blocks but the whole realization including memory elements and wiring area can be evaluated. It is possible to compare a conventional scanpath implementation of a FSM with a scanpath implementation based on a shiftregister realization. For the reasons stated above (reduced number multiplexers and wires), we expected to achieve better results than a +22% area increase. Unfortunately, we could not prove that assumption. It showed that the actual area of a standard cell

FSM	bits	SRs	Time	Jedi-MIS Lits	Area	Shiftreg-MIS Lits	Area	Lits-Chg. %/%
ex4	4	3-1	9.0	66/89	1048	81/97	1144	122/109
planet	6	2-2-1-1	83.2	564/697	8360	577/714	8664	102/102
shiftreg	3	3	1.1	9/11	11	2/2	2	22/18
scf*	7	3-3-1-1	146.8	829/1056	12840	757/991	11776	91/94
kirkman*	4	4	90.3	154/222	2544	170/248	2801	110/112
s1488*	6	2-2-1-1	57.2	543/667	8072	523/640	7688	96/96
s208**	5	5	39.8	49/71	824	42/65	736	86/92
shiftreg	3	3	2.0	9/9	112	2/2	32	22/22

Table 1. Results of multi-level implementations of shift register realizations

realization increased by approximately +35% when the shiftregister realization algorithm was applied.

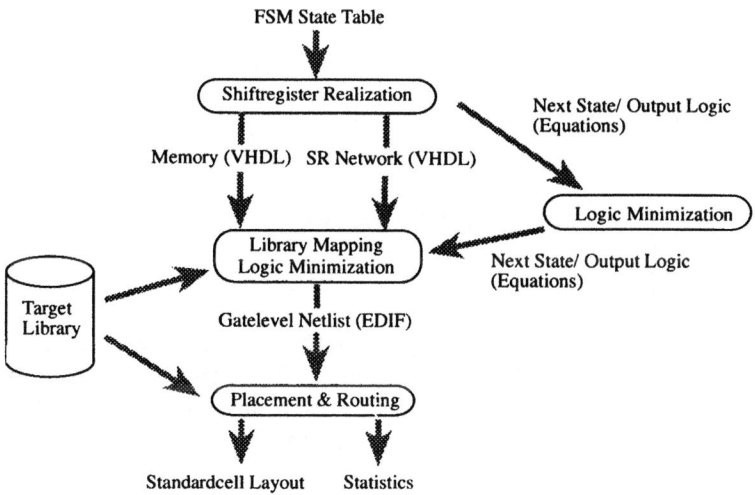

Fig. 4. SR realization:First Approach

A further analysis of the shiftregister realization reveals, that there is much unused optimization potential available:

1. In many cases the algorithm produces more shiftregister structures of equal quality, but always the first one is returned as result without further evaluation.
2. If a shiftregister structure is defined there is still some freedom to assign actual state codes. Only a random state assignment technique is being applied.

The shiftregister realization procedure treats the state transition functionality only and totally neglects the FSM's output logic. But output logic requires a big part of the whole logic area in most controllers.

This analysis of the shiftregister realization procedure is taken into account by a refined design flow, shown in Figure 5.

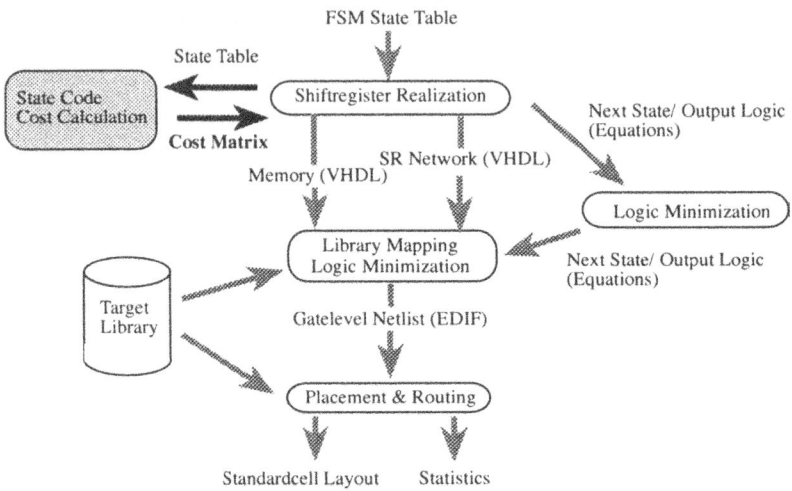

Fig. 5. Extended SR realization

In the refined shiftregister realization algorithm the impact of logic minimization (also of the output logic) is taken into account. Modern state assignment tools consider optimization procedures of lower levels of the design flow. In JEDI, for example, a cost matrix is calculated at the beginning of the algorithm, where a cost value is assigned to all pairs of states. High costs for a pair u, v denote a big logic area if very different binary codes are assigned to u and v. Thus this cost-matrix has to be considered for selecting shiftregister structures and offers a means for substituting randomly assigned (partial) state codes

The proposed framework approach enabled an easy integration of the evaluation of that cost-matrix within the shiftregister realization procedure. The costs are calculated by JEDI and fed into CAST.FSM. Currently the actual shiftregister realization code is being adapted and the coding of states is being improved. Some manual experiments showed promising results and motivated further work. Figure 6 compares final standardcell area of $s208^{**}$ (using an industrial library) followed by a scan path insertion and the corresponding layout achieved by the application of the refined SR realization procedure. Further result are available in [11].

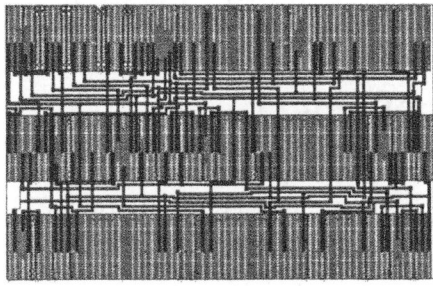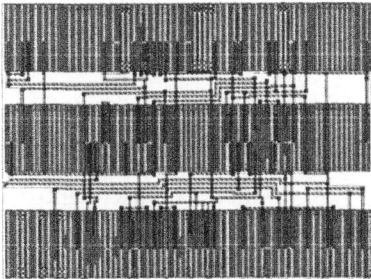

Fig. 6. Standardcell layouts of s208**: conventional scanpath (left) compared to the result of the SR-realization procedure (right)

5 Conclusion

The integration of CAST.FSM into a framework of interoperable VLSI design tools achieves powerful optimization capabilities for FSM designs. We presented a successful application of our framework targeting at testability and evaluated our results at the logic level. Then we also considered wiring aspects by producing final standard cell layouts. These results led to an refinement of the initial shiftregister realization procedure, which is subject of current work.

References

1. J. Fox, "A higher level of synthesis," *IEEE Spectrum*, vol. 30, no. 3, pp. 43–47, 1993.
2. F. Pichler and H.Schwaertzel, *CAST: Methods in Modelling*. Berlin: Springer, 1992.
3. T. Mueller-Wipperfuerth and M. Geiger, "Algebraic Decomposition of MCNC Benchmark FSMs for Logic Synthesis," in *EURO ASIC '91*, (Paris), pp. 146–151, 1991.
4. T. Mueller-W., J. Scharinger, and F. Pichler, "FSM Shift Register Realization for Improved Testability," in *Proceedings Eurocast 93*, Springer Verlag, 1993.
5. M. Mesarovic, D. Macko, and Y. Takahara, *Theory of Hierarchical, Multilevel, Systems*. New York and London: Academic Press, 1970.
6. D. Gajski, N. Dutt, A. Wu, and S. Lin, *High-Level Synthesis*. Boston: Kluwer, 1992.
7. Synopsys Inc., *Design Compiler Reference Manual*, Dec 1992. Version 3.0.
8. S. Yang, *Logic Synthesis and Oprimization Benchmarks User Guide*. MCNC, 1991.
9. K. McElvain, *IWLS'93 Benchmark Set*. MCNC, 1993. Version 4.0.
10. F. Pichler and H. Schwaertzel, *CAST - Computerunterstützte Systemtheorie*. Berlin: Springer, 1990.
11. T. Mueller-Wipperfuerth, *Finite State Machine Structuring with CAST.FSM* applied to VLSI Controller Design*. PhD thesis, University of Linz, 1994.

Using Logic Programming to Test Module Specifications in Early Stages of Software Development

Mario A. Nascimento*, Clenio F. Salviano** and Murat M. Tanik***

Department of Computer Science and Engineering, Southern Methodist University, Dallas, Texas, 75275-0122, USA. {mario, clenio}@seas.smu.edu

Abstract. To build a complete software system, it is widely recognized that a proper decomposition of the system into modules can facilitate the software construction process. Such decomposition can be accomplished by modeling software modules as Abstract Data Types (ADTs). ADTs can be formally specified by using Bartussek and Parnas' TRACE. We investigate the feasibility and practicality of translating TRACE specifications into Prolog programs, keeping a TRACE-like format, in such a way that can actually be executed. Besides exercising the specification before any implementation, we can run both the actual implementation and the TRACE specification and then verify whether the implementation matches the specification. Additionally, we can partially avoid the maintenance of test-case databases as the specification itself will provide output according to a given input. An experiment has been carried out.

1   Introduction

We consider two issues: software testing and software decomposition. In software testing it is common to assume that given an input, the corresponding output can be verified. Such verification either relies on an oracle or assumes that for every input the output is manually verifiable. The former might not be feasible and the latter usually restricts the universe of input data to a few special cases. A typical approach for such verification is to maintain a database of test-cases, with input-output pairs. This option is very costly and may be unreliable [14].

The process of decomposition in software development is widely used and needs no further justification. One approach to accomplish such decomposition is to use the principles of Information Hiding, Encapsulation and Data Abstraction [1]. Every module has all the information about its own behavior without needing to rely on information from any other module. Moreover, none of the

---

* On leave from CNPTIA–EMBRAPA, Campinas, Brazil (mario@cnptia.embrapa.br) and supported by CNPq (Process No. 260088/92.7,) Brasilia, Brazil.

** On leave from CTI, Campinas, Brazil (clenio@ic.cti.br) and supported by CNPq (Process No. 200801/87-2,) Brasilia, Brazil.

*** Current address: Department of Electrical and Computer Engineering, University of Texas at Austin, Texas.

modules should have any information on the internals of others. All modules should have an interface to communicate with other modules and all interactions among modules should be done only through these interfaces. This should prevent one module from inadvertently affecting the behavior of another module. In addition, every module can use the services of another one by calling it through an appropriate interface, thus enforcing reuse. However, to allow a productive use of such principle each module should have a well defined interface. The definition of the interface should be at a level that it is defined precisely, yet without suggesting any implementation approach.

Figure 1 shows a common way to build software modules. In this approach the module specifier is aware of the needs and writes the ADT's specification. We assume that such ADT's specification is done using the TRACE methodology [1, 6, 8]. Using the specification, as well as knowledge from the specifier, the implementor is able to write the actual code and test it, delivering the module implementation. Now consider one potential drawback of this approach. If the specification was wrong or incomplete, the actual code has to be rewritten (an expensive task) or "fixed." Indeed it may happen that during the test phase something said to be an error is discovered, when it actually is a symptom, i.e. a manifestation of a error that was in the specification [13]. In this case it is likely that the code is changed to overcome what should rather be a change in the specification.

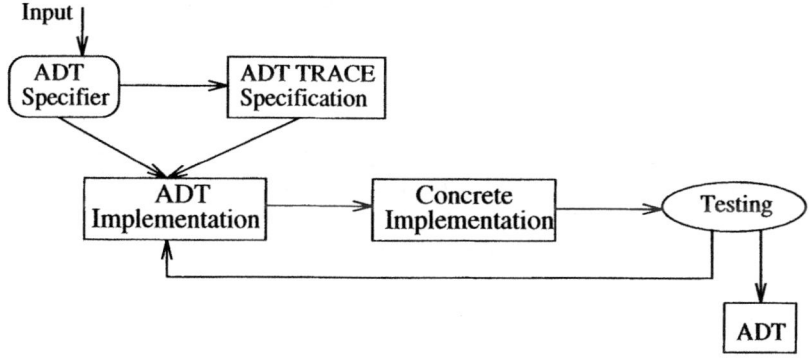

Fig. 1. A common approach for module development

In order to minimize the effects of such drawback we propose a different approach. The specification is translated into Prolog predicates, resulting in what we call Prolog-ed Trace (*PT*). This *PT* is executed on a set of input test-cases providing valuable and prompt feedback to the specifier. The specifier uses this feedback to correct, improve and/or extend the specification. This process is repeated until the specification reaches a stable status (see Figure 2.) This minimizes the chances of specifications errors being propagated into implementation.

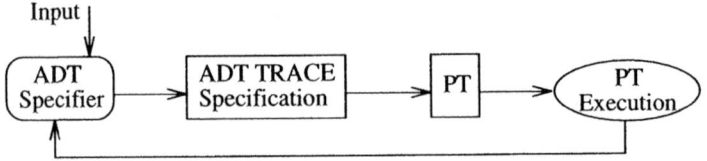

Fig. 2. Exercising the specification

We can couple the idea of exercising the specification with the common approach for module development. This results in the situation depicted in Figure 3. In this improved process, the actual implementation of the module is started, using knowledge obtained from the specifier and from the TRACE specification itself. Once implemented, the actual code is run with the same input test cases that the *PT* was tested on before. The results from both implementations should agree. Testing the actual implementation provides feedback to the implementor who can also provide further feedback to the specifier and possibly specification constraints could be relaxed or strengthened. After some iterations the process should converge to the actual ADT implementation.

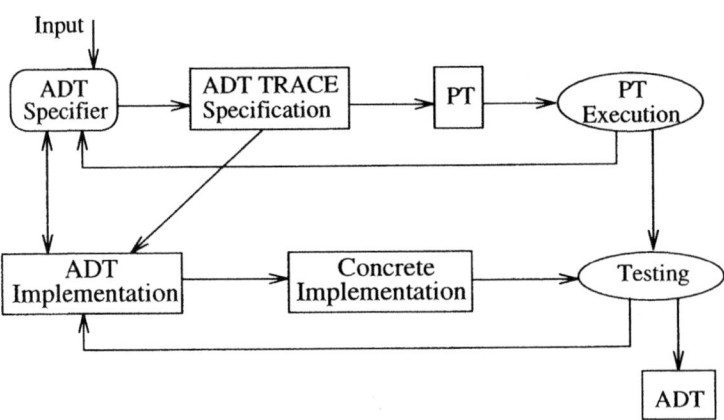

Fig. 3. The proposed approach

The literature presents some attempts to deal with this problem [5, 7, 9, 14, 17]. Some focused on the module implementation testing and others on the module specification. Most are concerned with testing the actual implementation against the specification. In contrast, we concentrate on exercising the specification before the actual implementation may take place. This agrees with one of the requirements (from a total of seven, presented in [11, page 13]) of the next-generation intelligent systems, which is the *"ability to set hypothesis and to test them."* We feel that the approach above presented allows the user to "play" with the specification without incurring in the cost of actually implementing it, hence avoid potential problems, for instancing in determining whether possible flaws are due to erroneous or incomplete specifications, or their implementations. Finally, our approach partially addresses the concern expressed in [11, page 11] in that *"... a user can concentrate on the specification of the system and the problem to be solved, rather than instructing a computer how to do what needs to be done."* Nonetheless, our approach allows the use of the executable specification to test the actual implementation as well.

Put in the context of Computer Aided Systems Technology, we consider the software development problem under the light of the three system engineering observations presented in [15]:

- *Specification:* Human being cannot specify correctly a reasonably sized system (or software module for that matter) at once;
- *Feeback:* Questions, answer and feedback about the system help in refining its specification/design and
- *Automation:* If convenient automated tools are provided to accelerate the feedback process, then the initial system design/specification can be enhanced more efficiently.

This paper addresses, in a practical way, the Automation Observation. We practically investigate the feasibility of using a tool (Prolog in this case) to automatize the process of gathering feedback about the software module, speeding it up and consequently helping in the refinement/correction of the specification. Finally it has been pointed out [10] that high quality software can be achieved by (1) computer aided specification of the problem and (2) translation of such a specification into an executable program. We address partially the second issue. Although we do not present such a translator, we use Prolog as an "executional" framework for one specification. We envision that the translation of TRACE specifications into Prolog predicates can be done if we constrain ourselves to a specific domain.

This paper is divided as follows. Section 2 presents a review of the TRACE specification methodology. Section 3 introduces the approach we propose to address the early exercise of a software module specification. Such an approach is based on using Prolog [3] and throughout this paper the reader is assumed to be familiar with the language. For that we use the unlimited integer stack as a supporting example. Section 4 compares our approach with related work. Finally, conclusions are drawn in section 5.

2 TRACE Specifications - A Review

Let us consider the issue of ADT modeling. Briefly, an ADT is an operand to which a set of operators can be applied. If every module is modeled as an abstract data type, and its interface is well defined, then the software decomposition process can potentially be achieved.

TRACE is a formal method to specify a software module as an ADT. Using TRACE, the module is an ADT which is designed as a "black box" in the sense that its internals are not detailed at all. The only way to access the ADT is through the use of its interface, i.e., its accessing functions: O-functions, which cause a change in the ADT's status; and V-functions which return information on the ADT's status without changing it. In what follows we briefly review the way TRACE specifications are written.

A TRACE specification consists of two main parts: Module Syntax and Module Semantics.

- In the Syntax part the name of the accessing functions and the type of its parameters are defined;
- In the Semantics part the accessing functions (both O-functions and V-functions) are defined. Remember that we are not maintaining details about how the ADT works internally.

A trace T is a sequence of calls to O-functions and/or V-functions. We say that a trace T causes an "error" if the ADT's behavior cannot be predicted using only the information given in the Semantics part. If a trace T causes no error it is called "valid." The Semantics part has three sections: Legality, Equivalence and Values.

- In the Legality section the sequences of traces that can be fed into the ADT without causing any "error" are declared. The empty TRACE is always legal;
- In the Equivalence section we can derive whether traces are equivalent. Two traces are said to be equivalent if their behavior is the same, and both have the same legality;
- Finally, the Value section, describe the information that should be returned upon using a V-function.

Below we present the TRACE specification (adapted from [1]) of an unlimited stack of integers.

SYNTAX:

PUSH:	integer, stack	$\rightarrow$ stack
POP:	stack	$\rightarrow$ stack
TOP:	stack	$\rightarrow$ integer
DEPTH:	stack	$\rightarrow$ integer

LEGALITY:

$L(T) \Rightarrow L(T.PUSH(a))$

$L(T.TOP) = L(T.POP)$

EQUIVALENCE:

T.DEPTH	$\equiv$	T
T.PUSH(a).POP	$\equiv$	T
L(T.TOP)	$\equiv$	T.TOP = T

VALUES:

$L(T) \Rightarrow V(T.PUSH(a).TOP) = a$

$$L(T) \qquad \Rightarrow \qquad V(T.PUSH(a).DEPTH) = 1 + V(T.DEPTH)$$
$$V(DEPTH) = 0$$

Let us make some observations on the example given:

- It is clear that no implementation approach is suggested;
- Using the Equivalence assertions a trace T can be reduced to a sequence of PUSHes only;
- The syntax of the overall specification seems to be very suitable to be implemented in a parser for automatic reading, provided that we constrain ourselves to a set of known operators.

3 The *PT* Approach

The main goal of the *PT* approach is to use the Prolog language and its inherent features to deal with TRACE specifications, thus addressing the issues presented in the Introduction. Although we do not present the theoretical framework underlying this research, there has been some more theoretically oriented research addressing Term Rewriting Systems which is the theoretical background for our work, and we refer the reader to [16, 4, 2, 17] among many others papers.

Therefore given a trace T, one should be able to:

- Verify whether it is valid or not (i.e. cause an "error");
- Reduce it to its Normal Reduced Form (NRF); we say that T1 is the NRF of T2 if T1 is the smallest trace equivalent to T2 and/or
- "Execute" it, thus exercising its behavior.

The approach we use to accomplish this can be written as follows:

(A) INITIALIZATION

- Read the input trace T using the format "$f_1.f_2.....f_n$", where each f_i is either a O-function or V-function;
- Initialize the normal equivalent trace N as empty;

(B) CHECK LEGALITY, EXECUTE AND REDUCE

- Extract each f_i from T, one at a time, and perform the following (when there is not a f_i to be extracted stop the processing):
- Append the f_i to the normal trace N that has been already built.
- Check the legality of using f_i at that position. If the legality check fails the trace is illegal and the process of execution is aborted otherwise continue;
- If f_i is a V-function perform its task using N (that is all knowledge the trace can provide so far) and delete f_i from N as V-functions do not change the ADT status;

- If f_i is a O-function check whether the new f_i (using the point of view of N) can be used to trigger some of the equivalence assertions. Once some equivalence assertions can be used, do so and further investigate (reducing for instance) N.

For instance, if the user gives the following trace input T = push(a).top.pop the process would be as described next:

(Read input)	T = push(a).top. pop
(Initialize the normal trace)	N = [ ]
(Extract each f_i)	f_1 = push(a)
(Check legality)	Is '[ ] + push(a)' legal ? Yes.
(Check type of function)	Is 'push(a)' a V-function ? No.
(Check equivalence)	Can we reduce '[push(a)]' ? No.
(Extract each f_i)	f_2 = top
(Check legality)	Is 'push(a)+ top' legal ? Yes.
(Check type of function)	Is 'top' a V-function ?
	Yes. Output 'a' (the previous "push".)
(Check equivalence)	This V-function is not added to N.
(Extract each f_i)	f_3 = pop
(Check legality)	Is 'push(a) + pop' legal ? Yes.
(Check type of function)	Is 'pop' a V-function ? No.
(Check equivalence)	Can we reduce 'push(a).pop' ? Yes. N = [ ].
(Extract each f_i)	No more input to read, STOP.

Therefore we can conclude that the trace is legal and its NRF is [ ]

3.1 The *PT* Operators

Here we present a set of operators that can, up to a certain extent, cover the specification written in TRACE style. The reader should be aware that although the translation from TRACE into Prolog is quite simple in this case, we still need to adapt the actual implementation in order to handle all semantic expressiveness of the original specification.

```
:- op(150, xfy, :).
:- op(200, xfy, implies_in).
:- op(200, xfx, equal).
:- op(200, fx, always).
:- op(200, xfy, not_equal_to).
:- op(200, xfx, is_equivalent_to).
:- op(250, xfx, returns).
```

Once given these operators the STACK specification presented previously can be re-written in Prolog as (parameter type checking has been omitted for the sake of simplicity):

```
/* Syntax */
o_functions([push(_), pop]).
v_functions([depth, top]).
/* Semantics */
/* Legality */
l(T)                          implies_in   l(T : [push(_)]).
T       not_equal_to []       implies_in   l(T : [pop]).
T       not_equal_to []       implies_in   l(T : [top]).
always l( T : [depth]).
/* Equivalence */
T : [depth]          is_equivalent_to      T.
T : [push(_),pop]    is_equivalent_to      T.
T : [top]            is_equivalent_to      T.
/* Values */
T : [push(Data),top] returns Data.
[] : [depth]         returns 0.
T : [push(_),depth]  returns (T : [depth] + 1).
```

Notice that we do not have an absolute translation of the original specification, rather we have a modified TRACE-like specification hence the name Prolog-ed TRACE. The modification appear in the third assertion of the Legality and it is introduced in order to avoid a potential looping, actually a positive loop, in the Prolog execution. Namely to verify whether a "top" is legal would require to verify whether a "pop" is legal instead, but then to verify the latter the former verification would be required again and so forth.

Nevertheless this stack specification is well-defined and well-founded. Notice that no additional information was given about the stack, it was only the case that some previous information was "split" for execution feasibility.

We have noticed that it is not very likely that one can come up with a general algorithm to translate a TRACE specification into Prolog predicates. This is mainly due to the fact that the TRACE assertion method is purely a methodology for specifications, there is not, as far as we could verify in the literature, a *language* for it. As long as the designer of the specification maintains the specification complete and consistent, including the definition of any operands/operators used, there is no problem whatsoever.

3.2 *PT* predicates

The basic idea, and fortunately a Prolog feature, that we are going to use is "pattern matching." At each step of the substeps of (B) presented above we should be able to match the current normal trace format with some of the legality assertions (otherwise it would not be legal); return a value (should this be the case) and reduce it.

One of the nice features provided by Prolog is that the *PT* predicates that are going to be introduced can be expanded easily to cover a broader class of specifications.

The predicates for the legality checking are the following:

```
legal([]).                     /* Legality verification */
legal(T) :-
        append(L, R, T),
        l(L) implies_in l(L : R),
        legal(L).
legal(T) :-
        append(L, R, T),
        L not_equal_to C implies_in l(L : R),
        L C.
legal(T) :-
        append(L, R, T),
        always l( _Any : R).
```

For equivalence assertions we used the following predicates:

```
equiv(Tin, Tout) :-            /* Equivalence verification */
        append(L, R, Tin),
        L : R is_equivalent_to L,
        Tout = L.
equiv(Tin, Tin).
```

Finally for the value assertions we need a little bit more work. Recall that the specification for "depth" require a numerical recursive computation. In this case we need to handle only the sum ("+") operator. Should some other be necessary we could handle it. Such handling should be done in a "demand" basis as it is very unlikely to be possible to cover, at once, all possible cases. Actually the user do not need to confine him/herself to known operators, he/she could provide new ones, given that they are "specifiable" at any time.

```
value(Tin, V) :-               /* V-function computation */
        append(L, R, Tin),
        L : R returns Return,
        (atomic(Return) ->
                V = Return
        ;       compute(Return, V)
        ).
compute(Exp, V) :-
        Exp = P1 : P2 + P3,
        append(P1, P2, L),
        value(L, R),
        V is R + P3.
value(Tin, Return) :-
        Tin returns Return.
```

This predicate, as the previous one, always has a return object. The reader should observe that this definition is far away from being comprehensive. We have defined strictly what was required by the specification we were dealing with. It should be reasonably easy to extend it to perform others computations as well as using other operators that the specification may require. Notice that this predicate **value/2** is performed only when a V-function is found, and from the Syntax part we know whether a given operator is a V-function or not.

3.3 The *PT* engine

Here we present what we call the *PT* engine that is responsible to use the predicates just introduced in order to honor queries from the user. The user can make use of the following main predicates:

- **legal/1**: To check whether a trace is legal or not;
- **reduce/2**: To reduce a trace, what implicitly invokes **legal/1**;
- **run/3**: To execute the trace, it implicitly invokes **reduce/2**.

In order to verify the legality of the input trace it should be reduced, but **legal/1** does not output such a reduction. To obtain the reduced trace the user should use the predicate **reduce/2**, naturally the reduction succeeds only if the input trace is legal. To "execute" the input trace the user should use **run/3**, which implicitly checks the input's legality and reduces it as well. The interrelations among these predicate are shown in Figure 4. The actual predicates follow.

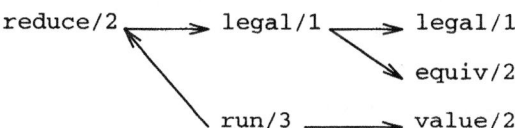

Fig. 4. Relationship between main predicates

```
legal([]).
legal(Tr) :-
      rec_legal(Tr, _Normal, []).
rec_legal([H | T], N, N1):-
      append(N1, [H], L),
      legal(L),
      equiv(L, N2),
      (T [] ->
         rec_legal(T, N, N2)
      ;  N = N2
      ).
reduce([], []).
```

```
reduce(Tin, Tout) :-
      rec_legal(Tin, Tout, []).
run([], [], []).
run([H | T], Normal, Output) :-
      run(H, T, [], Normal, [], Output).
run(H, T, N1, N, O1, O) :-
      append(N1, [H], L2),
      legal(L2),
      v_functions(Set_V),
      (member(H, Set_V) ->
         value(L2, V),
         O2 = [V | O1]
      ;  O2 = O1
      ),
      equiv(L2,L3),
      (T [] ->
         T = [NH | NT],
         run(NH, NT, L3, N, O2, O)
      ;  N = L3,
         O = O2
      ).
```

Notice that the predicate value/2 is called just if it is the case of a V-function, which is verified upon calling member/2 on the set of V-functions (Set_V). Also the reader should notice by now that the complete source code, with the exceptions of predicates append/3 and member/2, has been presented.

3.4   An example exercise session

In this section we give a brief taste of *PT*'s use. Consider that the user wants to investigate the specified stack using the following trace:

depth.push(a).push(b).pop.depth.top.pop.depth

If the user wants to check its legality he/she should execute:

```
:- legal([depth,push(a),push(b),pop,depth,top,pop,push(x),top]).
yes
```

Should the trace not be legal the answer would be **no**. If the user wants the normal form of such trace, he/she should do:

```
:- reduce([depth,push(a),push(b),pop,depth,top,pop,push(x),top],
Normal).
Normal = [push(x)]
```

Notice that the unbound variable Normal was bound to the reduced normal form of the input trace. Finally if the user wants to execute the given trace, he/she should use the predicate run/3 as follows:

```
:- run([depth,push(a),push(b),pop,depth,top,pop,push(x),top],
Normal,Output).
Normal = [push(x)]
Output =[0,1,a,x]
```

Again the unbound parameters were bound accordingly.

4 Related Work

In the approach presented by Macario and Camargo [9] there was a sample of traces to be used as input, where these traces were intended to be a representative subset of all possible traces. In fact, this sample should be of canonical traces of a given specification, which may be very hard, if not impossible to obtain beforehand. The idea is to use these traces samples and substitute them in the specification. Then, run the concrete implementation (written in C, in that particular research) of the specification and verify whether such execution was violating any assertions in the TRACE specification. They do not investigate the issue of whether the specification is correct/complete, unlike we do, rather they investigate the concrete implementation of the specification.

Strooper and Hoffman's [14] work do not use a formal specification as a starting basis as we do. They capture the behavior of the ADT into prose and then use this as a basis to write the Prolog code. Also they make use of a facility that most Prolog compilers provide, namely, the capability to call C functions from within the Prolog code. Thus, instead of running the specification they used Prolog to build test drivers for the C modules.

Wang and Parnas [17, page 16] describe a trace rewriting system and how such system can simulate a TRACE specification. Interestingly they rely on a "... reduction function r which maps any possible trace to its representative canonical trace." We could observe that despite a different denomination, our approach implicitly presents such a function r, this is the case when the input trace is reduced to what we call its normal form. They present a basis for rewriting systems, but without many arguments for its concrete implementation. In this paper we are primarily concerned with the practical aspects of specification exercise in software development. Nevertheless our approach may be considered a trace rewriting system as well.

In [7] it is also attempted to use Prolog to exercise TRACE specification. However that paper concentrates on the specification itself, providing for instance some heuristics for writing a TRACE specification. Using those heuristics can indeed help. However, looking at the classical stack example, it seems that the specification itself can be hard to understand when compared to the original specification [1] and the new tabular format [12]. As a minor drawback we may point out the fact that their Prolog interface (which is generated automatically) uses the Prolog's built-in predicates *assert* and *retract* that are often expensive to execute and can be understood only in terms of their side effects. Indeed they may yield some unexpected side effects, unless carefully used. In our approach this is not the case. We rely solely on a list (in the Prolog sense) which is at "execution" time reduced to the input's normal form. An important contribution from that particular research is that they build an interface for C programs, so that C programs can actually use the Prolog prototype (which in turn is the specification itself) to perform the functions they need. Hence, allowing a rapid prototyping approach.

According to Strooper and Hoffman [14, page 599] *"... oracles determine whether the outputs are correct given input/output pairs, whereas implementations generate correct outputs for given inputs."* The main difference from their research and the one presented in this paper is that they build an *oracle* whereas we build an *implementation*. Note, however, that both can be considered as executable specification. They argue convincingly that for testing the actual implementation, oracles are superior. However our main intention in this research is to allow the specification to be experimented before being implemented, thus recognizing the fact that the specification may need maintenance itself. Moreover an implementation might also be used as an oracle, in such a case there would be a necessary overhead to check whether the concrete implementation output matches the specification output, assuming it to be correct and executable.

In all these and our research the goal of reducing the maintenance of the databases for test-cases is accomplished. However a representative input set has to be maintained anyway. Furthermore, neither Macario and Camargo [9] nor Hoffman and Strooper [14] attack the problem of using the TRACE specification to reduce a trace input to a normal irreducible form, as we do, and as Wang and Parnas point out, it is a way to discover canonical traces.

Finally one main difference between our research and the ones previously presented is that we use the TRACE specification method, rewriting it in Prolog and then making it suitable for execution, whereas the most of the others use actual concrete implementation to be executed. Such TRACE specification may have to be rewritten in order to be feasibly implemented. Finally we are able to exercise the behavior of the ADT before it is actually implemented, as we can execute the specification itself.

5   Conclusions

We have investigated in this paper the issue of executing a TRACE-based software module specification. We have seen that the Prolog language has features that ease such task, and we have then proposed the use of a Prolog-ed TRACE (*PT*.) The way that the specification is given in *PT* seems to be reasonable, and future effort could be oriented to further improve the expressability of the "language" as well as give it more deductive power. For instance, in the original stack specification we can see that either a "pop" or a "top" is only allowed if the stack is not empty, this knowledge was explicitly given in the *PT* specification. However there should be paid some price for that. Indeed it is a tradeoff between an easier translation versus a closer (to TRACE) translation.

The approach we presented can be very useful for software testing. In the process of software decomposition one could:

- Specify each modules formally in TRACE, this should provide "for free" a nice documentation of the specifications;
- Implement a *PT* specification, without much difficulty given a TRACE specification of a module and
- run both the actual Software code and the *PT* specification in order to verify whether the implementation output matches the output obtained by the specification itself.

Another feature of a specification *PT* is that there is no more need to maintain a database of test-cases. As the requirements change, and they always do, the specification should also change and when it is run again, the *PT* specifications will already reflect the update. Thus granting the test cases database more reliability in the sense that it will always be following the actual software implementation. The gain is that when a database is maintained all test-cases should be verified, eliminate or changed and a whole new set of output, to be used as an oracle, should be re-generated. This is costly, and the approach we presented avoids this problem while maintaining regression testing viable. Lastly, future research is needed towards the automatic generation of test cases based on the specification itself.

Acknowledgment

The authors gratefully acknowledge Dr. Weidong Chen's (Southern Methodist University) invaluable comments and several references kindly provided by Dr. Dan Hoffman (University of Victoria.) We also thank the anonymous referees' comments, which helped to improve this paper.

References

1. W. Bartusseky and D.L. Parnas. Using assertion about trace to write abstract specification for software modules. In N. Gehani and A.D. McGettrick, editors, *Software Specification Techniques*, pages 111–130. Addison-Wesley, 1986.
2. C. Beirle and U. Pletat. Integrating logic programming and equational specification of abstract data types. In J. Grabowski, P. Lescanne, and W. Wechler, editors, *Proceedings of the International Workshop on Algebraic and Logic Programming*, number 343 in Lecture Notes in Computer Science, pages 71–82, Gaussig, GDR, November 1988. Springer-Verlag.
3. W.F. Clocksin and C.S. Mellish. *Programming in Prolog*. Springer-Verlag, Berlin, Heidelberg, third edition, 1987.

4. K. Drosten. Translating algebraic specifications to prolog programs: A comparative approach. In J. Grabowski, P. Lescanne, and W. Wechler, editors, *Proceedings of the International Workshop on Algebraic and Logic Programming*, number 343 in Lecture Notes in Computer Science, pages 137–146, Gaussig, GDR, November 1988. Springer-Verlag.

5. D. Hoffman. A CASE study in module testing. In *Proceedings of the Conference on Software Maintenance*, pages 100–105, Los Alamitos, USA, October 1989. IEEE Computer Society Press.

6. D. Hoffman and R. Snodgrass. Trace specifications: Methodology and models. *IEEE Transactions on Software Engineering*, 14(9):1243–1252, September 1988.

7. D. Hoffman and Y. Wang. Executables prototypes of trace specifications. In *Proceedings of the Canadian Information Processing Society*, pages 176–184, Edmonton, Canada, October 1987.

8. D.A. Lamb. *Software Engineering - Planning for Change*. Prentice-Hall, Englewood Clifs, USA, 1988.

9. C.G.N. Macario and F.B. Camargo. The component production line. In *Proceedings of the 2nd International Workshop on the Brazilian Software Plant Project*, pages 13–29, Campinas, Brazil, 1990. BB/EMBRAPA/CTI.

10. T.I. Ören. Artificial intelligence and quality assurance in computer aided systems theory. In F. Pichler and R. Moreno-Diaz, editors, *Proceedings of the International Workshop on Computer Aided Systems Theory – EUROCAST'89*, number 410 in Lecture Notes in Computer Science, pages 336–344, Las Palmas, Spain, March 1989. Springer-Verlag.

11. T.I. Ören. Computer aided systems technology: Its role in advanced computerization. In F. Pichler and R. Moreno-Diaz, editors, *Proceedings of the Third International Workshop on Computer Aided Systems Theory – EUROCAST'93*, number 763 in Lecture Notes in Computer Science, pages 11–20, Las Palmas, Spain, February 1993. Springer-Verlag.

12. D.L. Parnas and Y. Wang. The trace assertion method of module interface specification. CRL Report 244, McMaster University, May 1992.

13. G.G. Schulmeyer. The move toward zero defect software. In G.G. Schulmeyer and J.I. McManus, editors, *Handbook of Software Quality Assurance*, chapter 10, pages 189–224. Van Nostrand-Reinhold, New Yourk, USA, 1992.

14. P. Strooper and D. Hoffman. Prolog testing of C modules. In V. Sraswat and K. Ueda, editors, *Proceedings of the 1991 International Symposium on Logic Programming*, pages 596–608, San Diego, USA, October 1991. MIT Press.

15. M.M. Tanik and A. Ertas. Design as a basis for unification: Systems interface engineering. In *Proceedings of the 16th Annual ASME–ETCE Conference*, pages 113–114, Houston, USA, 1992. ASME. Symposium on Computer Applications and Design Abstractions (PD - Vol.43).

16. A. Togashi and S. Noguchi. A program transformation from equational programs into logic programs. *Journal of Logic Programming*, 4(2):85–103, June 1987.

17. Y. Wang and D.L. Parnas. Simulating the behavior of software modules by trace rewriting. In *Proceedings 15th International Conference on Software Engineering*, Baltimore, USA, 1993. IEEE Computer Society Press.

Systems Theory and Systems Implementation—Case of DSS

Yasuhiko TAKAHARA   Naoki SHIBA

Department of Management and Systems Engineering
Tokyo Institute of Technology,
2-12-1 Ookayama, Meguro-ku, Tokyo 152, JAPAN

Abstract. A DSS (decision support system) is a CBIS (computer based information system) designed to support a rational decision making for a semi-structured problem. In order to make a rational decision we need models for a problem. A DSS can be distinguished from other CBIS by the fact that a model plays an essential part in it.
Our group has been engaged in the project to develop an intelligent DSS for several years. In this paper we investigate our implementation of the model management system of our DSS which is called actDSS from the system theoretic view point. We discuss how a system technology DSS is implemented on a computer technology and how a computer technology is assisted by the systems theory.

1 Introduction

A DSS is a CBIS(computer based information system) designed to support a rational decision making for a semi-structured problem. In order to make a rational decision making of a problem we have to have a model of it. A DSS can be distinguished from other CBIS by the fact that a model plays an essential part in it. Researches on a DSS is mainly concerned with model management.

On the other hand a model is a principal object of the systems theory. In this sense, although a DSS is used for business activities rather than for scientific or engineering problems, it should be one of the central topics of CAST study.

Our group has been engaged in the project to develop an intelligent DSS for several years. In this paper we investigate our implementation of the model management system of our DSS which is called actDSS from the system theoretic view point.

2 actDSS and Model Management System

In this chapter we will survey briefly actDSS and a model management system in general for the sake of the succeeding discussions[1].

A DSS is usually considered to consist of three components, dialog system, MMS(model management system) and DBMS(date base management system). The dialog system is an interface between a user and the system. MMS provides models to the user according to his problem specification, gets data from DBMS and supports him in execution of the models, that is, simulation, sensitivity analysis, risk analysis, goal-seeking analysis and others so that he can get an insight into his problem.

Our DSS is designed to achieve the following three goals:

1. Realization of an intelligent assist.
2. Realization of a structural flexibility so that actDSS can be easily modified by the user's requests.
3. Realization of the formal approach in construction of a DSS.

In order to achieve these objectives actDSS is developed on Prolog or its extension.

Fig. 1 shows the hierarchy of the implemented structure of actDSS. As the figure shows, it is constructed on a UNIX work station.

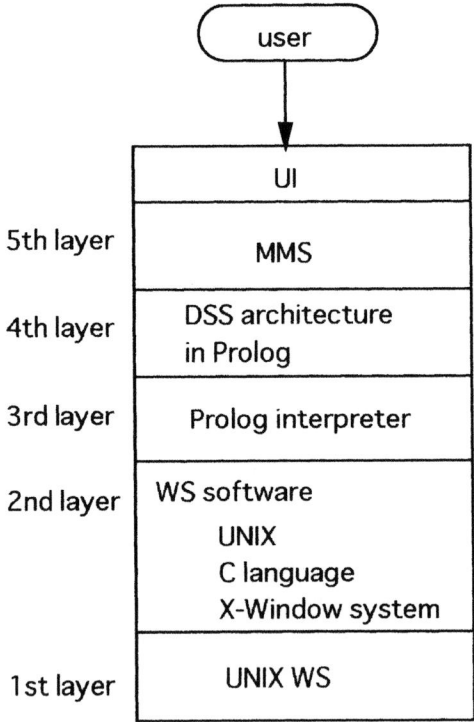

Fig. 1. Hierarchical structure

The Prolog interpreter is designed for our research purpose and developed on the UNIX environment. In particular, most of the basic predicates necessary for the system implementation are defined as subroutines of C and the Prolog program controls these predicates to realize necessary functions. The functional structure or the process structure of actDSS is expressed by the Prolog.

The MMS is specified in accordance to the architecture. The highest level is the user interface. The user supported by the interface makes a decision using models.

Fig. 2 shows the functional structure of actDSS. As the figure shows, actDSS is different from the conventional DSS in two aspects, existence of an operating system, which is called actOS, and existence of a model space.

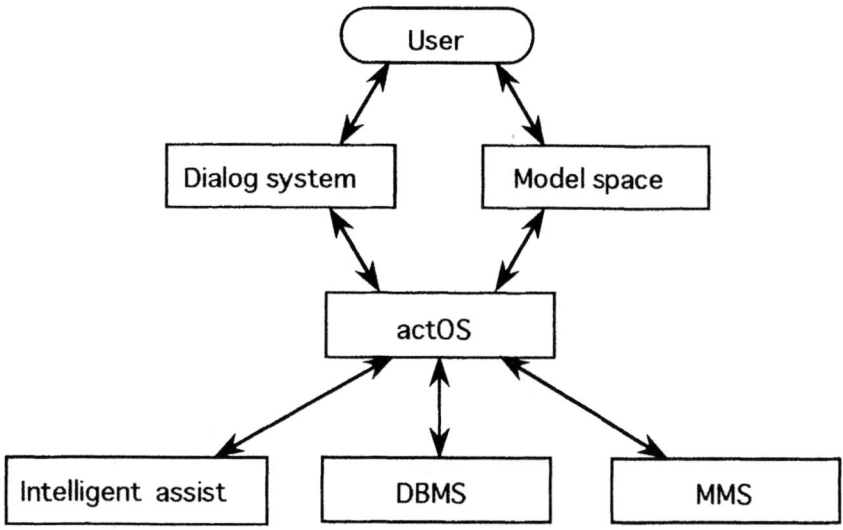

Fig. 2. Functional structure

The operating system which is written in Prolog manages the dialog system, MMS and DBMS. An intelligent behavior is emphasized for the management.

The model space is a main topic of this paper.

Fig. 3 shows the behavior process of actDSS. As the figure shows, actDSS works in the following steps:

1. The window manager defined as a predicate in actOS waits for an input from the user. User's inputs or requests are accepted through windows.
2. If the input is simple enough to be processed by the window manager, an output is given immediately by it.
3. If the window manager cannot respond to the input, which may be a complicated request such as model execution, it transfers the control to the core of actOS. This transfer is called interrupt.
4. When actOS is given the control, it processes the input, yields a response and returns the control to the window manager.

An important feature of actDSS is that actOS memorizes both the history of the user inputs and that of the system outputs as its state and processes an input depending on the state. This is the basis of the intelligent assist of actDSS.

An MMS is required to perform at least three functions:

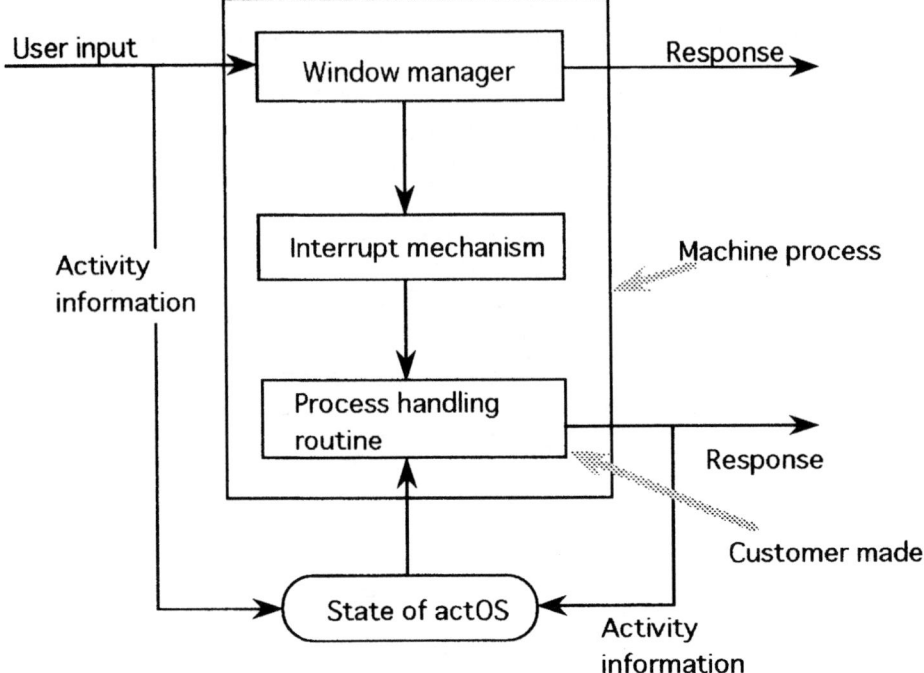

Fig. 3. Behavior process

1. Support of a model description language.
2. Support of model management.
3. Communication to and from DBMS.

A model description language is a language in which a user describes his model. Since it is assumed for DSS that the user is not a professional in computer science, the language must be easy to learn but strong enough to cover every business problem. Design of a model description language is one of the topics of the paper. Communication with DBMS means that the MMS gets necessary data from the DBMS to run a model or sends data to the DBMS to save them.

3 MMS and Systems Theory

In this chapter we will examine the relation between MMS and systems theory in two aspects, model description language and model management.

Since an MDL(model description language) is a programming language, it can be represented as a formal system, that is, its conventional representation is:

$$MDL =< N, T, S, \wp >$$

where N, T and S are sets of non-terminal symbols, terminal symbols and the initial symbols, respectively. $S \in N$ is assumed. $\wp$ is a set of production rules.

Formal languages are classified according to the types of $\wp$ and it is well known that conventional computer languages and basics of natural languages belong to a class of cfl(context free language). That is, each element of $\wp$ takes the following form:

$$X \to Y$$

where $X \in N$, $Y \in (N \cup T)^*$, $Y \neq \wedge$(null string). The MDL of actDSS is also designed as a cfl because it is simple but satisfactorily strong.

We can associate an input–output system with the MDL of the formal system in a natural way.

Let

$$P_{MDL} : T^* \to \{-1, 0, 1\}^*$$

where for $x \in T^*$ and $l(x) = n$ (length of x)

$$P_{MDL}(x)(k) = \begin{cases} 0 & \text{if } k < n-1 \\ 1 & \text{if } k = n-1 \text{ and } x \in MDL \\ -1 & \text{if } k = n-1 \text{ and } x \notin MDL \end{cases}$$

P_{MDL} is an input–output system which yields 0 until the end of the input x and yields 1 or -1 at the end depending on whether $x \in MDL$ or not.

The systems theory says that a cfl can be processed (for instance, compiled) by a pda (push down automaton)[2]. A state space representation or a computation process of the P_{MDL} is, therefore, given by a pda. Actually, the MMS of actDSS transforms a model by the MDL into a Prolog like language by a Prolog interpreter constructed as a pda[3].

In this paper an MDL is dealt with as an input–output system P_{MDL}.

It is natural that not every cfl should be desirable for an MDL of a DSS. actDSS requires its MDL to satisfy the following properties:

1. Typeless
2. Non-procedural
3. Category free

A term of an MDL is classified as a constant symbol, a variable symbol, a function symbol, an operation symbol or a predicate symbol. It often happens, however, in a DSS that the same variable symbol can take more than one types of value, depending on the situation. For example a variable can be a scalar or a vector. Therefore, it is not desirable that a user is required to declare the type of a variable. Typeless means that the user is not required to declare the type of a variable.

If a model consists of more than two sentences, the order of the sentences is critical to the meaning of the model in a usual computer language. It is clear, however, that a user can easily write or modify a model if the order of sentences is irrelevant to the meaning of the model. Non-procedural means the irrelevancy of the order of sentences in a model.

A model of actDSS is an input–output time system, whose variable is conventionally classified into a parameter, an input, an output or a state. This classification is important in practice as well as in theory. For instance, in order to execute a model we have to have data for input and state variables beforehand, which implies we should know which variable is an input or a state, or conversely, since we cannot assign a value to an output variable, the classification tells us which variable is in our direct control. It is cumbersome for a user to construct a model declaring the role of each variable. Furthermore, the role of a variable can change on the decision making process. Typically, an input may become an output or a state when the model is modified. Category free means that a user can build a model without paying attention to the roles of variable.

A model construction strategy is a very important factor for the design of an MMS. As mentioned above, since a DSS deals with a semi-structured problem, the structure of a model is not well understood at the beginning. Furthermore, usually, models of a DSS are rather complicated. Hence, it is not easy to build a total model all at once and even if we can do so, it is hard to understand the meaning of the result a complex model yields. Therefore, actDSS adopts the model integration approach for model construction[4].

Fig. 4 shows the procedure of MIA(model integration approach) of actDSS. A user is supposed to start from a analysis of his problem to determine variables and objective functions. Then, he selects a class of atomic models which can describe the relation of the conceptualized system and the objective function. An atomic model is assumed to be small and simple enough to be clearly understood about its behavior. It is also assumed that atomic models are built using domain knowledges and saved in the model base.

Then the user links and integrates hierarchically the atomic models to get a global model. Naturally he cannot build a desirable global model in a straight forward way.

The global model which is called c-model(complex model) is used to analyze the user's problem using meta models such as risk analysis model, goal-seeking model, optimization model and others.

If the analysis yields a satisfactory result, the decision making process is over. However, it is usual that the whole process is carried out in a iterative way.

Since a desirable c-model is built evolutionary, it is necessary to save intermediate stage of the process. A configuration model of Fig. 4 is the representation of an intermediate stage, which will be discussed in Chapter 4.

Let us formulate MIA. Following the systems theory[5] let a user model of a DSS be $< P, G >$ where

$$P : X \to Y$$

$$G : X \times Y \to R^n$$

P represents a process while G a goal where X, Y and R are an input set, an output set and the set of reals. R^n implies a multi-objective decision making.

Let $\{P_i : X_i \to Y_i | i \in I\}$ be the class of atomic models selected by the user. Let the desired model be

$$P_d : X_d \to Y_d$$

Let

$$X_i = M_i \times V_i \times U_i$$

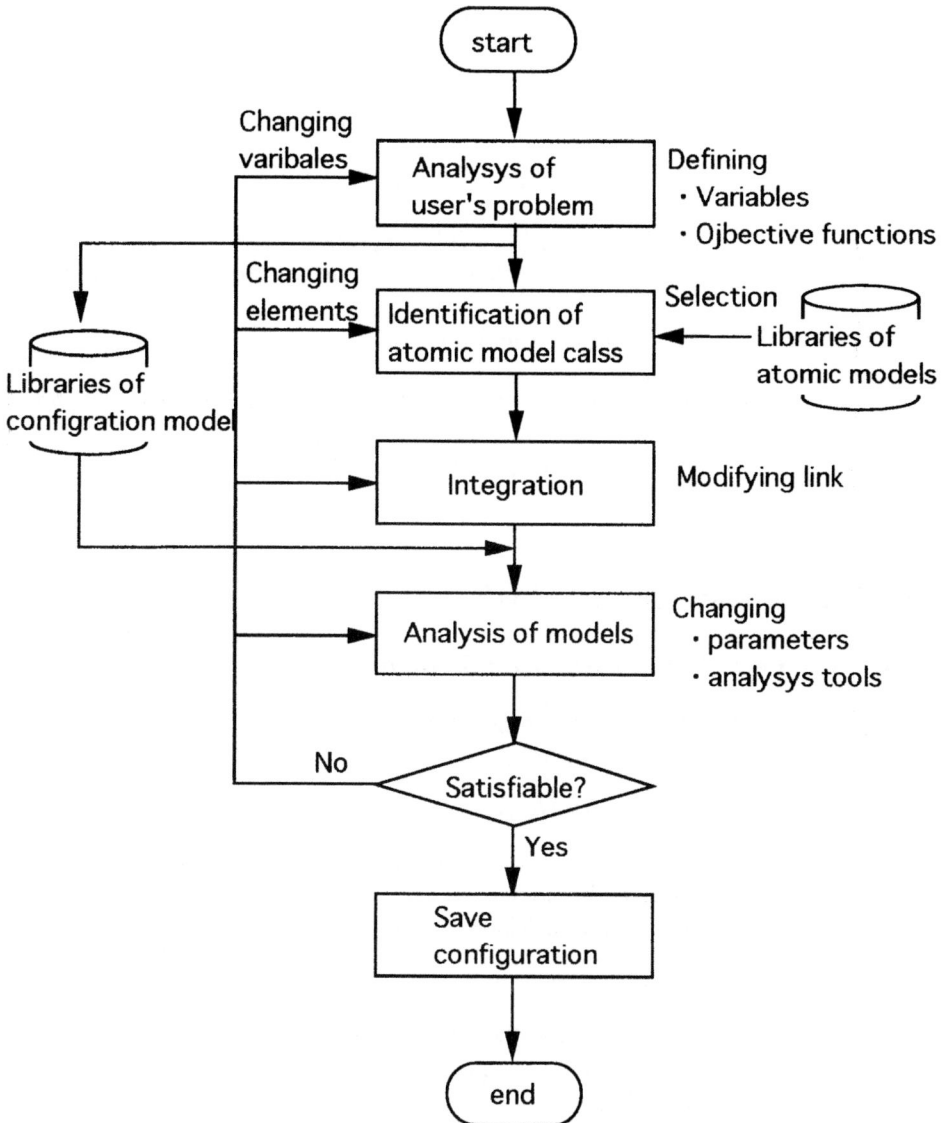

Fig. 4. MIA (Model integration approach)

where M_i, V_i and U_i represent a decision variable, a parameter and an interaction variable, respectively. How X_i is divided into M_i and U_i depends on the modeling situation.

Let

$$X = X_1 \times \cdots \times X_n$$

and

$$Y = Y_1 \times \cdots \times Y_n$$

Then, the interaction U_i is, in general, represented by a mapping K_i as:

$$K_i : X \times Y \to U_i$$

Conceptually, $\{< P_i, K_i > | i \in I\}$ determines a process

$$P^* : X \to Y$$

eliminating $\{U_i\}$ by substituting $\{K_i\}$ into $\{P_i\}$.

$K = \{K_i | i \in I\}$ and P^* are called link structure and generated system, respectively.

A desirable K should satisfy the following:

There exist onto maps $h_1 : X \to X_d$ and $h_2 : Y \to Y_d$ such that for any $x \in X$

$$h_2(P^*(x)) = P_d(h_1(x))$$

holds.

4 Implementation of MMS

4.1 A general framework of implementation

Let us start from formulation of a general framework of implementation[6]. Let an SI(system implementation) be:

$$SI = < Spec, SSR, AT, realization, evaluation >$$

where $Spec$, SSR and AT are sets of system specifications, state space representations and available technologies, respectively.

Following the formulation of Chapter 3, let $Spec$ be:

$$Spec = \{P_d : X_d \to Y_d\}$$

Since a model of actDSS is a discrete time system, its state space representation is an automaton type but is not necessary finite. Let $P : X \to Y$ be a time invariant process where

$$T = \{0, 1, 2, \ldots\}$$

$$X \subset \{x | x : T \to A\}$$

$$Y \subset \{y | y : T \to B\}$$

A and B are called input and output alphabets. Then, its automaton type state space representation is:

$$\varphi = < A, B, C, \delta, \lambda, c_0 >$$

where C is a state space, $c_0 \in C$ and

$$\delta : C \times A \to C$$

$$\lambda : C \times A \to B$$

$$P(x)(t) = \lambda(\delta(c_0, x|[0, t)), x(t)). \tag{1}$$

Then,

$$SSR = \{< A, B, C, \delta, \lambda, c_0 >\}.$$

When the relation (1) holds, we will write

$$P = Res(\varphi).$$

Since φ specifies the computation of $P(x)$ by a recursive process, φ is considered to represent the mechanism of the process P.

AT represents an environment of implementation and eventually specifies the range of realizable systems. Let

$$AT \subset SSR.$$

As shown later, AT will be specified depending on the implementation situation.
The above formulation implies

$$realization : Spec \to SSR$$

should satisfy the following conditions:

1. $realization(P_d) \in AT$
2. The following commutative diagram holds:

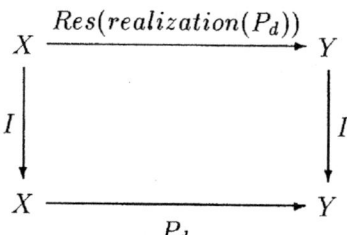

Finally, let

$$evaluation : SSR \to R^n$$

where R^n implies a multi-objective evaluation.

4.2 MDL

Let us consider the MDL of actDSS using the framework of SI mentioned above.

A specification of an MDL was given by P_{MDL} of Chapter 3. That is, $Spec$ is the set of P_{MDL}'s. As mentioned in the chapter, the computation process of P_{MDL} is given by a pda whose representation is different from $< A, B, C, \delta, \lambda, c_0 >$ but it is easily seen that if we combine the state of the stack and the state of the head to define the state space C, we can transform the pda into the automaton type representation. Then SSR is the set of pda's.

It should be noticed that a cfl needs a non-deterministic pda to be processed in general. We assume AT of MDL to be the set of deterministic pda's.

The requirements about MDL mentioned in Chapter 3 cannot be expressed in $Spec$ directly. They are meta specifications.

The requirements of typeless and non-categorical can be easily realized if we do not introduce any term to describe the type or the category into the MDL.

The requirement of procedural free is realized by:

1. The MDL is defined as a functional language as shown below.
2. The model by the MDL is then transformed into a Prolog type internal representation.
3. The internal representation is executed by the Prolog interpreter. The back-tracking function of the interpreter realizes the procedural free property.

Let us show the MDL of actDSS using two examples.

Fig. 5 illustrates a simple business model described by the MDL. As the example indicates, a model consists of the following functional expressions:

```
// s: sales
// tl: total liability
// se: stockholder's equity
// i: income
// lP0: liability payment
// fc: fixed cost
se=gal(se) + kK2 * i
tl=gal(tl) + kK1 * gal(s) - lP0
asset=tl + se
i=-fc + kK3 * gal(s) + kK4 * s
```

Fig. 5. Business model

$$y = \text{if} < boolean\ expression > \text{then } f(x_1) \text{ else } g(x_2)$$

or

$$y = f(x).$$

The right hand part of the equality symbol "=" will be called definition part.

The difference operation is expressed by "**gal**" function. Fig. 6 shows its internal representation. The Prolog interpreter saves the trajectory data of each variable and when it interprets gal(se, X1), it extracts the final value of the trajectory of **se** and assigns it to $X1$.

```
se(X0):-gal(se,X1),kK2(X2),i(X3),X0:=X1+X2*X3,!;
tl(X0):-gal(tl,X1),kK1(X2),x(X3),gal(s,X4),lP0(X5),
            X0:=X1+X2*X3*X4-X5,!;
asset(X0):-tl(X1),se(X2),X0:=X1+X2,!;
i(X0):-fc(X1),kK3(X2),x(X3),gal(s,X4),kK4(X5),s(X6),
            X0:=-X1+X2*X3*X4+X5*X4*X6,!;
```

Fig. 6. Internal representation of business model

When the evaluation of se(Yse) is done so that the value of Yse is given as **v**, the following rule is asserted at the head of the model:

$$se(v):-!.$$

This rule avoids recomputation of **se**.

Fig. 7 shows another example of a model by the MDL which simulates Turing machine. This example demonstrates the typeless property and the describing power of the MDL. As the comments show, **tape**, **pos**, **state** and **sym** represent the tape, the head position, the head state, the symbol under the head of the Turing machine, respectively. **Action** expresses the behavior of the head. If action = R or L, then the head moves right or left and if action = 0 or 1, then the symbol under the head is rewritten as 0 or 1, respectively.

The production rule of Turing machine is given by two functions rule1.a and rule2.a where

$$rule1.a : \{state\} \times \{sym\} \rightarrow \{action\}$$

$$rule2.a : \{state\} \times \{sym\} \rightarrow \{state\}$$

rule1.a and rule2.a are functions which can be easily defined on SS (spread sheet) of actDSS in a table form.

The variable **turing.mstop** specifies a stooping condition. The Prolog interpreter executes the model until the value of **turing.mstop** becomes 1.

Fig. 8 shows a final result of the execution where $2 + 1 = 3$ is demonstrated.

4.3 Implementation Structure

Since an element of SSR specifies the computation process of a model, it may well be supposed that implementation is completely given by SI. But it is not so.

```
//sym=symbol of tape under head
//pos=position of head on tape
//state=state of head
//action={R,L,0,1}
sym=project(gal(tape),gal(pos))
action=rule1.a(gal(state),sym)
tape=if action='right' or action='left' then gal(tape) else
                    replace(gal(tape),gal(pos),action)
@trjlast(tape)
pos=if action='right' then gal(pos)+1 else if action='left' then
                    gal(pos)-1 else gal(pos)
state=rule2.a(gal(state),sym)
turing.mstop=if sym=action and state=gal(state) then 1 else 0
initial(pos,1)
initial(tape,[[1,1,0,1]])
initial(sym,1)
initial(action,'right')
initial(state,'q.0')
initial(turing.mstop,0)
```

Fig. 7. Turing machine

In order to meet various requests by a user on execution another auxiliary representation is necessary, which is called IS(implementation structure). Notice, for instance, that SI specifies the computation process of the vector of variables but does not computation order of variables. These detailed information is important to realize a friendly and intelligent user environment. A concrete structure of IS depends on the execution requirements of a system. The IS of actDSS is:

$$IS = < input\ name\ list, state\ name\ list, output\ name\ list, structural\ model,$$

$$process\ model, root\ name\ list > .$$

It is called compilation to generate IS from an element of SSR in actDSS. That is,

$$compile : SSR \rightarrow \{IS\}.$$

A *process model* is the program consisting of Prolog type expressions given by transforming a model of the MDL. The other components of IS are also generated on compilation as shown below.

The compilation is done by the following steps. Suppose a model is given by

$$model = \{v_i = f_i(v_{i_1}, \ldots, v_{i_{n_i}}) | i = i, \ldots, l\}.$$

Step 1 When the *process model* is constructed, the following set of node-predicates is defined as *structural model*.

$$structural\ model = \{node(v_i, [v_{i_1}, \ldots, v_{i_{n_i}}]) | v_i = f_i(v_{i_1}, \ldots, v_{i_{n_i}}), i = 1, \ldots, l\}$$

We will write $CL_i = [v_{i_1}, \ldots, v_{i_{n_i}}]$.

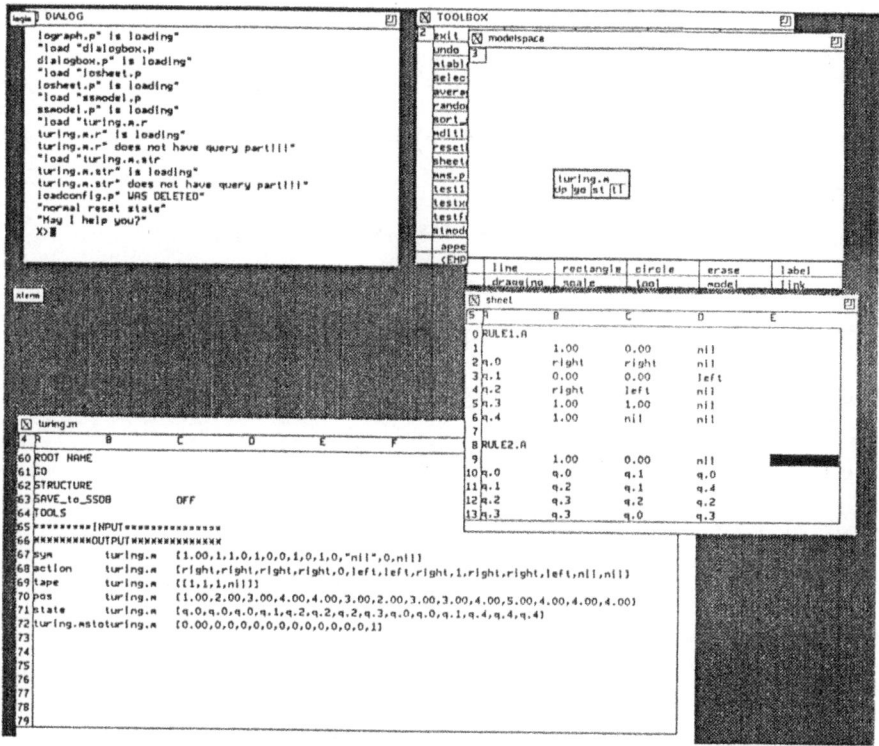

Fig. 8. Execution result of Turing machine

Step 2 Let *name list* be the set of all variables in the model. Then *root name list* is determined by the *structural model* and the *name list* as follows:

root name list =

$$\{v_i | v_i \in name\ list\ \&\ \neg(\exists node(v_k, CL_k) \in structural\ model)(v_i \in CL_k)\}$$

Step 3 Derive a tree structure representation of the *structural model* whose root nodes are give by the *root name list*.

Step 4 Let the *input name list* be as follows:

input name list = the set of leaf nodes of the tree derived in Step 3

Step 5 Let the *state name list* be as follows:

state name list =

the set of variables whose definition part contains *gal* functions.

Step 6 Let the *output name list* be as follows:

output name list = *name list* − *input name list* − *state name list*

Let us explain the above procedure by a simple example. Suppose the model is given as follows:

$$asset = tl + se$$

$$se = gal(se) + kK2 * i$$

Then

$$name\ list = [asset, tl, se, kK2, i]$$

$$structural\ model = \{node(asset, [tl, se]), node(se, [[se], kK2, i])\}$$

$$root\ name\ list = [asset]$$

The tree of the structural model is given by Fig. 9.

$$input\ name\ list = [tl, kK2, i]$$

$$state\ name\ list = [se]$$

$$output\ name\ list = [asset]$$

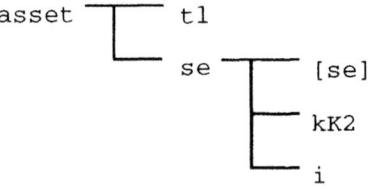

Fig. 9. Tree

[se] of the *structural model* means that se exists in the definition part as an argument of the *gal* function.

As the above procedure implies, it is a key condition for derivation of IS that the tree structure can be generated from the structural model, that is, the current MMS of actDSS can deal with only models whose structural models are loop free.

The tree representation has another important function to determine the computation sequence of variables as well as to categorize them. If we sort the variables of Fig. 9 using the depth-first strategy, we have the sequence

$$[se, asset]$$

for *state name list* ∪ *output name list* and if we compute the variables in this sequence, the Prolog interpreter can do with computation without backtracking. This fact is essential for determination of the execution sequence of atomic models of MIA which we will discuss in detail.

4.4 Model link and hierarchical structure

The MMS of actDSS adopts MIA. Although MIA is a very natural and convenient approach for a DSS, there is a big problem with it, that is, how to
implement a linking mechanism in a user friendly way. For the sake of the implementation the MMS of actDSS introduces a work place where a user can link
submodels flexibly, graphically and on real time. The work place is called model
space. Actually, the model space is used for all kinds of model management
including linking.

Fig. 10 illustrates an instance of the model space. In this instance there are
four atomic models, mesa1.m, mesa2.m, input.g and out.g. The first two models
are numeric dynamical models while the latter two models are for graphical input
and output operations. The variable pop of mesa2.m is linked to the variable
popl of mesa1.m in the way that the value of pop is always equal to that of popl.
This relation can be described symbolically as:

$$var(mesa2.m, pop) = var(mesa1.m, popl)$$

where var indicates variable. Similarly, mesa1.m and out.g are linked to input.g
and mesa2.m, respectively, as follows:

$$var(mesa1.m, rpopurl) = var(input.g, input.g).$$

$$var(out.g, out.g) = var(mesa2.m, pop)$$

Notice that a graphical input or output model $<name>$ .g has a default variable
name $<name>$ .g as shown in Fig. 10. How these links are realized will be
discussed below.

As the figure shows, two models mesa1.m and mesa2.m are integrated as
one model whose name is m.c. Furthermore, input.g, m.c and out.g are again
integrated as t.c. In this way the four models are integrated or conceptualized
in a hierarchical way. The conceptualization is realized by enclosing component
models with a rectangle and giving a name to it. The operation is carried out
by the pointing device of the system. The conceptualized model, m.c or t.c can
be treated as one atomic model, which we call c-model or complex model.

In order to realize links among variables we have to deal with variables of
models explicitly. We, therefore, display the *input name list*, the *state name list*
and the *output name list* of the IS of a model in a table form so that the user
can apply a link operation directly. The table form is realized on SS of actDSS.
The table is called IORep of the model.

Let the i-th atomic model be m_i and its process be

$$P_i : X_i \rightarrow W_i$$

where W_i consists of the state and the output while X_i is the input of m_i. We
assume that the initial state values are fixed.

Let

$$N_{x_i} = input\ name\ list\ of\ m_i$$

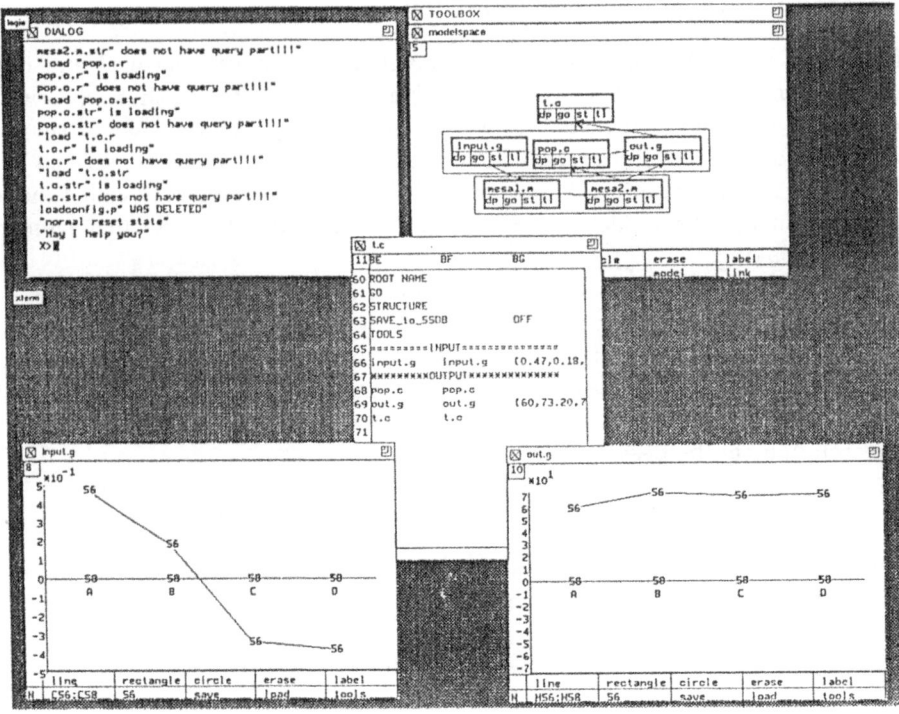

Fig. 10. Instance of model space

$$N_{w_i} = (state\ name\ list\ of\ m_i) \cup (output\ name\ list\ of\ m_i)$$

$$N_i = N_{x_i} \cup N_{w_i},$$

that is N_i is the *name list* of m_i.

Let

$$V_{ij} = \text{the set of values of the variable name } n_{ij} \in N_i$$

We will denote $V_{ij} = \varphi(n_{ij})$.

Let

$$V_i = \bigcup_j V_{ij}$$

and

$$N = \{var(m_i, n_{ij}) | i, j\}$$

Notice that $n_{ij} \in N_i$.

Let

$$F = \{f | f = \text{arithmetic formula on } N\}.$$

For instance, $f = var(mesa1.m, popl) * 100$.

Let
$$F_i^* = F \cup V_i$$
Then, the $IORep_i$ of m_i is
$$IORep_i \subset (F_i^* \times V_i)^{N_i}$$
or
$$((n_{i_1}, (f_{i_1}, v_{i_1})), \ldots, (n_{i_l}, (f_{i_l}, v_{i_l}))) \in IORep_i$$
where
$$f_{i_j} \in V_i \text{ if } n_{i_j} \in N_{w_i},$$
$$v_{ij} \in \varphi(n_{ij}) \text{ and } v_{ij} \text{ is the evaluation of } f_{ij} \text{ if } n_{i_j} \in N_{x_i} \text{ and}$$
$$(v_{i_1}, \ldots, v_{i_s}) \in X_i \Rightarrow (v_{i_s+1}, \ldots, v_{i_l}) = P_i(v_{i_1}, \ldots, v_{i_s}).$$

Links among models are specified by formulas in F. As Fig. 10 shows, the values are displayed on the left hand side of the $IORep_i$ while the formulas are hidden behind them. The formulas can be written on the $IORep_i$ by the keyboard because the $IORep_i$ is implemented on SS.

4.5 Complex model structure and execution order

When we execute an integrated model or c-model, there is a proper order of the execution sequence of component models due to their dependency relation on the link structure. In the case of Fig. 10 execution of the component models should be done in the order input.g → mesa1.m → mesa2.m → out.g. It is usually a heavy burden for a user to run an integrate model by executing each component model step by step manually in a correct order. Actually, it is not easy even to run such a simple model as that of Fig. 10 many times without making mistakes. We have to introduce some mechanism so that the system can support execution of a complex model. The MMS of actDSS supports it using the IS.

Let the dependency relations among component models be represented by node predicates as follows:

If a variable of m_i depends on a variable of m_j, let us write as
$$m_j \to m_i.$$
Then, a node predicate is given for each atomic model by
$$node(m_i, CL_i)$$
where $CL_i = \{m_j | m_j \to m_i\}$.

If a.c is a complex model and if it includes component models $m_1, \ldots, m_l$, then
$$node(a.c, [m_1, \ldots, m_l]).$$
The complex model structure is, then,

complex model structure of a.c =
$$\{node(a.c, [m_1, \ldots, m_l])\} \cup \{node(m_i, CL_i) | i = 1, \ldots, l\}.$$

In the case of Fig. 10

complex model structure of $m.c =$

$$\{node(mesa.2, [mesa1.m]), node(m.c, [mesa1.m, mesa2.m])\}.$$

When we have a *complex model structure* specified by node predicates, we can apply the categorization procedure of variables mentioned in Section 4.3 to the component models. Then, the tree representation of the *complex model structure* yields a correct execution order among the components as mentioned in the section. It should be clear that we have always a correct execution order regardless in which sequence the components and links are defined.

4.6 Task structure

When a user makes a decision using a DSS, he does not do so working with only one model or one complex model but with an environment consisting of graphs, data, meta models for analysis and others. Such an environment is called task structure.

The concept of task structure is essential for MIA which is carried out by evolutionary process. A user cannot build an ideal environment all at once. He must be allowed to follow a try and error process. In order to make a try and error process effective we have a mechanism that can save an intermediate stage from which the user can start for building a next higher level environment. A task structure can be saved and regenerated if the system memorizes the whole input sequence given by the user. But this approach is not feasible. Since the inputs to a DSS are complicated, saving of them consumes a huge memory space. Furthermore, even if we succeed in saving them, it may take us a unbearable long time to reach the final environment if we follow the sequence faithfully. Notice that a try and error process usually requires the regeneration operation many times. A practical solution to the problem is to save a specific aspect of the state of the system in accordance with saving purpose. This method, certainly, cannot reproduce the complete state but if an appropriate aspect of the state is chosen to be saved, it gives us a satisfactory solution.

The MMS of actDSS saves as the state of the system the task structure shown in Fig. 11, that is,

task structure $=$

$$< window, model\ process, meta\ model, state\ date, control\ flag > .$$

window is a structure which specifies what kind of windows are opened in the model space. It consists of SS type windows of *IORep*, model type window for the model space, graph type windows for graph and others and each component is composed of window structure(window id, type, dimension, position and others) and the data structure in the window.

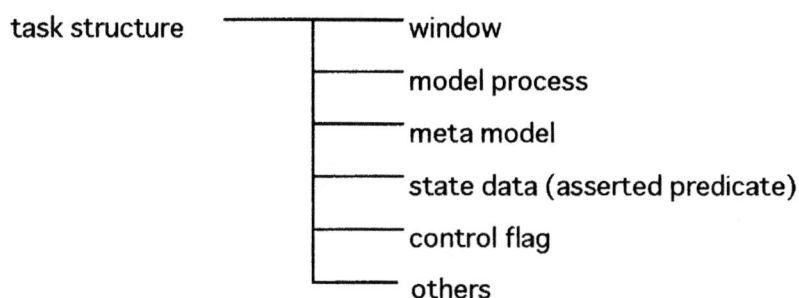

Fig. 11. Task structure

model process is a structure to specify the model process loaded on the model space.

meta model represents the meta models in the model space introduced to analyze models.

state date is a simplified version of an input sequence of the user. It consists of the asserted predicates by him.

control flag represents control status manipulated by the user input.

This task structure concept and its saving mechanism makes the MIA process shown in Fig. 4 feasible.

5 Object Oriented Approach and MMS

The MMS of actDSS is not designed by the object oriented approach to which many system scientists as well as computer scientists show keen interests as a new system design approach. But the present system is quite compatible with the philosophy of the approach[7]. We can reformulate the MMS in the approach.

Fig. 12 illustrates the reformulation. The objects of the approach are boxes in the model space. As mentioned in Chapter 4, the object is specified by the class

$$class =< input\ name\ list, state\ name\ list, output\ name\ list, process\ model,$$

$$structure\ model, message, message\ execution\ routine > .$$

As Fig. 10 shows, "go", "structure", and "save" of *IO Rep* are messages defined to the object. "go" implies execution of the model, "structure" display of the model structure in the tree form and "save" request to save current values of variables into the DB. The execution routines for these messages exist as methods in the class.

The relations among objects are given by the linking relations.

We can extend these correspondences to the concepts of abstraction, encapsulation, inheritance, abstract class and others. This compatibility may explain why the present implementation of the MMS is easy to be understood or conversely implies that the object oriented approach is a natural one.

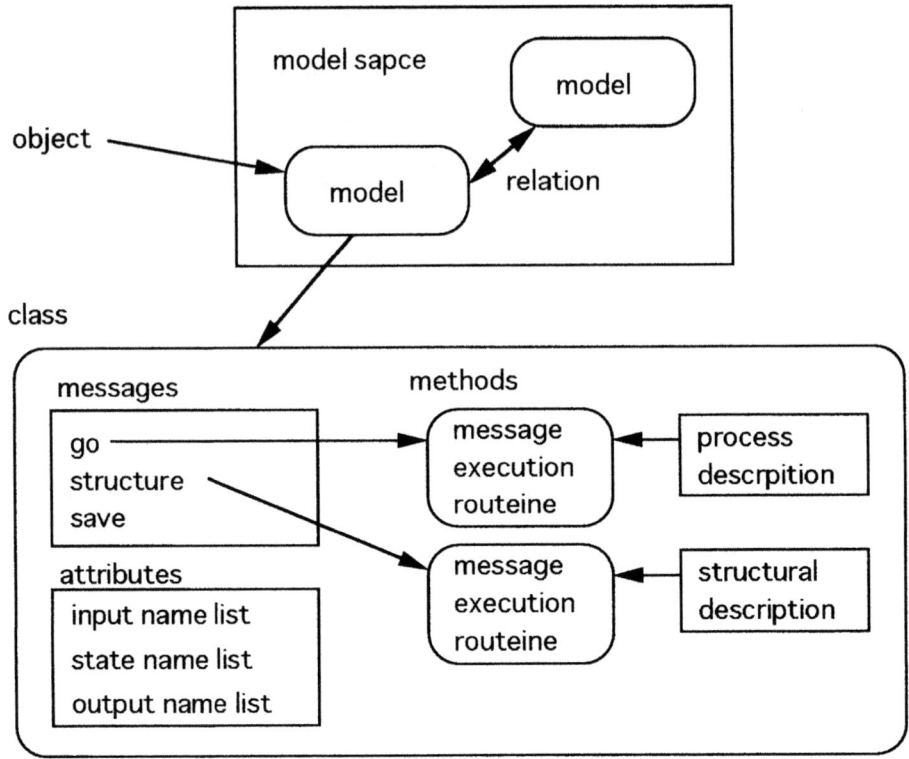

Fig. 12. Object oriented approach

6 Conclusion

In this paper we discussed how a system technology DSS is implemented on a computer technology. And at the same time this paper implies how a computer technology is assisted by the systems theory.

References

1. Takahara,Y., Iijima,J., Shiba,N., "A Model Management System and its Implementation", Systems Science, Vol.19, No.4, pp.17-33, 1993
2. Hopcroft,J.E., Ullman,J.D., *Introduction to Automata Theory, Languages, and Computation*, Reading, MA, Addison-Wesley, 1979
3. Takahara,Y., Iijima,J., Shiba,N., "DSS as a System", in preparation to Int. J. General Systems.
4. Dolk,D.R. and J.E.Kottemann,"Model Integration and Theory of Models",Decision Support Systems,9(1),51-63.

5. M. D. Mesarovic, Y. Takahara, *Abstract Systems Theory*, Lecture Notes in Control and Information Sciences, Springer-Verlag, Berlin, 1989.
6. Wymore,A.W, *Model-based systems engineering : an introduction to the mathematical theory of discrete systems and to the tricotyledon theory of system design*, CRC Press, Boca Raton, 1993
7. Halladay,S., Wiebel,M., "Object-Oriented Software Engineering", R & D Publication,Inc., Lawrence, Kansas, 1993

CASCADE:
A Computer-Aided Tool for Low Energy High-Performance Multi-DSP VLSI System Design

Gerard K. Yeh, Jim B. Burr, Kallol K. Bagchi and Allen M. Peterson

Space, Telecommunications, and Radioscience Laboratory
Department of Electrical Engineering
Stanford University
Stanford, CA 94305-4055 USA
gerard@nova.Stanford.EDU

Abstract: The design of an energy efficient digital signal processing system benefits from trade-offs of system complexity and performance during the algorithmic design and architectural design. CAST methods are useful during these early phases of design exploration. The traditional CAD tools are used in the final implementation phases. CASCADE is a design environment consisting of method banks of CAST and CAD tools. The system modeling and simulation method called OPERAS is described in context with the CASCADE.

1 Introduction

A CAST *(Computer Aided System Technology)* tool is a software which implements model-based activities in which a system theoretic modeling formalism is used [3]. The *system*, as described here, refers to sequential and parallel processors for digital signal processing *(DSP)* and digital image processing *(DIP)* applications [4, 5, 6]. Low power VLSI system design is an emerging area of electronic engineering. The limited power budget for systems such as a small remote sensing space probe or a portable video camera place a constraint on the available power for signal processing. These applications can benefit from CAST by exploring algorithms and architectures for best energy and performance trade-offs.

Computer aided design *(CAD)* systems, on the other hand, support the design and implementation of processing units with the help of cell libraries and logic synthesis tools. This paper's contribution is based upon deriving elements of a computer aided tool-set called CASCADE *(CAST and CAD Environment)*. This environment for DSP system design is a combination of CAST tools based upon mathematical transformations and CAD tools based upon integrated circuit design. CASCADE embodies the interactions and interrelations of CAST and CAD tools. As DSP systems increase in complexity, the role of combined CAST and CAD tools becomes more important in system modeling, design and implementation.

Software implementations of system theory based concepts are common among many engineering disciplines. These CAST/CAD systems have a collection of interactive methods [4] for modeling, synthesis and analysis. Both analysis and synthesis methods are included in such tools as aids to the iterative design process.

Trends for the integrated signal processing system indicate that single chip or multi-chip module (MCM) implementation of multi-DSP's will be important in the future. Applications such as video conferencing, multi-media, document processing, computer graphics and SAR, use large-scale parallel algorithms which are highly compute-intensive in nature [10]. Scalable multicomputer systems that can be integrated in one or several VLSI chips are attractive candidates for large-scale energy efficient DSP implementations. Early in the DSP system design, the detailed algorithmic design is driven by hardware complexity and system theory with the aid of software tools. For example, the theory of discrete-time signals and systems is closely applied to the system design and realization. Important issues such as numerical stability, quantization and signal sampling affect the system performance and implementation efficiency. Also, transformation of a algorithm can reduce computation and power.

CASCADE uses C, C++ and Unix as the programming environment. Tools from industry and academia including a new system simulation tool are integrated under the umbrella of object-oriented design from algorithm specifications to VLSI implementations. The system simulation tool is called OPERAS *(Object-Oriented Power Estimating Restructurable Architecture Simulator)*. In section 2, the design stages and relevant DSP system design tools are identified. Object-oriented system modeling and simulation tools and techniques are described in section 3. In section 4, signal processing system designs using different method banks are illustrated from the algorithmic development to architectural implementation.

2 Design Stages

Researchers have previously outlined the essential characteristics of CAD and CAST tools [1, 2, 26]. The interactivity is considered essential to the design of such tools. Object-oriented (OO) and knowledge-based techniques are also used. The CASCADE tool set is based on OO-principles and covers the design spectrum from algorithm to system implementation. The final circuit and physical design phases are conducted using proven CAD tools.

The various stages in the design of such systems are depicted in Figure 1. Tools are needed in each stage to support the design process and some examples of these tools are shown including tools in the public domain. A designer starts the design process by selecting a basic algorithm, and this algorithm is transformed into different forms for various types of implementation. Architectural mapping refers to the process of mapping an algorithm to the underlying system architecture. In the simple case of standard DSP processors, this step represents the code compilation process. Once the system (hardware and software) is specified, one needs to model and simulate the system architecture. When a suitable system solution is found, one can use many of the existing CAD tools to design the integrated circuits. The algorithmic and architectural design stages are identified as within the domain of the CAST. Tools spanning the algorithmic and architectural mapping exist within the CAST domain. OPERAS is a tool that is used to rigorously evaluate each system architecture in terms of area, power and performance using a VLSI design knowledge database.

In the present context, many of the existing tools either do not address the entire system design flow, or there are not adequate for handling problems at all levels. The tool set contains elements that can be refined to form a global interactive method bank. The strength of the tool set lies in providing various method banks and a platform for comparing solutions obtained from the method banks. This scheme allows flexibility in the sense that one can iterate within the earlier stages of design and make further design trade-offs.

2.1 Algorithm Transformation and Architectural Mapping

The design from an application description to an efficient realization requires several steps and tools at each step. This includes parallel algorithm design, conversion to a flow-graph such as a synchronous data-flow graph (SDF), mathematical transformations, data gathering and analysis. Mapping includes code generation, operation scheduling on a systolic array, shared memory or distributed memory processors. The design process proceeds in a hierarchical manner from the algorithm specification to implementation.

Algorithm transformations and mappings often require considerations of scheduling heuristics. Algorithms are represented by flow-graphs such as a dependency graph *(DG)*, and the flow-graphs are ideal in depicting the data dependencies. Details such as interprocess communication (IPC) and architectural characteristics can be included in the flow graph to arrive at multiprocessor solutions [12, 13]. Several methods have been developed that transform flow-graphs from the algorithmic descriptions to forms suitable for multi-DSP [20] or array [11] implementations. The area, energy, and performance estimators that can yield realistic estimates are useful to evaluate these implementation alternatives. These practical estimation models drive the selection process of the parallel architecture.

Ptolemy is an example of an object-oriented work environment for DSP algorithm and system design. Several systolic array tools such as Diastol exist to transform an algorithm in SDF/DDF graph notations to find an efficient schedule of computation [20, 11].

2.2 Architecture Simulation and Evaluation

A combined system simulation of both DSP processors and interconnection networks is needed for evaluation and verification. A DSP architectural simulator provides experimental facilities for design of data-flow, instruction cache, data caches, execution units, instruction set and others [10]. This simulator in combination with the interconnection network simulator realizes powerful modeling capabilities. A network simulator simulates a large network structure and provides a flexible platform for experimenting with network topologies such as mesh, array, hypercube and shared memory.

412

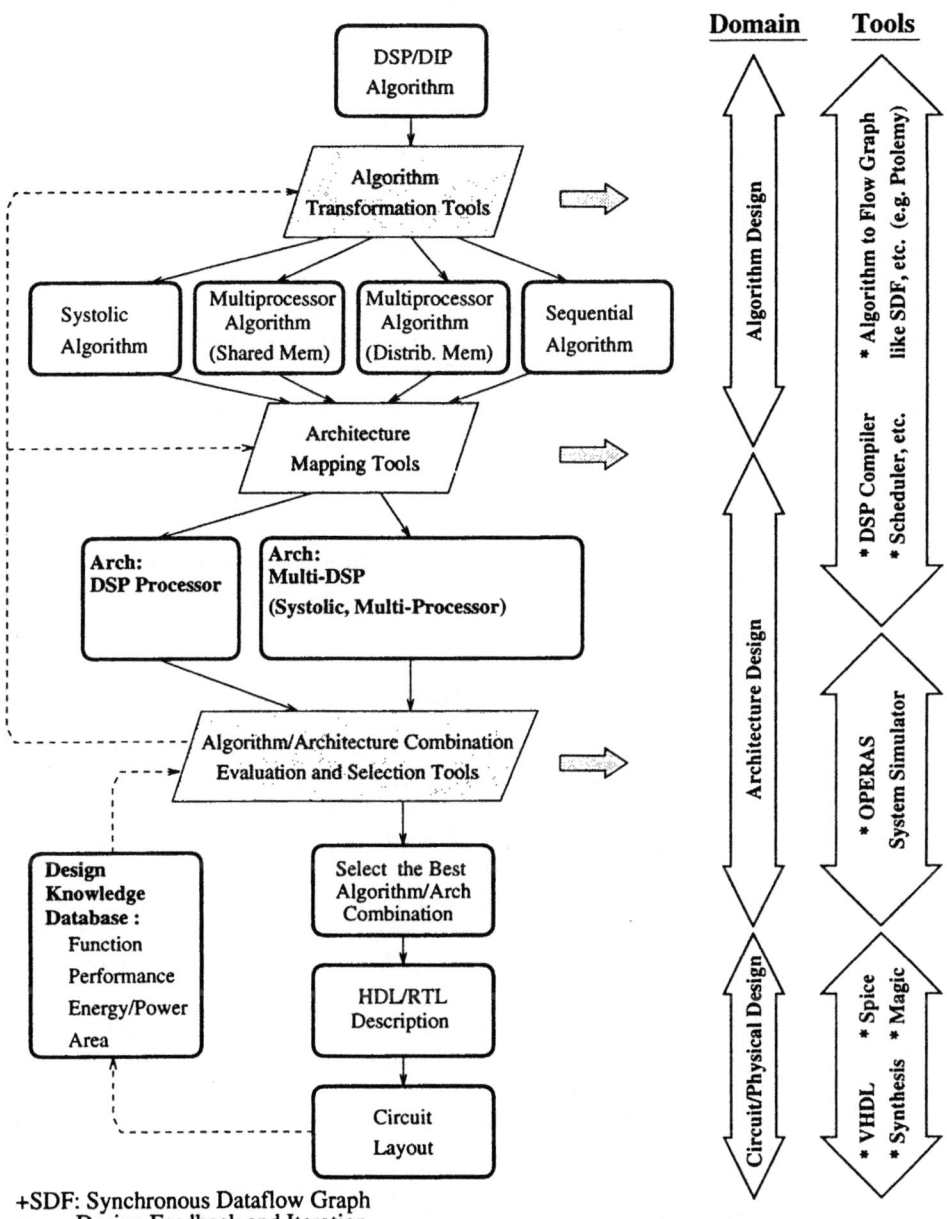

+SDF: Synchronous Dataflow Graph
◄- - - Design Feedback and Iteration

Figure 1: The Overall Design Scheme of CASCADE

OPERAS is an architectural simulation tool devoted to the system design as discussed above. Architectures are modeled in a hierarchical manner in OPERAS. For example, any DSP processor can be simulated first at the PMS level. The total instruction counts, memory and I/O references are measured and time estimates are obtained. In the next stage, actual instructions are simulated, and execution times of each instruction are measured in terms of clock cycles. Register-transfer level (*RTL*) and circuit level simulations verify the details of chip design. OPERAS simulates components at various levels of hierarchy and combines these components to model different solutions. This object-oriented and heterogeneous simulation process allows a trade-off between accuracy and speed of the simulation. For example, a DSP array processor can be integrated with models of other components such as memory and I/O to realize a DSP system. A particular processor element within the processor array is modeled in detailed RTL model. Other instances of the processor elements in the processor array is modeled by a high-level behavioral model.

For simulating the whole parallel architecture, the processor simulator and the network simulator are combined. The network simulator provides detailed performance results such as bandwidth and latency within the network. The features of different system architectures and network topologies can be measured and compared at this stage.

2.3   Power and Energy Modeling

Traditionally, the optimal processing speed and the implementation cost of a system have been the main goals for system design. Recently, power consumption has become an equally important concern for some systems. To achieve low power implementation, a key approach is to minimize energy per operation or per sample. A holistic minimization approach to power is needed because energy consumption is ubiquitously distributed throughout the system in the forms of storage, communication and computation [19]. In addition to device and circuit technologies, circuit packaging and system interconnect techniques have an influence on the overall system power. The amount of each energy type is also strongly determined by the algorithm, architecture and the interactions between the two. For example, many signal processing systems' design goals can be summed up as minimizing energy consumption while satisfying the system performance requirements. Therefore, energy efficiency becomes one of the design parameters.

For example, an image compression system must produce decoded images with acceptable fidelity while minimizing the energy consumption. To achieve such a goal, a combined investigation of algorithm and architecture solutions is needed. One trade-off in the low power design of a signal processing system occurs between signal accuracy and module power/area. For example, by reducing the signal accuracy, a filter implementation is smaller and consumes less power, but the corresponding quantization noise affects the system fidelity. It has been shown in reference [15], for example, that different implementations of a digital filter have significantly different sensitivities to quantization error. In particular, the sensitivity of the roots of a polynomial to the accuracy of its coefficients increases with the order of the polynomial. Other discrete-time digital implementation issues such as dynamic range and scaling also have effects such as numerical overflow and limit cycle oscillation [14]. The energy driven analysis and design of the whole system, consisting of algorithm, architecture, circuits and devices, needs an accurate energy estimating system simulator. Such a simulator can provide functional simulation of the effects of quantization and limit cycles. Energy, performance and silicon area are estimated to allow for trade-off and exploration of whole design space.

Finally, a performance matrix including power parameters needs to be developed for rigorous comparison of the VLSI-implementable parallel solutions. An interactive platform which can aid in comparing and selecting the right solution from the design space has been proposed [12, 13]. Each solution can be refined with the help of the corresponding interactive method. Several solutions derived from different methods can be compared against each other, and the trade-offs are made. These kinds of method banks are unique and useful for DSP system designers.

3 Object-Oriented System Modeling

An object-oriented approach to CAST and CAD design was proposed [7, 8], and has many advantages [22]. Object-oriented programming languages offer many facilities such as encapsulation, modularity, flexibility, code re-use, portability, maintainability, hierarchical and incremental modeling [8, 9]. Re-use of code, object inheritance mechanisms and object code linking in particular, have been shown to be desirable features [7]. OPERAS, an object-oriented system architectural simulator, is discussed next.

In OPERAS, the user describes the system design by defining a group of modules. Each module has the interface specification, functional description, performance information and energy estimation function. Furthermore, each module can also contain instantiations of other modules interconnected together. The system is represented as a module at the root level of the hierarchical tree. The root module definition consists of other components called cells, and each cell is just an instantiation of a module definition. Each component cell, in turn, consists of still other cells. The relation of instantiation is a descendant relationship in the hierarchical tree structure. The modules' definitions are taken from the user input file or from existing libraries of module definitions. The definitions of the library modules are stored in library files to be linked via the linker. The interconnections of modules and the instantiations in a hierarchical manner allow one to describe a complex system [8].

3.1 OPERAS Based System Modeling

Using the OPERAS based system modeling, the discrete system theory based concepts can be related to the performance of the actual system implementation. performance. The system theoretic transformations can also be evaluated and compared. Specifically, criteria such as power, area, latency and throughput can be rigorously compared between implementations. Subjective criteria such as acceptability of processed image may also be evaluated.

In OPERAS, the user works with two fundamental object classes: *Net* and *Module*. Through these two object classes, the user describes the whole system in a hierarchical and modular manner. The detailed behavior of a module is specified by the user using C++, and it is encapsulated within the module. Net objects are used to interconnect the modules. The event-driven simulation is conducted through the system of interconnected modules and nets.

Each module represents an implementation of a system component. For example, in a DIP system, a module may represent a 2-D image filtering operation implemented on a DSP processor, and another module describes a systolic array implementation. The behavioral descriptions of the filter module are derived from the output of the upstream tools as shown in Figure 1. A user can iterate with different algorithmic and architectural combinations as well as experiments with various hardware components to find the ideal trade-off of performance, area and power.

By placing the existing design in a library, the user can build a library of modules embodying the various system theoretic concepts. In OPERAS, new design is inherited using the OO mechanism from an existing module or a module within the module library. From the original design, only aspects that are different in the inherited design are changed. For example, a new adder can be inherited from an old adder design. The energy estimation function may change, while the rest of adder remains unchanged. Fast prototyping of system design is possible by re-using existing modules.

3.2 OPERAS Implementation

The user describes the system design by defining a group of modules using the OPERAS input language *(OSDL)*. The input description is preprocessed by the OSDL preprocessor which automatically generates the C++ code. The C++ statements are compiled to generate the module object code. The object linker is invoked to link the object codes with the module libraries and the OPERAS simulation library. With the availability of dynamic linking facility, one can have an interactive environment where a user can incorporate new design without recompilation of the simulator.

The user can direct the simulation by providing the test vectors and initial contents to the memory modules. The information is stored on the host machine file system. When the simulator starts to run, the information can be read by the internal memory module. Similarly, the input test vector is used during the simulation to test the design. The simulation results can be displayed on a waveform display program, and the processed and intermediate data can be stored in files for other analyses. Reference [23] contains detailed description of the OPERAS for interested readers.

4 Method Banks with Examples

In this section, the sequential, systolic and multi-processor method banks will be described using several examples. A simple finite impulse response *(FIR)* filter suitable for sequential single-chip DSP is discussed first. Secondly, motion estimation for real-time video compression application will be explained to illustrate the systolic and multi-processor method banks.

In the video image compression applications, image correlation is often used to take advantage of temporal redundancy of video sequences. The image is broken into small blocks, and the previous or next image frames are searched for a matching block. Compression is achieved when a good match is found, and the motion vector of the image block is sent instead of the entire block. The large image size and large processing requirement make parallel and systolic system implementations attractive.

Using the block-based mean absolute difference *(MAD)* as the error function has been found to produce acceptable results without using the high power and large area multiplication based correlation function. The MAD error function is shown in equation (1). Along with equation (2) and equation (3), these equations define the algorithm known as the block-based motion estimation of video image sequences [16, 17]. M and N are the image block sizes, and u is the minimum error corresponding to the best matched block in the search area, and v is the corresponding motion vector.

$$e(i, j) = \sum_{m=1}^{M} \sum_{n=1}^{N} |v(m, n) - u(m + i, n + j)| \qquad (1)$$

$$u = \min_{(i,j)}\{e(i, j)\} \qquad (2)$$

$$v = (m, n)|_u \qquad (3)$$

In section 4.2, the canonical design procedure for array processors, as proposed by Kung [11] will be introduced. The procedure will be applied to the motion estimation algorithm to illustrate the elements of the systolic method bank. In particular, the contributions of the OPERAS based system modeling, simulation and evaluation are demonstrated.

4.1 FIR Filter Design Using Sequential Method Bank

Different VLSI components have been simulated and stored in the OPERAS design library. Since the simulation can be conducted at different levels of detail, a procedure is developed to successively model the system from high level algorithm to detailed micro-architecture using models of increasing details.

A 256-taps FIR filter running on TMS320C25 is considered as an example. The architecture was simulated with OPERAS as discussed above. At the highest level, the initial estimate of the performance of the system is obtained by counting the number of cycles needed to compute one output sample of the filter, and then multiplying the cycle count with the cycle time from the data book [21]. At the next level of modeling, details of individual instructions from the TMS320C25 data book are used as input to the instruction level simulation. Finally, the detailed block diagram of the data-path is modeled using OPERAS. In this simulation, the detailed model of the data-path is built using the VLSI modules such as multiply-accumulators (*MAC*), adders, shifters and latches. The FIR program is interpreted and then simulated on the model of the data-path. The detailed energy estimate of the system is formed by adding the energy of each of the activated modules (e.g. adder, MAC, etc.) at each cycle. Finally, the VLSI area estimates of each module are generated based upon the actual VLSI design within our research group. For example, when running the 256-taps FIR program, the TMS320C25 data-path implemented in $1.8\mu m$ technology and operated at $5V$ have an estimated area of $5.688mm^2$ and an estimated power of $88.8mW$. By varying the supply voltage and transistor feature size, one can relate these technological parameters to the overall system performance.

4.2   Method Bank Using Systolic Solutions

In this section, the use of a method bank is illustrated with the help of the motion estimation algorithm discussed in the beginning of Section 4. Figure 2 shows a step-by-step methodology for the design of an array processor system. The steps two, three and four of this methodology are based on Kung's canonic mapping methodology [11].

Low power design is strongly influenced by the selection of the algorithm. As mentioned above, when the multiplication within the correlation function is substituted with the mean absolute difference, the area, power and latency of the system is reduced because adders are used instead of the larger multipliers. Although the MAD based search does not produce as high image quality as the traditional multiplication based search, in view of the overall system requirements, the trade-off for simpler MAD search is chosen.

The steps two, three and four have been described in detail in the book [11]. In each step of the array processor design, several possible system designs are generated. For example, given an algorithm several dependency graphs *(DG)* can be generated, and each DG can be projected into several signal flow graphs *(SFG)*. Furthermore, each SFG implies a different specification for the processing elements. The systolic array design space consists of the various design alternatives. In this approach, the cost function is rigorously defined in terms of power, area and latency. The designer can make other trade-offs based on system issues such as fault tolerance. Please note that other mapping scheme such as heterogeneous DG are not discussed here. Interested readers can refer to the book by Kung [11].

In step five, the system designer models and simulates the system using the OPERAS whereby one can conduct experiments such as number format versus output signal to noise ratio utilizing realistic input data. Compiled simulator from C++ code runs rapidly and efficiently. Furthermore, since the models representing VLSI circuit models are used, the level of modeling can be detailed. The detailed delay modeling and functional simulation will provide the designer with the ability

1. Algorithmic Design

Suitable algorithms identified.
The system designer expresses the algorithms in terms of regular indexed expressions.
Some classes of algorithm can only be represented as Heterogeneous DG's. (see [11])

2. DG Generation

Several Dependency Graphs (DG) can be generated for each algorithm.
DG has a strong effect on the final solution, so iteration in DG design may be necessary.

3. SFG Design

From each of the DG's, several Signal Flow Graph (SFG)
can be generated based on different projections.
I/O and memory bandwidth requirements are established for the system.

4. Array Processor Design

The SFG from the 3rd stage is used to specified the physical design
of the Parallel Array Processors.
The datapath structure is specified and pipelined to meet the computational requirement.
The control structure (e.g. SIMD, systolic, pipeline, etc.) is specified.

5. OPERAS based System Modeling and Simulation

Model each specified array processor system solutions using the new or existing components.
Verification of each architecture and algorithm solutions are conducted through simulation.

6. High Level Estimates of System Cost Functions

Area, Power and Performance are estimated based on realistic input data.
Designer may optimized design and explore design space by changing different components.

Figure 2: A Methodology for Array Processor System Design

to meet the requirements of the system. In step six, high level estimates of the system area and power are generated, and they can combine with the latency to form a system cost function. The appropriate system cost function is determined by the designer, weighted by the power, area and latency. Literatures exist to discuss the issue of VLSI system cost functions [24, 25].

Figure 4.2 shows one pass through the CAST design flow. The basic sequential algorithm of the motion estimation is shown in Figure 4.2a, and a sequential implementation is shown in Figure 4.2b as a C program. This program can be compiled by a DSP C compiler to generate an executable code for a standard single chip DSP processor. Other tools such as Ptolemy contains a compiler that can compile an algorithmic description in the synchronous data-flow (SDF) domain into executable code for a programmable DSP [20]. Given a DSP executable code and an OPERAS based model of the DSP chip, one can simulate the performance and the energy consumption of the system solution. Since the video compression system has a real time requirement, the sequential solution is not considered further.

For an array implementation of the motion estimation algorithm, Kung's method is used, and one scaled down solution is chosen from existing work [18] to illuminate the design process with CASCADE. The algorithm is expressed as a DG in Figure 4.2c, and other possible DG's can lead to very different solutions. The DG diagram shows a block size (N) of three with $p = 2$ and $q = 2$. The dotted line in DG is the time scheduling of the data flow, and the data enter each of the nodes at the specified time following the direction of arrow. In other words, the dotted line represents a wave front of data flowing into the system. Each node specifies the functionality that will be performed: AD node performs absolute difference, and ADD node is an adder, and the VEC node outputs the motion vector when a smaller average absolute difference is found.

Figure 4.2d shows a SFG mapping of the DG, and the SFG is generated by projecting the DG along the m dimension. Shift registers are required to feed the processor elements according to the time schedule in the DG. The overall system implementation will not only include the processing elements (e.g. AD, ACC and VEC) shown in SFG, but the shift registers and controls to feed the systolic array are also required for the complete system.

Given the SFG diagram, the processor functionalities are defined and mapped to hardware. The model of the systolic portion in Figure 4.2d is implemented using the OPERAS input description [23] as shown in Figure 4.2e. The module defines input and output interfaces and instantiates each of the processing elements. Finally, the module itself will model the function of the VEC element by looking for a new minimum MAD value. When a new minimum is found, a signal is sent to the system controller. With this module definition and with realistic modeling of the VLSI cells, the simulation can be conducted to verify the system functionality and performance. A typical simulation waveform is shown in Figure 4.2f. Based upon the simulation history and the VLSI component model, a table of area and power estimate is generated. At this stage, one can conduct experiments such as reducing the internal fixed point number representations to reduce area and power. Rigorous comparisons with other systolic solutions can also be done.

4.3   Method Bank Using Multiprocessor Solutions

For multiprocessor solutions, another method bank needs to be developed. For example, the key elements of shared memory multiprocessor solutions are: [13, 20]

- the software needed to detect and decompose a given algorithm for mapping to a multi-processor. The algorithm is transformed into a data-flow graph. The nature of the graph may vary depending on the grainuarity of decomposition whether it is coarse or fine grained.

- the scheduler and clustering softwares are used to map the data-flow graph onto a multiprocessor architecture, taking into consideration factors such as node processing time, host architecture, interprocessor communication models and precedence constraints among the nodes of the graph.

Comparable method banks for distributed memory systems can also be designed and implemented.

420

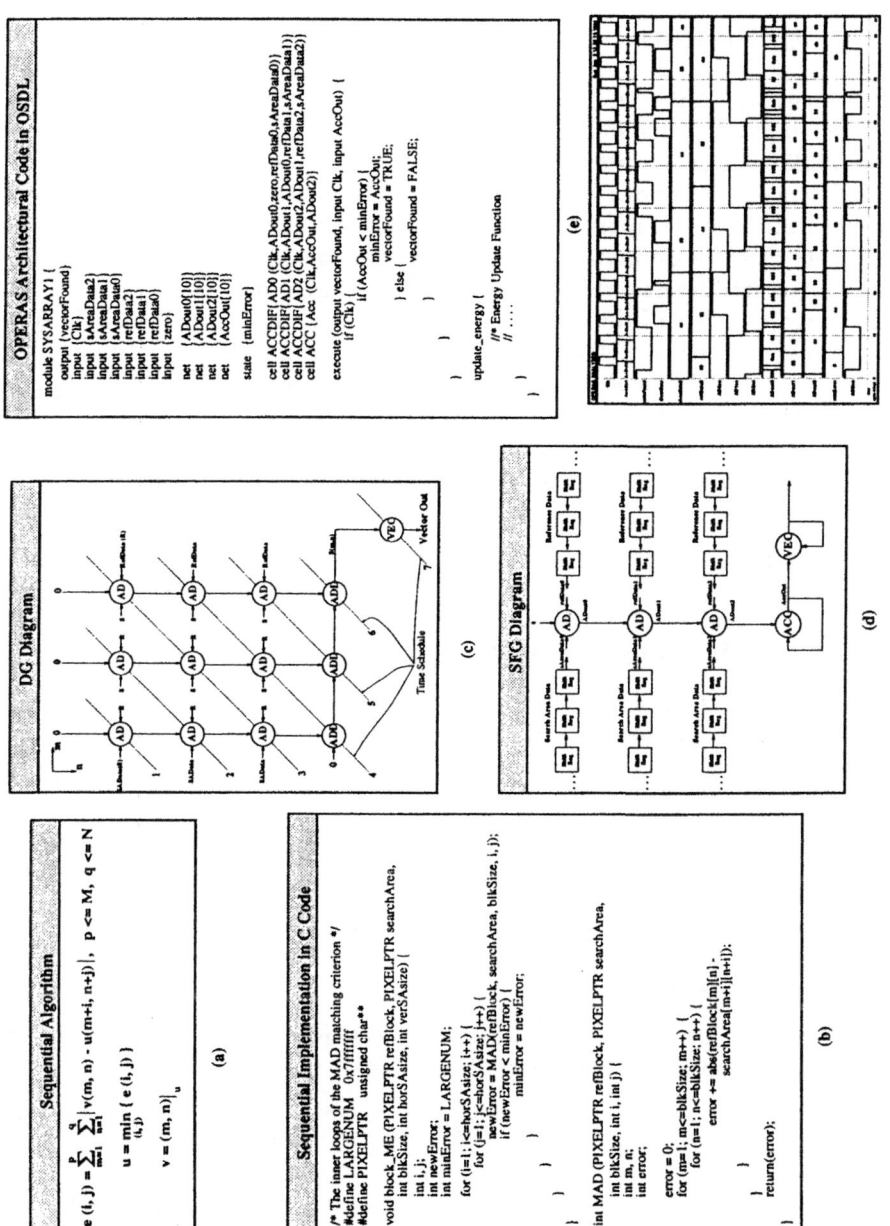

Figure 3: An Motion Estimation Design Example

5 Conclusions

The elements of CASCADE, a CAST-CAD tool set for the design of energy efficient VLSI signal processing systems is presented. The system design from the algorithm to the VLSI solution proceeds hierarchically through several steps using the existing method banks. Iterations of the design are conducted to find the best system implementation based upon performance, area and power estimates. The early availability of these estimates allows a designer to make trade-offs in algorithmic and architectural designs which in turn leads to efficient designs.

A design example of the image correlation system illustrates the whole system design cycle in which the algorithm transformation and architectural mapping are applied. The array implementations can be rigorously compared using OPERAS to arrive at the best array solution. The complete CASCADE environment is a global method bank containing several interactive method banks. CASCADE is effective toward designing the energy efficient VLSI signal processing system.

References

[1] F. Pichler and R. Moreno–Diaz (Eds.) *Computer Aided Systems Theory - EU-ROCAST '89*, Springer Verlag, 1989.

[2] F. Pichler and R. Moreno–Diaz (Eds.) *Computer Aided Systems Theory - EU-ROCAST '91*, Springer Verlag, 1991.

[3] T. Ören. *Artificial Intelligence and Quality Assurance in Computer-Aided Systems Theory,* LNCS 410, pp. 336-344, Springer Verlag, 1989.

[4] H. Hellwagner. *CAST.FOURIER–An Interactive Method Bank for Generalized Spectral Techniques,* LNCS 410, pp.355-366, Springer Verlag, 1989.

[5] M. Geiger. *CAST.FSM Applied to VLSI Synthesis* LNCS 585, pp.422-441, Springer Verlag, 1991.

[6] F. Bretschneider et al., *Infrastructure for Complex Systems-CAD Frameworks* LNCS 410, pp.125-132, Springer Verlag, 1989.

[7] R. Mittelman. *Object-oriented Design of CAST Systems* LNCS 410, pp. 69-75, Springer Verlag, 1989.

[8] B. Zeigler. *Hierarchical, Modular, Discrete Event Simulation in an Object Oriented Environment,* Simulation Journal, Vol. 49, No. 5 , pp. 219-230, 1987.

[9] G. Booch. *Object-Oriented Design With Applications,* Benjamin/Cummings Pub. Co., 1991.

[10] K. Guttag. *The MVP, Hot Chips Symposium,* 1993.

[11] S. Y. Kung. *VLSI Array Processor,* Prentice-Hall, 1987.

[12] A. F. Nielsen, P. M. R. Jensen, K. Bagchi and O. Olsen. *Comparing Transformation Schemes For VLSI Array Processor Design* , Proc. VLSI '91, Edinburgh, North-Holland, pp. 367-376.

[13] P. Koch, K. Bagchi and K. Hermansen. *Some Experiments on Realistic and Efficient Scheduling of DSP Algorithms onto Multiprocessor Architectures*, Proc. of the 26th IEEE Asilomar Conf. on Signals, Systems and Computers, California, IEEE Press, 1992.

[14] A. V. Oppenheim and R. W. Schafer. *Discrete-Time Signal Processing*, Prentice Hall, 1989.

[15] L. B. Jackson. *Digital Filters and Signal Processing*, Vol. 5, pp. 181-194, Kluwer Academic Publishers, 1986.

[16] A. K. Jain. *Fundamentals of Digital Image Processing*, Prentice Hall, 1986.

[17] A. N. Netravali and B. G. Haskell. *Digital Pictures*, Plenum Press, 1988.

[18] T. Komarek and P. Pirsch. *Array Architectures for Block Matching Algorithms*, IEEE Transactions on Circuits and Systems, pp. 1301-1308, October, 1989.

[19] J. Burr and A. Peterson. *Energy Considerations in Multi-chip Module-based Multiprocessors*, IEEE Intl. Conf. on Computer Design, pp. 593-600, 1991.

[20] E. A. Lee. *A design lab for statistical signal processing*, International Conference on Acoustics, Speech, and Signal Processing, pp. 81-4, March 1992.

[21] *Second-Generation TMS320 User's Guide*, Texas Instruments, 1987.

[22] D. Jordan. *Implementation Benefits of C++ Language Mechanisms*, CACM, pp. 61-64, September 1990.

[23] G. Yeh and K. Bagchi and J. Burr and A. Peterson. *OPERAS - An Object-Oriented Signal Processing System Architecture Simulator*, 27th Annual Simulation Symposium, pp. 198-207, April 1994.

[24] C. D. Thompson. *Area-Time Complexity for VLSI*, Artificial Intelligence and Information Control Systems of Robots, pp. 373-382, Elsevier Science Publishers, 1984.

[25] C. D. Thompson and P. Raghavan. *On Estimating the Performance of VLSI Circuits*, 1984 Conference on Advanced Research in VLSI, MIT, pp. 34-44, January, 1984.

[26] S. W. Director, editor. *Proceedings IEEE, Special Issue on VLSI CAD*, IEEE Press, February, 1990.

Illustrating Constraint Programming Systems in Logistic Planning

Jean-Michel Thizy

Systems Science Programme, University of Ottawa, On, Canada K1N6N5 thizy@csi.uottawa.ca
supported by Department of National Defence contract W0141-95-AA01, NSERC Grant OGP 0042197

Abstract. Logistic systems analysts use a wide array of modeling systems such as algebraic and logic programs. Two simple examples show how they mesh as Constraint Logic programs, emphasizing the visual style of formulation. A facility location problem is solved in Prolog. Its standard mathematical programming formulation is then represented in CLP($\Re$) to reach an optimal solution faster. The examples are chosen to raise systemic issues such as model re-use and adaptability that are at the heart of logistic analysis and planning.

Introduction

Systems technology is usually expected to provide a method for the coordination of heterogeneous components. Equally challenging is the coordination of distinct modes of representation of a particular system, illustrated here by the meshing of logical and mathematical programming that arises in logistic planning. For problem solving, artificial intelligence techniques promoted functional and declarative languages such as Lisp, while operations researchers honed mathematical programming packages[1]. After these methods had long coexisted, (Blair, 86; Jeroslow 88, 89; Hooker 88, 90) revived numerical optimization methods for logic inference, while logic programs were enriched by systems of constraints, in particular, (linear) inequalities. The rapprochement lead to a novel constraint logic programming (Cohen, 90; Jaffar 87).

The following section analyzes specific features of three landmark modeling languages, summarizing two main families: those offering great compatibility with tabular database systems and those featuring rigorous mathematical transcriptions. Brief logistic examples emphasize program layouts which determine largely the quality of the interface between the modeler and the computer-assisted solution process. In Section 3, the classic Uncapacitated Facility Location Problem (UFLP) is given a logic formulation in Prolog (Colmeraurer, 72). In Section 4, a constraint logic program in CLP($\Re$) solves this problem and computational advantages are presented in Section 5. The conclusion emphasizes the need for systems modeling in logistic planning.

---

[1] A short list of standard software for optimization includes: Apex (Krabek, 80), CPLEX (Bixby, 90), FMPS and Sprint (Dickson, 73), KORBX (Cheng, 89), LINDO (Schrage, 91), Minos (Murtagh, 78), MPS (IBM, 76), MPS III/Whizard (Ketron, 75), OSL (IBM, 91) Sciconic (Beale, 85), Tempo (Carstens, 78), Umpire (Forrest, 74), XMP (Marsten, 81) and Zoom (Singhal, 89).

1 Modeling Systems for Mathematical Programming

The first logistic example below displays a simple, yet not entirely symmetric figure (it is not a complete bipartite graph). It is inspired by an aggregate production planning model, presented by (Bowman, 56), where the demand b_t for an item in period t ($\leq T$) must be met at a minimum cost from available sources a_t in the T periods. Products can be stored at a cost proportional to the number of items and the storage duration, but no backlog is allowed. Its traditional mathematical notation is:

$$\min h \sum_{t=1}^{T}\sum_{\tau=t}^{T} (\tau - t)\, x_{t\tau}$$

s.t.

(1) $\sum_{\tau=1}^{T} x_{t\tau} \leq a_t$ for all $t = 1, ..., T$

(2) $\sum_{t=1}^{\tau} x_{t\tau} = b_\tau$ for all $\tau = 1, ..., T$

$x_{t\tau} \geq 0$ for all $t, \tau = 1, ..., T,$

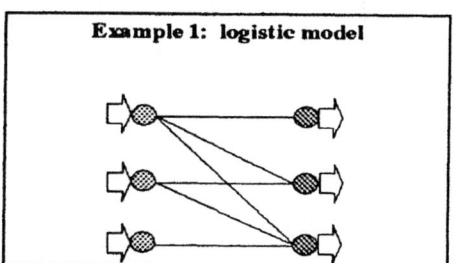

Example 1: logistic model

where $x_{t\tau}$ is the quantity stored from period t to period τ, and h is the unit inventory cost. Although the problem can be solved by a single pass procedure (Johnson, 74), it can be considered as a building block of a more complex logistic problem.

1.1 PAM, a large-scale tabular system for mathematical programming

Originally designed as user interfaces for procedural modeling languages[2], table-oriented systems for mathematical optimization have long supported large-scale, logistic applications on mainframe systems, and they were adapted to micro-computers recently. Table 1 formulates the inventory problem (1)-(2) in PAM (Welch, 87), where highlighted lines are discussed. PAM works best with rectangular structures; Table 1 provides a counter-example requiring some artifact such as two column indexes T and T1 over the same domain PERIODS; in other models, simple non-rectangular (multi-period or echelon) structures can be represented with some icons. Although the focus of this section is on flexible mathematical programming systems developed more recently, tabular systems possess database features unmatched by their successors:
- each variable and constraint index is recognized by its name, instead of simply its position; thus x[1, 2] is different from x[2, 1], while x[FROM=1, TO=2] is the same as x[TO=2, FROM=1] (this feature prefigures typing polymorphic messages.)
- data domains can be supersets of the model domains: the model generator performs its own projections and selections, whereas algebraic models need to declare explicit subsets to formulate problems of reduced sized, using the original data.

Other tabular modeling systems such as MathPro (Hirschfeld, 89) and PLATOFORM (Palmer, 84) have an analogous syntax; MIMI/LP (Baker, 86) has comparable features.

---

[2] such as Apex/MRG (Krabek, 80), CFMS (Honeywell, 1982), DATAFORM (Ketron, 87), SESAME and DATAMAT (Orchard-Hays, 78abc), GAMMA (Sperry, 77), MaGen (Haverly Systems, 77), MGG (Simons, 87), MGRW (IBM, 72), MODELER (Burroughs, 80), OMNI/PDS (Haverly Systems, 76), and RPMS (Bonner & Moore, 79).

```
COLUMNS         STOCK
   L            X
   T            PERIODS
   T1           PERIODS

MATRIX          STOCK          RTYPE          RHS
   COST         HCOST          <>
   SUPPLY       /1             <              AVAILABL
   DEMAND       /1             =              REQUIRED

ROWS            L              T              T1
   SUPPLY       S              PERIODS
   DEMAND       D                             PERIODS

DATA            TABLE          STUB           HEAD
   HCOST        RATES          T              T1
   AVAILABL     RATES          T              UNITS
   REQUIRED     RATES          UNITS          T1

DOMAINS
                PERIODS
                  1
                  2
                  3
DATA TABLES
     RATES
       1        1      2       3      UNITS
       2       1E-15  10      20      2
       3              1E-15   10      1
     UNITS                    1E-15   3
                1      2       2
```

Table 1: An Inventory Problem in PAM

1.2 GAMS and AMPL, algebraic modeling systems for mathematical programming

Table 2 formulates the problem in GAMS (Brooke, 88), probably the best-known among mathematical programming systems offering a syntax close to conventional algebraic notation. Yet, GAMS does not represent indices, but only sets, as tabular systems do, hence the need of an alias set T1 to represent Constraints (1). Without an alias set, the declaration SUPPLY(1) would represent the constraint: $\Sigma_t x_{tt} \leq a_1$. In the same spirit, GAMS allows only global set manipulation (such as lags: T-1, T-2, etc.), making constraints (2) difficult to model. An awkward solution is to use the function ORD to index each element of a set. Even in simpler cases, set operators such as lags must be applied cautiously when representing boundary conditions (e.g., t=0). These properties are found in many other systems such as LINGO (Cunningham, 89).

The same problem is modeled with AMPL (Fourer, 90) in Table 3. Indices simplify the model considerably as local variables evaluated before constraint interpretation. However, AMPL models can become unwieldy to represent complex constraints or data which GAMS and AMPL manipulate in imperative style. For example, in Problem (1)-(2), the cost of inventory h could be influenced by the bill of materials featured in Table 4, which AMPL models by the traditional programming statements

```
SETS
     T    periods    /1, 2, 3/
     ALIAS (T,T1);
PARAMETERS
     A(T)   supply   /1 2
                      2 1
                      3 3/

     B(T)   demand   /1 1
                      2 2
                      3 2/
SCALAR H     inventory cost (dollars per unit per period)    /10/ ;
PARAMETER C(T,T1)    inventory cost between periods T and T1 ;
          C(T,T1)$(ORD(T1) GE ORD(T)) = H * ( ORD(T1) - ORD(T) ) ;

VARIABLES
     X(T,T1)      inventory quantities from period t to period t1;
     Z            total inventory costs;
  POSITIVE VARIABLE X,Y;

EQUATIONS
     COST           objective function
     SUPPLY(T)      available supply in period t
     DEMAND(T)      satisfy demand in period t;

  COST ..      Z  =E=  SUM((T,T1), C(T,T1)*X(T,T1)) ;
  SUPPLY(T) ..    SUM(T1, X(T,T1)) =L=  A(T) ;
  DEMAND(T1) ..  SUM(T$(ORD(T1) GE ORD(T)), X(T,T1)) =G=  B(T1) ;
```

Table 2: An Inventory Problem in GAMS

of Table 5. Note the abundance of simple logical conditions (if), calling for a logic programming syntax such as the declarative Prolog program of Table 6. Even though parameters and variables are governed by a fairly uniform syntax, they are distinct and so are their related constraints, with no provision for submodel or sequential solution.

```
set periods  := 1 .. T;
param T >0 integer;
param h
param c{t,t1}                # inventory cost between periods t and t1 ;
      c(t,t1) {if t1 >= t}  = h * ( t1 - t ) ;
var x{periods,periods} >= 0;

minimize cost:  sum{t in periods} sum{t1 in periods}   c[t,t1]*x[t,t1];
subject to supply{t in periods}: sum{t1 in periods} x[t,t1] <=   a[t];
subject to demand{t in periods}: sum{t1=t..T} x[t,t1] = b[t];

data;
param T := 3 ;                # horizon
param h := 10 ;               # inventory cost
param : a  :=    1 2
                 2 1
                 3 3 ;        # supplies
param : b  :=    1 1
                 2 2
                 3 2 ;        # demands
end;
```

Table 3: An Inventory Problem in AMPL

1.3 Toward logic programming: disjunctive mathematical programming

Logic programming was advocated early to formulate linear programs (Kallio, 79). It is even more appropriate to formulate disjunctive programs (Balas, 85), in which constraints can be connected not only by the logic operator AND, but also by the disjunction OR. Most modeling languages, except Ketron's MPSIII (1992), can only formulate disjunctions, e.g. in preemptive scheduling:

(3) $t_1 - t_2 \leq 2$ OR $t_2 - t_1 \leq 2$,

by transforming them into a restricted form, i.e., zero-one requirements, as in AMPL: `var x >= 0 <=1 integer;` such transcription may not be strictly possible (e.g. with complementary slackness conditions), or may hamper both the formulation and its ensuing solution method. Introducing logical constructs in the formulation vastly expands the scope of mathematical programming systems and their syntactic requirements. For example, let $\{P_k\}_k$ represent a family of constraints. A

Product	Part	#
bicycle	frame	1
	wheel	2
frame	saddle	1
wheel	flange	2
	hub	1

Table 4: Bill of materials to determine unit inventory cost

condition $(P_1 \vee P_2) \wedge (P_3 \vee P_4)$ can be modeled by simple extensions of most mathematical programming systems, but statements such as

$(\bigwedge_{i \in I} P_i) \wedge \bigvee_{j \in J} (P_j \text{ and} \bigvee_{k \in K_j} P_k)$, where

the sets I, J and K_j may overlap, require a new syntax as found in logic programs, the most popular of which is Prolog. One approach is to formulate mathematical

```
set product;
set assemble within product cross product;

param max_level := (sum {product} 1) - 1;

set level {n in 1..max_level} :=
    if n==1
    then assemble
    else level[n-1] union {i in product, j
            in product: exists {k in product}
            ( (i,k) in level[n-1] and
              (k,j) in level[n-1] ) };

set parts := {level[max_level]};
```

Table 5: Exploring a bill of materials in AMPL

programs, using the syntax of Prolog, as in the system implemented by McKinnon (88), where the disjunction (3) can be written: `at_least( 1,{ `$t_1-t_2\leq2$`, `$t_2-t_1 \leq2$` })`. An interpreter rewrites, or embeds, the model as a mixed integer program, an implementation that opens up an array of alternatives. At the semantic level, such an immediate transcription only allows the formulation of *static* constraints (Jaffar, 92), e.g. $t_1-t_2\leq2$. In contrast, a delayed embedding permits a greater use of the symbolic manipulation of Prolog, e.g., to process *dynamic* templates that generate families of constraints recursively. At the operational level, the transcription transfers the solution process to a compiled model, enhancing its speed. The price to pay is a fixed embedding and an optimization algorithm over which the modeler has no control. In contrast, logic declarations in the original formulation can specify computing procedures; thus, at the extreme, the logic programs themselves can be used not only to formulate, but also to specify the optimization process, thereby integrating data, specification and solution control. This alternative is explored in the following sections, where Prolog is chosen not only for its data and constraint structuring convenience, illustrated by the disjunction (3) and Table 6, but mostly for its power to let modelers investigate different formulations and evaluate their solution processes.

```
/* Bill of Materials :   Product   ,    Part    ,   Part_Qty   )   */
      assemble          ( bicycle   ,   frame    ,      1       ).
      assemble          ( bicycle   ,   wheel    ,      2       ).
      assemble          ( frame     ,   saddle   ,      1       ).
      assemble          ( wheel     ,   flange   ,      2       ).
      assemble          ( wheel     ,    hub     ,      1       ).

/* find how many subparts form a given product */
require(Product,   Part   , Qty):- assemble(Product, Part, Qty).

require(Product, Subpart, Qty):- assemble(Product, Part, Part_Qty),
                                 require(Part, Subpart, Subpart_Qty),
                                 Qty = Part_Qty * Subpart_Qty.
?- require(bicycle, Part , Qty);
```

Table 6: Exploring a bill of materials in Prolog

2 Logic Programming

The Uncapacitated Facility Location Problem (UFLP) (Balinski, 63), consists of selecting some facilities as sources to keep open among a finite set J of potential sites in order to minimize the cost of serving a given finite set of destinations (clients). A cost c_{ij} is incurred to satisfy the totality of the demand of Client i from Location j, if a facility is kept open there, and a cost f_j is incurred to maintain a facility open at Location j. UFLP can be formulated as a logic program. Let the predicate $Open_j$ denote opening Facility j; each customer i must be served from at least one facility j as noted by the predicate $Serve_{ij}$. A propositional logic representation of UFLP is:

(4)   for each $i \in I$: $\bigvee_{j \in J}$ $Serve_{ij}$

and

(5)   for each $i \in I$, $j \in J$: $Open_j$ if $Serve_{ij}$.

In addition, the costs of all true predicates must be cumulated, and the cheapest overall cost is selected (much as expert systems select highest confidence levels). Although Clauses (4) are not definite, the satisfaction of (4) and (5) can be controlled explicitly by a set of Horn clauses, as proposed by (Clark, 77; Kowalski, 79). Accordingly, Table 7 contains a fairly natural, but rather inefficient implementation ("generate and test"), using additional variables such as $Serve_i$ to emulate the disjunction (4), and predicates such as $supply_j$ to control the conjunctions of clauses. Faster "test and generate" implementations perform feasibility tests before fully constructing a candidate solution. Such test constraints can often be expressed algebraically. Unfortunately, standard Prolog can verify algebraic expressions only after the associated variables have been assigned values. Thus, implicit testing of these constraints without numerical evaluation requires insight not revealed explicitly by the formulation. For example, in Table 8, the enumeration of solutions can be dramatically reduced by the following cost property. Denote by J1 the set of the sites selected in a given candidate solution; then the site j(i) to serve customer i can be determined as $j(i) = \arg\min_{j \in J1} \{c_{ij}\}$, and no other route from J1 to i need to be enumerated for this candidate solution. (Van Hentenryck, 89) proposes further improvements to the enumeration, using some ideas presented in (Khumawala, 73), which could speed the programs markedly. They are

```
route_ij( 1 , 1 , 0) .                  /* in full-scale application */
route_ij( 1 , 2 , 9) .                  /* facts would be read as data */
route_ij( 2 , 1 , 7) .
route_ij( 2 , 2 , 0) .
facility_j( 1 , 10 ) .                  /* indices I,J are used instead */
facility_j( 2 ,  8 ) .                  /* of lists for simplicity */

demand( 0, [], 0 ) .
demand( I, [Serve_i|Serve], Sum_Cost ) :- I>0, I_1 = I - 1 ,
                                demand( I_1, Serve, Sum_i_Cost ),
                                route_ij( I, Serve_i, Cost_i),
                                Sum_Cost = Sum_i_Cost + Cost_i.
supply( 0,      _ ,       0    ).
supply( J, Serve, Sum_Cost )       :-     J>0, J_1 = J - 1 ,
                                supply_j( J, Serve, Fixedcost ),
                                supply( J_1, Serve, Sum_j_Cost ),
                                Sum_Cost = Sum_j_Cost + FixedCost.
supply_j( _,       []    ,     0    ) .
supply_j( J,    [J|_]  , Fixedcost) :-     facility_j( J, FixedCost ).
supply_j( J, [K|Serve], Fixedcost) :- J<>K,supply_j( J, Serve, Fixedcost ).

?- demand( 2, Serve, Demand_cost ),     /* displays every solution cost */
   supply( 2, Serve, Supply_cost),      /* use predicate assert to find */
   Cost = Supply_cost + Demand_cost;                /* the minimum cost */
```

Table 7: A simple Prolog program to generate and test UFLP solutions

not implemented in this paper which emphasizes *explicit* modeling, not combinatorial insight. Similar improvements to the CLP($\Re$) program of Table 10 are not undertaken, to stress the relative merits of the *primary* features of each modeling approach.

```
route_ij( 1 , 1 , 0) .
route_ij( 1 , 2 , 9) .
route_ij( 2 , 1 , 7) .
route_ij( 2 , 2 , 0) .
facility_j( 1 , 10 ) .
facility_j( 2 ,  8 ) .

supply( 0,           []    ,       0    ) .
supply( J, [Fixedcost|Y], Sum_Cost ) :- J>0, J_1 = J-1,
                                facility_j( J, Fixedcost),
                                supply( J_1, Y, Sum_j_Cost),
                                Sum_Cost = Sum_j_Cost + Fixedcost.
demand( 0, _, _,     0    ) .
demand( I, J, Y, Sum_Cost ) :-     I>0, I_1 = I - 1 ,
                                demand_i( I, J, Y, Cost_1),
                                demand( I_1, J, Y, Sum_i_Cost ),
                                Sum_Cost = Sum_i_Cost + Cost_i .
demand_i( _, 0,   []   ,   9999 ).
demand_i( I, J, [Yj|Y], Cost_i ):-  J>0, J_1 = J - 1,
                                demand_i( I, J_1, Y, Cost_j ),
                                route_ij( I, J, Cost ),
                                demand_ij( Yj, CostCost_j, Cost_i).
demand_ij( 0,     _   , Cost_j, Cost_j ).
demand_ij( J, Cost, Cost_j, Cost   ) :- J>0, Cost <= Cost_j .
demand_ij( J, Cost, Cost_j, Cost_j ) :- J>0, Cost > Cost_j.

?- supply( 2 , Y , Supply_Cost ) ,
   demand( 2 , 2 , Y , Demand_Cost ) ,
   Cost = Supply_Cost + Demand_Cost ;
```

Table 8: A simple Prolog program to test and generate UFLP solutions

3 Constraint Logic Programming

The constraint logic program CLP($\Re$) (Jaffar, 92) recognizes as privileged predicates the constraints formed by arithmetic expressions over the real numbers and relations such as equality and inequalities ($\leq$, $\neq$, $\geq$). Therefore, CLP($\Re$) can solve linear programs as systems of inequalities, for instance those of Program (1)-(2) highlighted in Table 9. While Table 9 departs from mathematical programming systems as well as Prolog, it bears some similarity with the tabular models of Section 1.1. A procedural reinterpretation reveals also the classical forward solution pass (Johnson, 74).

```
lp_t(_, 0,     []    ,      0      ,    0    ,       []       , []       ).
lp_t(T, Z, [Xtz|Xt], (T-Z)*Xtz+CXt, Xtz+EzXt, [Xtz+EtXz|Xt_plus_EtX], [EtXz| EtX]):-
               Z>0,                     /* Z denotes tau,  t the t-th constraint */
               Xtz>=0,                  /* plus ls for vector addition */
               lp_t(T, Z-1, Xt, CXt, EzXt, Xt_plus_EtX, EtX).

lp(0,   []  ,    0   ,    EtX    ,   []   ).
lp(T, [Xt|X], CXt+CX, Xt_plus_EtX, [At|A]):- T>0,           /* Xt is the vector Xt. */
               lp_t(T, T, Xt, CXt, EzXt, Xt_plus_EtX, [0|EtX]),
               lp(T-1, X, CX, EtX, A),   /* E is the sum for all indices */
               At>=EzXt.                 /* Ez is up to index tau-1 only */

solution(Opt):-   lp(3, X, Opt/10, [1,2,2], [2,1,3]) .
```

Table 9: A CLP($\Re$) program to solve the inventory problem (1)-(2)

Direct formulation of propositional calculus

More circuitously, CLP($\Re$) can represent propositional calculus as a 0,1 linear program. Consider for example the formulation (4)-(5), recalling that A if B is equivalent to not B or A. Let x_{ij} denote the truth valuation of Serve$_{ij}$, y_j that of Open$_j$, then (4) and (5) are transformed into the well-known formulation of UFLP:

(6) $\sum_j x_{ij} \geq 1$ for all i
(7) $x_{ij} \leq y_j$ for all i and j
(8) $0 \leq x_{ij}$ for all i and j
(9) $y_j = 0$ or 1 for all j,   with an objective function:

(10)   min $z = \sum_j c_{ij} x_{ij} + \sum_j f_j y_j$

Table 10 represents (6)-(10) in CLP($\Re$). The optimal solution is found by enumerating the variables y in a standard branch-and-bound using Prolog's backtracking; at each node, the value of the local linear relaxation is calculated by solving the primal-dual system for (6)-(10):

$\sum_i\sum_j c_{ij} x_{ij} + \sum_j f_j y_j \geq \sum_i \sum_j u_{ij} - \sum_j w_j$,
$\sum_j x_{ij} = 1$ $\Big\}$ for all i
$x_{ij} \leq y_j$ $\Big\}$
$0 \leq x_{ij}$ for all j
$u_i - v_{ij} \leq c_{ij}$
y_j 0 or 1 $\Big\}$
$\sum_i v_{ij} \leq f_j$ for all j.
$v_i \geq 0$

Table 10: A CLP($\Re$) program to solve the UFLP

```
/*------------------------------------ Linear Program ------------------------------------*/
objY(    [],     [], F, FY, FYj) .
objY([Yj|Y], [Fj|F], FY+Fj*Yj, [0|EVj]):-             objY(Y, F, FY, EVj).

constraints_j(       [],           [], Ui,      [], 0              ).
constraints_j([Xij|Xi], [Yj|Y], Ui, [Vij|Vi], Xij+EjXi, [Vij+EiVj|Vi_plusEiv], [EiVj|EiV],[Cij|Ci]):-
                                                        /* Primal     */
            Xij>=0,                                     /* Constraints */
            Xij<=Yj,                                    /*             */
            Vij>=0,                                     /*    Dual     */
            Ui-Vij<=Cij,                                /* Constraints */
            constraints_j(Xi, Y, Ui, Vi, EjXi, CXi, Vi_plusEiV, EiV, Ci).

constraints(    [],   Y,      [],        [], F, FY,       EiV   ):- objY(Y, F, FY, EiV).
constraints([Xi|x], Y, [Ui|U], [Vi|V], F, CXi+CX, Vi_plusEiV, Ui+EiU, [Ci|C]):-
            constraints(X, Y, U, F, CX, EiV, EiU, C),
            constraints_j(Xi, Y, Ui, 1, CXi, Vi_plusEiV, EiV, Ci ) .

branch_penalty_j( Yj ,   Fj ,   EVj ,   Yj*(Fj-EVj)           ):-   ground(Yj),   zero_one(Yj).
branch_penalty_j( Yj ,   Fj ,   EVj ,              0           ):-   nonground(Yj),   Yj<=I, EVj<=Fj.

branch_penalty(     [],       [],        [], 0               ):- branch_penalty_j(Yj, Fj, EVj, Penalty_j),
branch_penalty([Yj|Y], [Fj|F], [EVj|EV], Penalty+Penalty_j):-     branch_penalty_j(Yj, Fj, EVj, Penalty_j),
                                                                 branch_penalty(Y, F, EV, Penalty)·

/*------------------------------------ Branch (-and-bound) ------------------------------------*/

zero_one(0).
zero_one(1).

not_binary(Yj):-   nonground(Yj).
not_binary(Yj):-   ground(Yj),   Yj>0, Yj<1.

free_fractionalY(       [],           [],       [],     ).
free_fractionalY([IpYj|IpY], [IpYj|Y], [_|F]):-   ground(IpYj),   free_fractionalY(IpY, Y, F).
free_fractionalY([IpYj|IpY], [_|Y],   [_|F]):-   nonground(IpYj),   free_fractionalY(IpY, Y, F).

branch(     []    ,    []    ,  IpY, _, _, IpObj, IpObj):-   ground(IpObj),
                                                            write(IpObj) , writeln(IpY).
branch([Yj|Y], [IpYj|IpY], IpY, F, C, LpObj, InodeObj):-   ground(Yj),
                                                          zero_one(Yj),
                                                          branch(Y, IpY, IpY_, F, C, LpObj, IpObj).
branch([Yj|_], [IpYj|_], IpY, F, C, LpObj, IpObj):-   not_binary(Yj),
                                                      zero_one(IpYj),
                                                      node(Y, IpY, IpY, F, C, IncumbentLpObj+Penalty, IpObj).

node(Y, IpY, IpY, F, C, IncumbentLpObj+Penalty, IpObj):-free_fractionalY(IpY, Y, F),
                                                        constraints(X, Y, U, V, F, IncumbentLpObj+Penalty, EV, Penalty, C),
                                                        branch_penalty( Y, F, EV, IncumbentLpObj),
                                                        branch(Y, IpY, IpY, F, C, IncumbentLpObj+Penalty, IpObj).

?-      F=[10,   8],
        C=[[0,   9], [7, 0]],
        node(Y, IpY, IpY, F, c, LpObj, IpObj).
```

Table 10 displays the basic elements of a program at the cost of some inefficiency, e.g.
- Some recursive calls to the procedure branch could be eliminated.
- The search tree is not pruned by any bounding scheme.
- The data is simply given in the query, and the objective values are printed as the enumeration proceeds. Input and output could naturally be placed on files.
- For 0-1 variables, the branch-and-bound can be streamlined by systematically branching on every variable, using CLP($\Re$)'s fast reoptimization capabilities.

Table 10 emphasizes another strength of CLP($\Re$): branch-and-bound is declared as a *structure* (and not only as a procedure as in conventional mathematical programming). Even if it appears unlikely that the symbolic simplex algorithm of CLP($\Re$) can compete with standard resolution of Prolog, the computational results will show that it may be beneficial to provide such alternative control strategies. CLP($\Re$) could also specify standard integer programming methods, e.g., implicit enumeration (Balas, 65) or Lagrangian relaxation (Geoffrion, 1974). It could even emulate parallel constraint programming (Saraswat, 93) and thereby *formulate* parallel solution of integer programs (Zenios 89, 90), an approach that has been characterized mostly by empirical computational work (Boehning, 91; Cannon, 90; Dutta, 83; Mohamed, 92 ; Pruul, 88).

4 Computational results

Two Prolog and one CLP($\Re$) programs displayed in Tables 7, 8 and 10 are compared. The standard test problems possess as many potential sites as clients; distribution costs are proportional to distances between major U.S. cities (Karg, 64). Problems with 2 to 10 cities are tested with CLP($\Re$), Version 1, under Sun OS Release 4.1 (generic). Computational times in centiseconds are given in Table 11. Not only is the CLP($\Re$) program fastest, but its CPU time increases nearly linearly as a function of problem size. The efficiency of the programs of Tables 7 and 8 could be improved commensurately. CLP($\Re$) is primarily suited to manipulate constraints, hence, it could accommodate the cutting planes of UFLP proposed by Guignard (80), Cornuéjols (82), Cho (83). Already, this aptitude is partially exploited in Table 10, since the constraints (7) are often considered to be cutting planes of a weaker formulation (Morris, 78).

	CPU (cs), program displayed in Table		
Problem Size	7	8	10
2	2	4	4
3	30	14	6
4	376	36	6
5	5988	96	16
6	113132	256	24
7	2206074	646	44
8	-	1572	60
9		3806	82
10		9048	116

Table 11: Computational results to solve UFLP

5 Systems modelling: linkage and reuse

CLP($\Re$) is designed to solve disjunctive programs, i.e. when predicates are continuous (linear) constraints, subsuming the linear program of Table 10 as well as the propositional calculus of (6)-(10). However, we are interested in extending the results of Section 4 by imbedding disjunctive programs into mixed integer programs under the control of CLP($\Re$), instead of letting them be encoded and solved statically, as described in Section 1.3 . The embedding of propositional disjunctions performed in Constraint (6) can be extended (Balas, 85; Jeroslow, 89). Its principle is illustrated by the objective function (10), which, despite its casualness, is not patterned after (6). To highlight the technique itself, first create the intermediate variables $\gamma_j = \Sigma_i c_{ij} x_{ij}$; then the objective is to minimize $z \geq \Sigma_j \gamma_j + \phi_j$ where ϕ_j denotes the cost of opening Facility j. Thus, the proposition Open_j, becomes in fact $\text{Open}_j(\phi_j)$, standing for $\phi_j \geq f_j$, whereas its negation can be characterized as $\phi_j \leq 0$. Disjunctive linear inequalities are modelled by splitting their variables, e.g., $\phi_j = \phi_j^1 + \phi_j^2$ and representing $\text{Open}_j(\phi_j)$ as $\phi_j^1 \geq f_j y_j$. In the special case of (10), $\phi_j \leq 0$ is embedded as $\phi_j^2 \leq 0$ $(1-y_j)$ and thus ϕ_j^2 vanishes. Since $\phi_j = \phi_j^1$, we simply write $\phi_j \geq f_j y_j$ and get the simple form of (10).

The same technique yields a variant of UFLP: suppose the demand of a client can be met from different locations. Let $\xi_{ij} \geq 0$ be the proportion of demand i satisfied by Facility j and Serve_{ij} be the constraint: $\xi_{ij} > 0$. Conditions (4)-(5) are replaced by:

(4') $\Sigma_i \xi_{ij} = 1$ for all j .

(5')   for each $j \in J$: $\text{Open}_j \bigvee$ (for each $i \in I$: $\xi_{ij} \leq 0$ for each i).

in which Open_j is ($\xi_{ij} \leq 1$ for each i). Using the same splitting method, (5') becomes:

(11) $\xi_{ij}^1 \leq y_j$ for all i,j,
(12) $\xi_{ij}^2 \leq 0$ for all i,j,
(13) $\xi_{ij} = \xi_{ij}^1 + \xi_{ij}^2$ for all i,j.

Since in Constraint (12), $\xi_{ij}^2 = 0$, Constraint (13) yields $\xi_{ij} = \xi_{ij}^1$ and Constraint (11) simplifies as $\xi_{ij} \leq y_j$ for all i,j. Replacing ξ by x yields a system identical to (6)-(10), illustrating a simple case of model re-use toward a slightly different logistic case.

A last variant of UFLP illustrates model linkage. The Capacitated Facility Location Problem CFLP (Efroymson, 66) specifies the demand $d_i > 0$ of Client i and capacity $s_j > 0$ of Facility j, if it is opened. To model CFLP, add to Program (6)-(10) the constraints:

(14) $\Sigma_i d_i x_{ij} \leq s_j$ for all j .

However, the linear relaxation does not display the computational qualities of (6)-(10): the optimal value y has many fractional coordinates, causing a long branch-and-bound. To obtain a model closer to disjunctive form, i.e. one with a sharper linear relaxation, the conjunction of constraints (14) is distributed over the disjunction contained in (5'):

(15) for each $j \in J$: $(\Sigma_i d_i x_{ij} \leq s_j \bigwedge$ for each $i \in I$: $x_{ij} \leq 1) \bigvee (\Sigma_i d_i x_{ij} \leq s_j \bigwedge$ for each $i \in I$: $x_{ij} \leq 0)$

The rightmost term can be simplified as:

(16)   for each $j \in J$: $(\Sigma_i d_i x_{ij} \leq s_j \bigwedge$ for each $i \in I$: $x_{ij} \leq 1) \bigvee$ (for each $i \in I$: $x_{ij} \leq 0)$

which, using the same variable splittings and their subsequent cancellation, yields a strong linear formulation of CPLP derived from (6)-(10) by adding the constraint:

(17) $\quad \Sigma_i \, d_i \, x_{ij} \le s_j \, y_j \qquad$ for all j.

One would hope that a model such as UFLP would form a core around which logic conditions could be transcribed either to fit a particular variant of a model such as CFLP, or an algorithmic specification (as in the declarative branch-and-bound of Table 10). Unfortunately, neither Prolog nor CLP($\Re$) offer full modularity: although each procedure is quite autonomous (e.g. branch and constraints in Table 10), the terms of the predicates are not. For example, the lists of variables of Table 9 can be modified with difficulty. Similarly, it would be difficult to adapt Table 10 to CFLP.

Logic programming and mathematical programming systems are both criticized as slow by some practitioners, who acknowledge them at least as convenient tools for rapid prototyping, yielding programs to be re-written when faster solution is required. However, supporters claim that logic programming sets necessary foundations for concurrent processing (Conery, 87; Foster, 90; Gregory, 87; Hogger, 82; Shapiro, 87).

Mathematical programming systems have been advocated for economic studies, where models are often used for only one analysis (Brooke, 88) because they save programmer's time to write an imperative program, and they link data representation and solution tightly. In this context, faster development can be expected from the affinity of CLP with database query languages, even if, with a dry calculation of CPU time, large-scale CLP are slower than traditional imperative languages.

Acknowledgments: thanks to J. Tomlin and R. Fourer for bibliographic references.

Bibliography

Baker, T.W. and D.J. Biddel, "A Hierarchical/Relational Approach in Modeling", *Chesapeake Decision Sciences, Inc.*, presented at ORSA/TIMS Joint National Meeting, Miami Beach, 1986.

Balas, E., "An Additive Algorithm for Solving Linear Programs with Zero-One Variables", *Operations Research* 13, pp. 517-546, 1965.

Balas, E., "Disjunctive Programming and a Hierarchy of Relaxations for Discrete Optimization Problems", *SIAM Journal of Algebraic and Discrete Methods*, 6, pp. 466-486, 1985.

Balinski, M.L. and P. Wolfe, "On Benders Decomposition and a Plant Location Problem", *Mathematica*, ARO-27, 1963.

Beale, E.M.L., "Integer Programming", *Computational Mathematical Programming*, K. Schittkowski, ed., Springer-Verlag, pp. 1-24, 1985.

Bixby, R.E., "Implementing the Simplex Method: The Initial Basis," *ORSA Journal on Computing*, 4, pp. 267-284, 1990.

Blair, C., R.G. Jeroslow and J. Lowe, "Some Results and Experiments in Programming Techniques for Propositional Logic", *Computers and OR*, 13(5), 1986.

Boehning, R.L., R.M. Butler and B.E. Gillet, "A Parellel Integer Linear Programming Algorithm", *European Journal of Operational Research*, 34(3), pp. 393-398, 1988.

435

Bonner & Moore Management Science , "RPMS, The Refinery and Petrochemical Modeling System - A System Description", Houston, 1979.

Bowman, E.H., "Production Scheduling by the Transportation Method of Linear Programming", *Operations Research*, 4, 1956.

Brooke A., D. Kendrick and A. Meeraus, *GAMS, A User's Guide*, The Scientific Press, Redwood City, Ca., 1988.

Burroughs Corporation, "Model Development Language and Report Writer (MODELER), User's manuel", Detroit Mich, 1980.

Cannon, T.L. and K.L. Hoffman, "Large-Scale 0-1 Linear Programming on Distributed Workstations", *Annals of Operations Research*, 22(3), pp. 181-217, 1990.

Carstens, D., "Parallel processing for large scale linear programming and other applications programs", presented at ORSA/TIMS joint national meeting, Los Angeles (1978) and IX International Symposium on Mathematical Programming, Montreal 1979.

Cheng, Y.C., D.J. Houck, J.-M. Liu, M.S. Meketon, L. Slutsman, R.J. Vanderbei, P. Wang, "The AT&T Korbx System", *AT&T Technical Journal*, 68(1), pp. 7-19, 1989.

Cho, D. C., E. L. Johnson, M. W. Padberg and M. R. Rao, "On the Uncapacitated Plant Location Problem I: Valid Inequalities and Facets", *Mathematics of Operations Research*, 8(4), pp. 579-589, 1983.

Cho, D. C., M. W. Padberg and M. R. Rao, "On the Uncapacitated Plant Location Problem II: Facets and Lifting Theorems", *Mathematics of Operations Research*, 8(4) pp. 590-612, 1983.

Clark, K.L. and S.-A. Tärnlund, A First Order Theory of Data and Programs, *Proceedings of IFIP*, North Holland, pp. 939-944, 1977.

Cohen, J., "Constraint Logic Programming Languages", *Communication of the ACM*, 33(7), pp. 52-68, 1990.

Colmerauer, A., Kanoui, H., Roussel, P. and Pasero, R. Un système de communication homme-machine en français. Rapport préliminaire. Groupe d'Intelligence Artificielle, Université d'Aix-Marseille, 1972.

Conery, J.S., *Parallel execution of logic programs*, Kluwer Academic Publishers, Boston, 1987.

Cornuéjols, G. and J.-M. Thizy, "Some Facets of the Simple Plant Location Polytope", *Mathematical Programming*, 23(1), pp. 50-74, 1982.

Cunningham, K. and L. Schrage, "The LINGO Modeling Language", Technical Report, University of Chicago, 1989.

Dickson, J.C., "On keeping both storage and I/O requirements low in linear programming", paper presented to the VII International Symposium on Mathematical Programming, Stanford, CA, 1973.

Dutta, A., H. J. Siegel and A. E. Whinston, "On the Application of Parallel Architectures to a Class of Operations Research Problems", *RAIRO Operations Research*, 17(4), pp. 317-341, 1983.

Dutta, M. and S. Shen, "Parallel Computer Architectures for Combinatorially Hard Problems", Graduate School of Management, The University of Rochester, ca. 1983.

Efroymson, M.A. and T.L. Ray, "A Branch-and-Bound Algorithm for Plant Location", *Operations Research*, 14, pp. 361-368, 1966.

Forrest, J.J.H., J.P.H. Hirst and J.A. Tomlin, "Practical Solution of Large Mixed Integer Programming Problems with Umpire", *Management Science 20*, pp. 736-773, 1974.

Foster, I., *Systems programming in parallel logic languages*, Prentice Hall, New York, 1990.

Fourer, R., D.M. Gay, and B.W. Kernighan, "A Modeling Language for Mathematical Programming", *Management Science*, pp. 519-554, 1990.

Geoffrion, A.M., "Lagrangian Relaxation for Integer Programming", *Mathematical Programming Study*, 2, pp. 82-114, 1974.

Gregory, S., *Parallel Logic Programming in PARLOG - The Language and Its Implementation*, Addison-Wesley, 1987.

Guignard, M., "Fractional vertices, Cuts and Facets of the Simple Plant Location Problem", *Mathematical Programming Study*, 12, pp. 150-162, 1980.

Haverly Systems, Inc., MaGen, Denville, NJ, 1977.

Haverly Systems, Inc., Omni Linear Programming System: User and Operating Manual, 1st ed. Denville, NJ, 1976.

Hirshfeld, D.S., "Mathpro Usage Guide", Mathpro Inc., 1989.

Hogger, C. (1980). "Concurrent logic programming", in *Logic Programming*, APIC Studies in Data Processing No. 16, K.L. Clark and S.-A. Tärnlund eds., Academic Press, London, pp. 212-228, 1982.

Hooker, J.N., "Resolution vs. Cutting Plane Solution of Inference Problems: Some Computational Experience", *Operations Research Letters*, 7(1), pp. 1-7, 1988.

Hooker, J.N. and C. Fedjki, "Branch-and-Cut Solution of Inference Problems In Propositional Logic", *Annals of Mathematics and AI*, 1, pp. 123-139, 1990.

IBM World Trade Corporation, "Matrix Generator and Report Writer (MGRW) Program Reference Manual", New York, 1972.

IBM World Trade Corporation, "IBM Mathematical Programming System Extended/370 (MPSX/370) Program Reference Manual", 2nd ed., New York and Paris, 1976.

IBM Corporation, "Optimization Subroutine Library: Guide and Reference", Research Triangle Park, NC, 1991.

Jaffar, J. and J.L. Lassez, CLP($\Re$), "Constraint Logic Programming", *Proceedings of the Conference on Principles of Programming Languages*, Munich, 1987.

Jaffar, J., S. Michaylov, P. Stuckey and R. Yap, "The CLP($\Re$) Language and System", *ACM Transactions on Programming Languages* 14(3), pp. 339-395, 1992.

Jeroslow, R.G., "Computation-oriented Reductions of Predicate to Propositional logic", *Decision Support Systems*, 4(2), pp. 183-197, 1988.

Jeroslow, R.G., "Logic-based Decision Support; Mixed Integer Model Formulation", *Annals of Discrete Mathematics*, 40, 1989.

Johnson, L.A. and D.C. Montgomery, *Operations Research in Production Planning, Scheduling and Inventory Control*, John Wiley and Sons, New York, 1974.

Kallio, K., "On Designing LP Interface Structures", *Proceedings of BIFOA Symposium*, Szyperski, N. and Grochla, E., eds., Sijthoff & Noordhoff, 1979.

Karg, R.L. and G.L. Thompson, "A Heuristic Approach to Solving Traveling Salesman problems", *Management Science*, 10, pp. 225-248, 1964.

Ketron, Inc., "Dataform Users Manual", Arlington, Va, 1987.

Ketron, Inc., "MPSIII User Manual", Arlington, Va, 1992.

Khumawala, B.M., "An Efficient Heuristic Procedure for the Uncapacitated Warehouse Location Problem", *Naval Research Logistics Quaterly*, 20, 109-121, 1973.

Kowalski R., *Logic for Problem Solving*, New York: Elsevier North Holland, 1979.

Krabek, C.B., R.J. Sjoquist and D.C. Sommer, "The APEX Systems: Past and Future", *SIGMAP Bull*, 29, pp. 3-23, 1980.

Marsten, R.E., "The Design of the XMP Linear Programming Library", *ACM Transactions on Mathematical Software* 7(4), pp. 481-497, 1981.

McKinnon, K.I.M., "Constructing Integer Programming Models in the Predicate Calculus", *Annals of Operations Research*, 1988.

Mohamed, R.A.K., "Parallel Branch and Bound for Mixed Integer Programming", Technical Report 92-CSE-10, Department of Computer Science and Engineering, Southern Methodist University, 1992.

Morris, J.G., "On the extent to which certain Fixed Charge Depot Location Problems can be solved by LP", *Journal of the Operational Research Society*, 29, pp. 71-76, 1978.

Murtagh, B.A. and M.A. Saunders, "Large-Scale Linearly Constrained Optimization", *Mathematical Programming* 14, pp. 41-72, 1978.

Orchard-Hays, W., "History of Mathematical Programming Systems" *In Design and Implementation of Optimization Software*, Harvey J. Greenberg, ed., Sijthoff & Noordhoff (Alphen aan den Rijn, The Netherlands, pp. 1-26, 1978.

Orchard-Hays, W., "Scope of Mathematical Programming Software" Ibid., pp. 27-40.

Orchard-Hays, W., "Anatomy of a Mathematical Programming System" Ibid., pp. 41-102.

Palmer, K., *A Model Management Framework for Mathematical Programming*, Wiley, New York, 1984.

Pruul, E.A., G.L. Nemhauser and R.A. Rushmeier, "Branch-and-Bound and Parallel Computation: A Historical Note", *Operations Research Letters*, 7(2), 1988.

Saraswat, V.A. *Concurrent Constraint Programming Languages*, MIT Press, Cambridge, Ma., 1993.

Schrage, L. *LINDO, An Optimization Modeling System*, The Scientific Press, Redwood City, Ca., 1991.

Shapiro, E., *Concurrent Prolog - Collected Papers*, The MIT Press Series in Logic Programming, Cambridge, Ma., 1987.

Simons, R.V., "Mathematical Programming Modeling Using MGG", IMAJ, *Math Management* 1, pp. 267-276, 1987.

Sperry Univac Computer Systems, "GAMMA 3.4 Programmer Reference", St Paul, Mn, 1977.

Van Hentenryck, P. *Constraint Satisfaction in Logic Programming*, MIT Press, Cambridge, Ma., 1989.

Welch, J.S., Jr., "PAM - A Practioner's Approach to Modeling", *Management Science*, 33(5), 1987.

Zenios, S.A., S. Nielsen and M. Pinar, "On the use of advanced architecture computers via high-level modeling languages, in *The Impact of Recent Computing Advances on Operations Research*, pp. 507-518, Operations Research Series 9, Elsevier Science, 1989.

Zenios, S.A., "Integrating Network Optimization Capabilities into a High-Level Modeling Language, *ACM Transactions on Mathematical Software*, 16(2), pp. 113-142, 1990.

Author Index

Lecture Notes in Computer Science

For information about Vols. 1–1029

please contact your bookseller or Springer-Verlag